Organic Synthesis in Water

Organic Synthesis
in Water

Edited by

Paul A. Grieco

Department of Chemistry and Biochemistry
Montana State University

SPRINGER SCIENCE+BUSINESS MEDIA, LLC

First edition 1998

© 1998 Springer Science+Business Media New York
Originally published by Blackie Academic & Professional in 1998

ISBN 978-94-010-6077-6   ISBN 978-94-011-4950-1 (eBook)
DOI 10.1007/978-94-011-4950-1

A catalogue record for this book is available from the British Library

Library of Congress Catalog Card Number: 97–74074

⊚ Printed on acid-free paper, manufactured in accordance with
ANSI/NISO Z39.48-1992 (Permanence of Paper)

Contents

3 Claisen rearrangements in aqueous solution 82
J.J. Gajewski

4 Carbonyl additions and organometallic chemistry in water 102
A. Lubineau, J. Augé and Y. Queneau

5 Aqueous transition-metal catalysis 141
I.P. Beletskaya and A.V. Cheprakov

6 Oxidations and reductions in water   223
F. Fringuelli, O. Piermatti and F. Pizzo

7 Base-catalyzed aldol- and Michael-type condensations in aqueous media   250
F. Fringuelli, O. Piermatti and F. Pizzo

Contributors

J. Augé Département de Chimie, Université de Cergy-Pontoise, 5 Mail Gay-Lussac, Neuville 95031 Cergy-Pontoise, France

I.P. Beletskaya Department of Chemistry, Moscow State University, 119899 Moscow, Russia

A.V. Cheprakov Department of Chemistry, Moscow State University, 119899 Moscow, Russia

F. Fringuelli Laboratorio di Chimica Organica, Dipartimento di Chimica, Università di Perugia, Via Elce di Sotto 8, I-06123 Perugia, Italy

J.J. Gajewski Department of Chemistry, Chemistry Building, Indiana University, Bloomington, Indiana 47405-4001, USA

P.P. Garner Department of Chemistry, Case Western Reserve University, 10900 Euclid Avenue, Cleveland, Ohio 44106-7078, USA

P.A. Grieco Department of Chemistry and Biochemistry, Montana State University, Bozeman, Montana 59717-3400, USA

S. Kobayashi Department of Applied Chemistry, Faculty of Science, Science University of Tokyo (SUT), Kagurazaka 1-3, Shinjuku-ku, Tokyo 162, Japan

A. Lubineau Laboratoire de Chimie Organique Multifonctionnelle, URA CNRS 462, Institut de Chimie Moleculaire d'Orsay, Université Paris-Sud, Bât. 420, F-91405 Orsay, France

D.T. Parker Novartis Pharmaceuticals, 556 Morris Avenue, Summit, New Jersey 07901-1398, USA

O. Piermatti Laboratorio di Chimica Organica, Dipartimento di Chimica, Università di Perugia, Via Elce di Sotto 8, I-06123 Perugia, Italy

F. Pizzo   Laboratorio di Chimica Organica, Dipartimento di
                      Chimica, Università di Perugia, Via Elce di Sotto 8,
                      I-06123 Perugia, Italy

Y. Queneau   Unité mixte CNRS-Beghin-Say, c/o Eridania Beghin-Say,
                      BP 2132, F-69603 Villeurbanne, France

Preface

The use of water as a medium for promoting organic reactions has, for the most part, been non-existent, despite the fact that water has served, and continues to serve, as the solvent in which the vast majority of biochemical processes take place. Chemists have only recently come to appreciate the enormous potential that water holds for those engaged in synthetic organic and organometallic chemistry, in part because of water's unique enthalpic and entropic properties.

In this volume, an international team of authors, each taking advantage of the unique properties of water for carrying out organic transformations, is brought together in order to provide a timely and concise overview of current research. The chapters focus on the practical use of water in synthetic organic chemistry, with special emphasis on Diels–Alder reactions, Claisen rearrangements, organometallic chemistry, transition-metal catalysis, oxidations and reductions.

I am grateful to my colleagues, J. Augé, I. Beletskaya, A. Cheprakov, F. Fringuelli, J. Gajewski, P. Garner, S. Kobayashi, A. Lubineau, D. Parker, O. Piermatti, F. Pizzo and Y. Queneau, for participating in this venture. Their knowledge and experience have been invaluable in putting this volume together.

P.A. Grieco

1 Diels–Alder reactions in aqueous media

P.P. GARNER

1.1 Introduction

Prior to Breslow's pioneering work in the early 1980s, the use of water as a solvent for the Diels–Alder reaction was a fairly rare occurrence. It is perhaps fitting that the earliest known example of such a reaction was actually reported by Diels and Alder themselves in 1931 [1]. This cycloaddition was performed by adding furan (**1.1**) to a solution made up by dissolving maleic anhydride in hot water:

1.1   **1.2**   **1.3**

With vigorous shaking, the oily diene eventually dissolved and a crystalline adduct, identified as the diacid **1.3**, was isolated from the reaction mixture. This is the product that would be expected if the reaction had occurred via maleic acid (**1.2**). The endo-selectivity of this reaction was notable in comparison with the analogous Diels–Alder reaction of **1.1** with maleic anhydride in organic solvents [2].

The first practical application of water as a solvent for the Diels–Alder reaction may be attributed to Hopff and Rautenstrauch, who noted that cycloaddition products could be 'obtained in a smooth and non objectionable manner by carrying out the reaction in aqueous dispersion' [3]. Although this patent is often cited, its scope is not generally appreciated. Numerous examples were provided, along with practical experimental details, for a variety of dienes and dienophile combinations. The representative reaction of cyclopentadiene (**1.4**) with N-s-butylmaleimide (**1.5**) to give a quantitative yield of product **1.6** would certainly not be objectionable to most synthetic chemists:

In light of what is now known about the aqueous Diels–Alder reaction, it is significant that these workers: (i) recognized the need to promote the water solubility of the reaction components; (ii) noted that the resulting aqueous reactions occurred at lower temperatures; and (iii) observed that it was some-

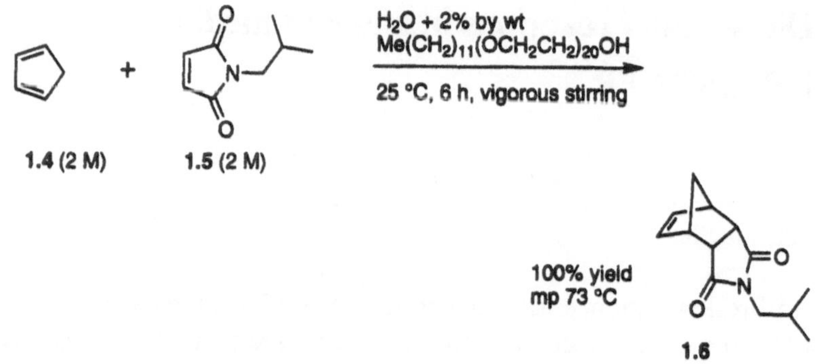

1.4 (2 M)
1.5 (2 M)
H_2O + 2% by wt
$Me(CH_2)_{11}(OCH_2CH_2)_{20}OH$
25 °C, 6 h, vigorous stirring
100% yield
mp 73 °C
1.6

times preferable to add inorganic salts to the reaction mixture. A few years later, Lane and Parker filed a patent that described similar aqueous Diels–Alder reactions of fumaric acid and a variety of dienes [4]. These early examples presaged the beneficial effect of water on the Diels–Alder reaction that would be 'rediscovered' in the early 1980s.

Finally, it is interesting to speculate that Nature may actually utilize aqueous Diels–Alder chemistry for the biosynthesis of the endiandric acids. These compounds are produced in the leaves of *Endiandra introrsa* (Lauraceae) – presumably in the vicinity of a largely aqueous cellular medium. The biosynthesis of endiandric acid A (**1.10**) proposed by Black's group [5] involves two sequential electrocyclization reactions (**1.7**→**1.8**→**1.9**) followed by an intramolecular Diels–Alder reaction to give **1.10**:

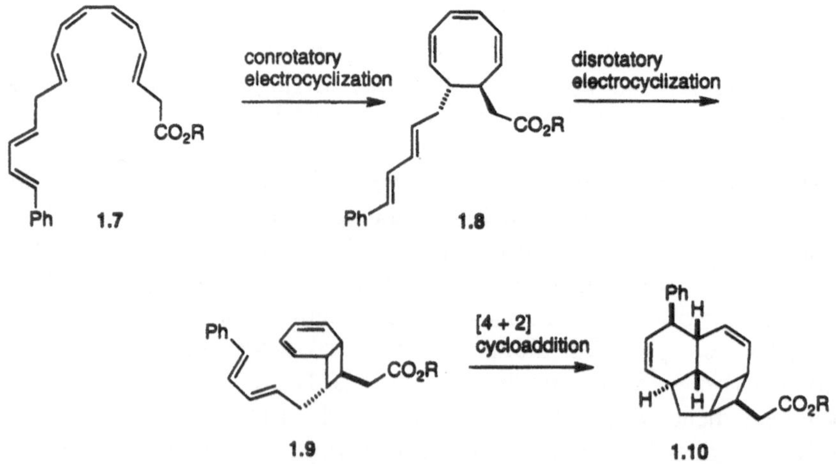

conrotatory electrocyclization
disrotatory electrocyclization
[4 + 2] cycloaddition
1.7
1.8
1.9
1.10

To be sure, one cannot say what role (if any) membranes play in the biosynthesis of these lipophilic compounds. However, since the endiandric acids are produced in racemic form, their biosynthesis does not seem to involve a chiral enzymatic environment.

1.2 The effect of water on Diels–Alder reactivity

1.2.1 Hydrophobic rate acceleration

Even though others had conducted Diels–Alder reactions in aqueous media, it was Breslow and his co-workers who first explored the extraordinary effect that water can have on the rate of Diels–Alder reactions in a quantitative manner [6]. They approached this mechanistic problem by comparing the rates of the Diels–Alder reactions shown in Scheme 1.1 in isooctane, methanol, and water (Table 1.1). All three reactions proceeded significantly faster in water than in the nonpolar organic solvent isooctane. That this rate enhancement was not due to a solvent polarity effect was suggested by the observation (Table 1.1) that reactions between cyclopentadiene (CPD) (2.1) and acrylonitrile (2.2) or between 2.1 and methyl vinyl ketone (MVK) (2.5) were only moderately accelerated in methanol. This is consistent with studies correlating the rate of Diels–Alder reactions with solvophobic parameters [7, 8]. On the other hand, the reaction between 9-hydroxymethylanthracene (2.8) and N-ethylmaleimide (2.9) was actually retarded in methanol relative to isooctane because methanol disrupted hydrogen-bonded association of the diene and dienophile. This was taken as good evidence that an additional factor must be responsible for the rate acceleration in water. Breslow proposed that water was accelerating these Diels–Alder reactions by promoting hydrophobic packing of the two reactants.

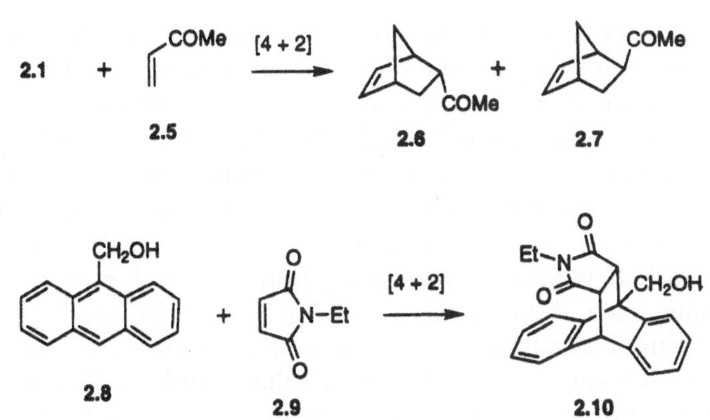

Scheme 1.1

Table 1.1 Rate constants for Diels-Alder reactions. Reprinted with permission from reference [6]. Copyright 1980 American Chemical Society

Solvent	Additional component	$k_2 \times 10^5$ $(M^{-1} s^{-1})^a$
(a) Cyclopentadiene + butenone, 20°C		
isooctane[b]		5.94 ± 0.3
MeOH		75.5
H_2O		4400 ± 70
H_2O	LiCl (4.86 M)	10 800
H_2O	$C(NH_2)_3{}^+Cl^-$ (4.86 M)	4 300
H_2O	β-cyclodextrin (10 mM)[c, f]	10 900
H_2O	α-cyclodextrin (10 mM)[c, f]	2 610
(b) Cyclopentadiene + acrylonitrile, 30°C		
isooctane[b]		1.9
MeOH		4.0
H_2O		59.3
H_2O	β-cyclodextrin (10 mM)[d, f]	537
H_2O	α-cyclodextrin (5 mM)[d, f]	47.9
(c) Anthracene-9-carbinol + N-ethylmaleimide, 45°C		
isooctane[b]		796 ± 71
1-butanol		666 ± 23
MeOH		344 ± 25
CH_3CN		107 ± 8
H_2O		$22\,600 \pm 700$
H_2O	β-cyclodextrin (10 mM)[e]	13 800

[a]Second-order rate constants. All data are the result of at least three runs at a given set of concentrations; error limits are given for cases in which triplicate runs were performed at more than one dienophile concentration.
[b]2,2,4-Trimethylpentane, >99% pure.
[c]Initial conditions: cyclopentadiene (0.4 mM), butenone (10 mM).
[d]Initial conditions: Cyclopentadiene (0.4 mM), acrylonitrile (200 mM).
[e]Initial conditions: anthacene carbinol (0.03 mM), N-ethylmaleimide (1.0 mM).
[f]The reactions were actually performed with 1 mol% of methanol and small amounts of HCl or formate buffer present; these had no effect on the reaction rate in pure water.

Evidence that a hydrophobic effect was involved came from a series of experiments that measured the rates of these reactions in the presence of additives known to increase or decrease the hydrophobic effect [9–11]. A prohydrophobic ('salting-out') agent such as LiCl enhances the hydrophobic effect by causing free water molecules to collapse around its ions (solvent electrostriction), thereby making it necessary to disrupt these stable hydration shells before a cavity can be opened for the solute (Diels–Alder reactants in this case). It is believed that antihydrophobic ('salting-in') agents like guanidinium salts decrease the hydrophobic effect by acting as a bridge between water molecules and nonpolar solutes. When the reaction between **2.8** and **2.9** was run in a 4.86 M aqueous solution of LiCl, the rate increased 2.5-fold. On the other hand, a two-thirds reduction in rate was observed when a 2.0 M aqueous solution of guanidinium perchlorate was used instead.

1.2.2 Enforced hydrophobic interactions

Blokzijl and Engberts have offered an alternative explanation for the enhanced rate of Diels–Alder reactions in water and highly aqueous binary mixtures [12–14]. Their data consisted of kinetically determined activation parameters (Gibbs energies of activation) for a series of simple Diels–Alder reactions, and thermodynamic parameters for transfer of reactants and activated complex from alcohol to water–alcohol mixtures and pure water (Gibbs energies of transfer). A plot of Gibbs energies of transfer for the initial state [2.1 + 2.5] and activated complex over the whole mole fraction range from pure 1-propanol to pure water (Figure 1.1) reveals that the rate acceleration in water relative to 1-propanol is due mainly to destabilization of the initial state. Stabilization of the transition state relative to the initial state was proposed to be a natural consequence of the reduction of the hydrophobic surface area as the reaction proceeds to the top of its reaction coordinate. This effect was termed 'enforced hydrophobic interaction' and is mechanistically distinct from explanations involving hydrophobic packing of reactants. Breslow's recent quantitative work probing the effect of (antihydrophobic) alcohol cosolvents on the rates of aqueous Diels–Alder reactions also

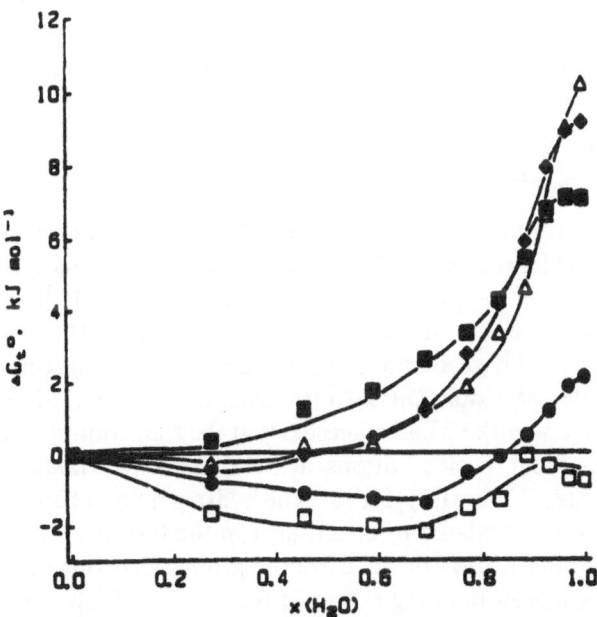

Figure 1.1 Relative standard chemical potentials of CPD ($\blacksquare$), MVK ($\bullet$), initial state (MVK + CPD) ($\blacklozenge$), activated complex ($\square$), and endo product ($\triangle$) (kJ mol^{-1}, standard state is 1 mol dm^{-3}) for the reaction of CPD with MVK in aqueous mixtures of 1-propanol as a function of the mole fraction of water at 298 K. Reprinted with permission from reference [13]. Copyright 1992 American Chemical Society.

appears to be consistent with differential solvation of reactants and transition states [15].

Although the rate of these Diels–Alder reactions was found to decrease steadily in binary solvents with increasing percentage of simple alcohols (above 5%), these workers also made the interesting observation that poly-hydroxylic additives such as glycerol and even D-glucose have the reverse effect! The rate of the reaction between **2.1** and **2.5** was found to increase with increasing concentration of glycerol (three hydroxyls per molecule). This rate acceleration was even more pronounced with D-glucose (five hydroxyls per molecule). Lubineau and co-workers have followed up on this effect of free sugars on the aqueous Diels–Alder reaction [16]. They found that, at a given molar concentration, the rate enhancements due to sugar additives increase in the order ribose (four hydroxyls per molecule) < glucose (five hydroxyls per molecule) < sucrose (eight hydroxyls per molecule), suggesting an additive effect. The activation parameters for these reactions suggest that glucose may be acting as a prohydrophobic or 'structure-making' agent much like LiCl. The inhibitory effect of α-cyclodextrin (18 hydroxyls per molecule) observed by Breslow may at first seem contradictory to this hypothesis, but is recon-ciled by the fact that cyclic dextrins are unique in that they possess a hydrophobic binding pocket. To the extent that one reaction partner gets 'trapped' within the α-cyclodextrin cavity, it will be effectively insulated from the other and the bimolecular rate will be reduced. Support for this interpre-tation comes from Lubineau's observation that the acyclic polysaccharide dextran 9000 was, in fact, a rate-enhancing additive.

1.2. 3   Hydrogen-bonding effects

If enforced hydrophobic interactions were the sole cause of rate accelerations for the Diels–Alder reaction of **2.1** + **2.5** in water, then the plot of Gibbs ener-gies of transfer for the product(s) **2.6/2.7** should resemble that of the acti-vated complex, which it does not (see Figure 1.1). This apparent inconsistency is nicely accommodated if one considers the effects of hydrogen bonding in the aqueous Diels–Alder reaction. Computational work by Jorgensen and co-workers has suggested that the transition states for this and related Diels–Alder cycloadditions are stabilized by enhanced hydrogen bonding to water (Figure1.2) [17, 18]. The hydrogen bond between water and the carbonyl was calculated to be stronger in the transition state because of the enhanced polarization of the carbonyl group. Experimental support for differential hydrogen-bonding effects in the aqueous Diels–Alder transition state was recently reported by Engberts and co-workers [19]. They compared the Gibbs free energies of the initial state and activated complex for the reaction between **2.1** and **2.5** in water (hydrophobic + hydrogen-bonding effects) and trifluoroethanol (hydrogen-bonding effects only) and found that hydrogen bonding stabilizes the activated complex more than the initial state.

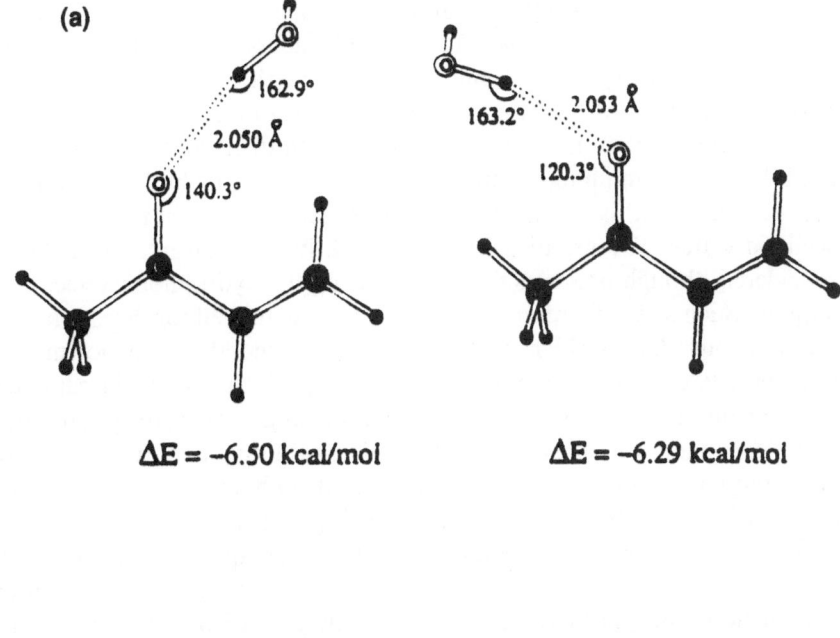

(a)

162.9°

2.050 Å

140.3°

$\Delta E = -6.50$ kcal/mol

163.2°   2.053 Å

120.3°

$\Delta E = -6.29$ kcal/mol

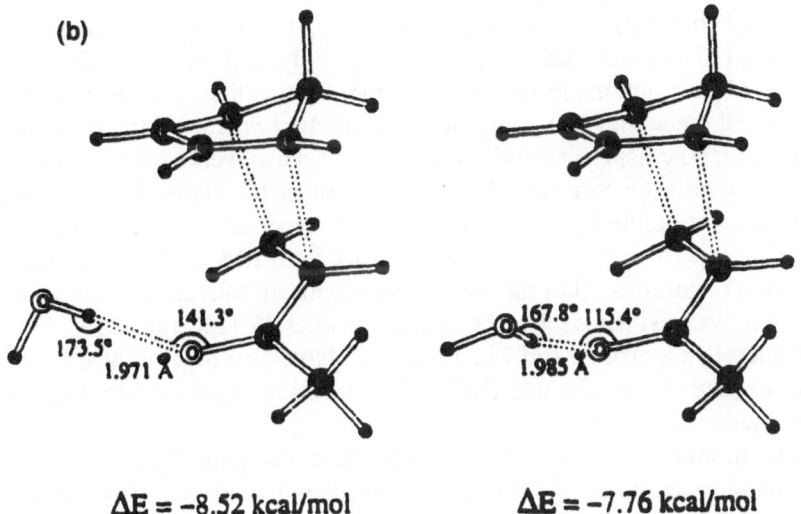

(b)

141.3°
173.5°
1.971 Å

$\Delta E = -8.52$ kcal/mol

167.8° 115.4°
1.985 Å

$\Delta E = -7.76$ kcal/mol

Figure 1.2 (a) Results of 6–31G* calculations for complexes of MVK and a water molecule. Six intermolecular degrees of freedom were optimized. The two lowest-energy structures that were found are illustrated. ΔE is the binding energy for the complex relative to the separated molecules. (b) The 6–31G* results for complexes of the CPD–MVK transition state with a water molecule, as in (a). Reprinted with permission from reference [17]. Copyright 1994 American Chemical Society.

1.2.4   Structure-facilitated hydrophobic effects

As we have just seen, the hydrophobic effect can lead to quite impressive rate accelerations in intermolecular Diels–Alder reactions. This effect (and thus the reaction velocity) can be enhanced even further by the use of certain 'prohydrophobic' or 'salting-out' agents. However, in order for the hydrophobic effect to manifest itself at all, both diene and dienophile must exhibit at least a minimum solubility in water. If one or both reaction partners do not meet this minimum solubility requirement, then no reaction will occur in water. The use of solubilizing additives or solvents must then be considered, though one runs the risk of losing the hydrophobic effect in the bargain. Such a set of circumstances does not augur well for development of the aqueous Diels–Alder reaction into a general synthetic method. Fortunately, one can overcome the solubility problem and at the same time enhance the hydrophobic effect by incorporating a prohydrophobic functional group into the substrate structure. The resulting 'hybrid' molecule will be amphiphilic, which could also lead to a rate enhancement if the resulting preorganization of the reactants in water was productive.

Grieco and co-workers demonstrated that this approach can make the aqueous Diels–Alder reaction generally useful for synthesis [20]. They found that diene carboxylates reacted with a variety of dienophiles in water at ambient temperature to give the expected cycloadducts in good to excellent yields (Table 1.2) [21]. In these reactions, the enhanced reaction rates observed in aqueous media stand in sharp contrast to the rates of analogous Diels–Alder reactions in organic solvents. The sodium salts were generally employed since they are conveniently generated *in situ* by treatment of the parent dienoic acid with sodium hydroxide. However, similar results were obtained with the corresponding ammonium carboxylates. With the benzoquinone dienophile **2.12**, the initially formed *cis*-fused adducts epimerized to the more stable *trans*-fused systems **2.13** and **2.15** under these aqueous reaction conditions. That the diene component can tolerate substantial modification was evidenced by the reactions of **2.26** [22] and **2.29** [23] with methacrolein (entries 6 and 7 in Table 1.2). Note also that, in the cases where it is an issue, these aqueous Diels–Alder reactions proceed with high regio- and endo-selectivities.

The monosodium diene phosphonate **2.31**, conveniently prepared by an Arbuzov reaction on the corresponding dienyl bromide, also participates effectively in aqueous Diels–Alder reactions (Table 1.3) [24]. The reactivity of this diene is comparable to that of the diene carboxylates. One of the potential advantages of the monosodium phosphonate headgroup is its inherent ability to buffer the reaction medium when sensitive functionalities are involved. The Grieco group also explored the use of dienylammonium chlorides in aqueous Diels–Alder reactions (Table 1.4). Not surprisingly, the pentadienylammonium salt **2.34** was less reactive than the hexadienyl

Table 1.2 Reactions of diene carboxylates with dienophiles in water[a]

Entry	Diene carboxylate	Dienophile	Reaction time (h)	Product(s)	Yield (%)
1	**2.11** (n = 1) **2.14** (n = 2)	**2.12**	1 1.5	**2.13** (n = 1) **2.15** (n = 2)	73 86
2	**2.14**	**2.16**	5	**2.17** (18:1) **2.18**	67
3	**2.14**	**2.19**	9 (0°C)	**2.20** (15:1) **2.21**	51
4	**2.14**	**2.22**	24	**2.23**	60
5	**2.14**	**2.24**	2	**2.25**	94
6	**2.26**	**2.16**	1	**2.27** (17:1) **2.28**	73
7	$(CH_2)_3NHCO(CH_2)_2CO_2Na$ **2.29**	**2.16**	15	$(CH_2)_3NHCO(CH_2)_2CO_2H$ **2.30**	79

[a]All reactions employed a five-fold excess of diene (2.0 M) and, with the exception of entry 4, were performed at 25°C. With the exception of **2.25**, all cycloadducts were characterized as their methyl esters (CH_2N_2 treatment).
Taken from references [21] and [22].

Table 1.3 Diels–Alder reactions of diene phosphonates in water[a]

Entry	Diene phosphonate	Dienophile	Reaction time (h)	Product(s)	Yield (%)
1	2.31	2.12	1.5	2.32	89
2	2.31	2.16	6	2.33	75

[a]Both reactions employed a five-fold excess of diene (2.0 M) and were performed at 25°C. The products were characterized as their methyl phosphonates (CH_2N_2 treatment).
Taken from reference [24].

Table 1.4 Diels–Alder reactions of diene ammonium salts in water[a]

Entry	Diene carboxylate	Dienophile	Reaction time (h)	Product(s)	Yield (%)
1	2.34	2.12	32	2.35	98
2	2.36	2.12	2	2.37	99
3[b]	2.38	2.12	11	2.39	62

[a]All reactions employed a five-fold excess of diene (2.0 M) and were performed at 25°C.
[b]The crude reaction mixture was directly acetylated (Ac_2O, Et_3N).
Taken from reference [23].

homolog **2.36** owing to inductive deactivation of the diene moiety in the former case. An apparent limitation of these primary ammonium salts is their incompatibility with easily polymerizable dienophiles (acroleins, acrylates, etc.) owing to the acidic nature of the reaction medium. This was, in fact, the impetus for preparing the succinic acid conjugate (see Table 1.2, entry 7), which reacts with such dienophiles normally as its carboxylate salt **2.29**.

The work of Keana and co-workers must also be mentioned in the context of structural modification of the diene and dienophile components in aqueous Diels–Alder reactions [25]. They found that the surfactant dienes **2.40** and **2.43** react very cleanly with hydrophilic dienophiles **2.41** and **2.45** in water to give the corresponding adducts, which no longer exhibit detergent-like behavior:

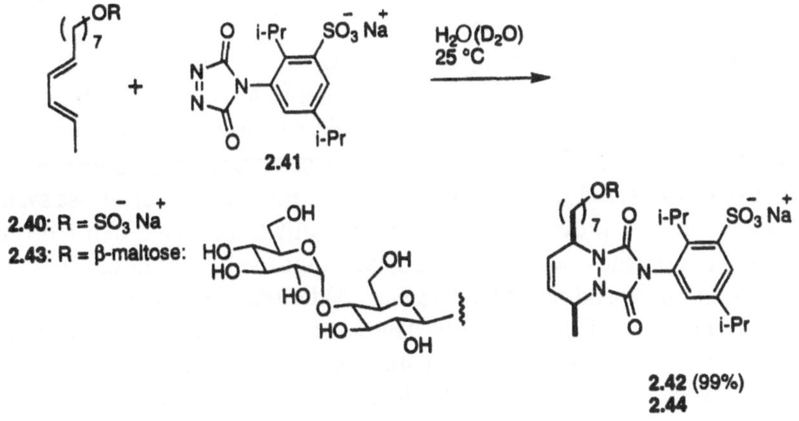

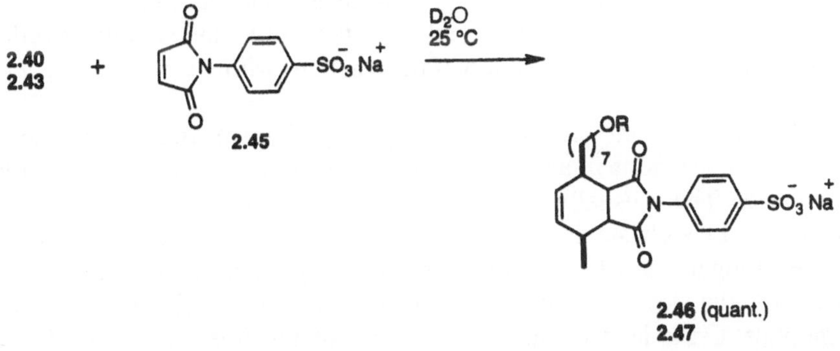

It was noted that the rates of these Diels–Alder reactions were qualitatively faster than analogous reactions in organic solvents. The use of the dienyl maltoside **2.43** in aqueous Diels–Alder reactions presaged Lubineau's dienyl glycosides, which will be discussed later in this chapter in the context of diastereoselective Diels–Alder reactions.

1.2.5   Miscellaneous effects

(a) 'Preorganizing' additives.   Breslow's original paper also reported the effects of cyclodextrin additives on the rates of these aqueous Diels–Alder reactions. The basis for these experiments was the hypothesis that the activated complex of diene and dienophile for reactions such as [2.1 + 2.2] and [2.1 + 2.5] should fit into the hydrophobic cavity of β-cyclodextrin (cycloheptaamylose) and that this would result in a rate enhancement. Indeed, β-cyclodextrin led to an increase in the rate of the first reaction equal to that observed with the prohydrophobic salt LiCl. A β-cyclodextrin-mediated rate acceleration was also observed for the second reaction but not for the reaction [2.8 + 2.9] since the diene 2.8 is too large for the cyclodextrin cavity. The fact that all three of these Diels–Alder reactions were decelerated by α-cyclodextrin (cycloheptaamylose) – whose smaller cavity cannot accommodate even the activated complexes of the first two reactions – was taken as further support for their hypothesis.

Sangwan and Schneider have studied the effect of cyclodextrins on a number of aqueous Diels–Alder reactions between acrylate, fumarate and maleate derivatives of varying hydrophobicities and (mainly) cyclopentadiene [26]. No simple correlation between substrate hydrophobicity and cyclodextrin-catalyzed rate enhancement was found. However, those systems that did respond to β-cyclodextrin catalysis exhibited enzyme-like saturation kinetics. This led these workers to conclude that the hydrophobic effect can, in fact, be counterproductive to the Diels–Alder reaction if it leads to unproductive orientation of the reactants. The same can be said about the effect of amphiphiles (detergents capable of micellization) on aqueous Diels–Alder reactions since sodium dodecylsulfate (SDS) decelerated the reaction between cyclopentadiene and methyl acrylate. Those cases in the literature claiming micellar catalysis of the aqueous Diels–Alder reaction may simply be benefiting from the solubilizing effect of the amphiphilic additives rather than any *bona fide* preorganization of the reactants within a micelle [27, 28].

(b) Lewis-acid catalysis.   It has long been known that conventional Diels–Alder reactions in aprotic organic solvents are subject to catalysis by Lewis acids. This is usually accomplished by complexation of a cationic metal species to a Lewis basic site on an activating group attached to the dienophile. The resulting increase in rate and endo-selectivity can be attributed to differences in the lowest unoccupied molecular orbital (LUMO) of the dienophile–Lewis acid complex versus that of the free dienophile. If one considers the rate-accelerating effect of hydrogen bonding on the aqueous Diels–Alder reaction to be the result of a Brønsted acid–base interaction, then Lewis-acid catalysis in water should be possible. However, in order to achieve effective Lewis-acid catalysis in water: (i) the change in free energy upon metal complexation with the substrate must compensate for the (energetically

unfavorable) dehydration events preceding it; (ii) Lewis-acid-catalyzed side-reactions with water must be minimized; and (iii) the Lewis acid should dissociate once the Diels–Alder reaction has occurred to allow catalyst turnover.

Otto and Engberts recently reported catalysis of the Diels–Alder reaction between cyclopentadiene (**2.1**) and the enone **2.48** by Lewis acids in water [29]:

reaction medium	k_2 (M^{-1}s^{-1})
acetonitrile	1.32×10^{-5}
water	4.02×10^{-3}
0.010 M Cu(NO$_3$)$_2$ in water	3.25

Thus, the reaction in aqueous Cu(NO$_3$)$_2$ solution proceeded about 800 times faster than in water alone and 250 000 times faster than in acetonitrile. This Lewis-acid-catalyzed aqueous Diels–Alder reaction presumably occurs via a transition state such as **2.50** with bidentate complexation to the metal as shown. In a related study, Kobayashi reported that scandium triflate can be used as a water-tolerant Lewis-acid catalyst for a Diels–Alder reaction in (9:1) tetrahydrofuran–water (THF–H$_2$O) [30], though this mixed solvent system cannot take advantage of the hydrophobic effect observed in pure water or highly aqueous mixtures.

(c) 'Water-like' reaction media.   Although the focus of this chapter is the Diels–Alder reaction in aqueous media, it is instructive to round out this section with some discussion of Diels–Alder chemistry in highly polar, 'water-like' solvent systems. Thus, Breslow and Guo reported that reaction between **2.1** and **2.5** proceeds about four times faster in formamide and six times faster in ethylene glycol than it does in methanol [31]. Quantitatively similar results were reported by Liotta and co-workers for Diels–Alder reactions in ethylene glycol (a six-fold rate enhancement over reaction in methanol) [32]. The source of these rate accelerations in water-like solvents was proposed to be due to a 'solvophobic effect' analogous to (though of smaller magnitude than) the hydrophobic effect seen in water. Breslow also found that β-cyclodextrin produces an additional rate acceleration and kinetic saturation in these 'water-like' solvents much like the corresponding

cyclodextrin-catalyzed reaction in water. Additional evidence for this idea came from the fact that 'salting-out' additives such as LiCl also led to a rate increase in 'water-like' solvents. However, traditional 'antihydrophobic' additives like urea also speeded up rather than retarded the Diels–Alder reaction in these 'water-like' solvents. Of all the additives tested, only quaternary ammonium salts slowed the reaction down – presumably by acting as 'pseudo-detergents'. Solvophobic effects may also be responsible for the Diels–Alder rate enhancements reported by Jaeger in liquid ethylammonium nitrate (m.p. ca. 12°C) + lithium iodide [33] and by Smith in buffered aqueous ethanol [34].

1.3 Regioselectivity and endo/exo-selectivity

1.3.1   Control of regiochemistry

The regiochemical outcome of aqueous Diels–Alder reactions is generally under frontier molecular orbital (FMO) control, paralleling the Diels–Alder regioselectivity observed in organic solvents. Exceptions to this rule can arise due to unique structural features that respond differently to aqueous versus organic media. For example, Casetta et al. reported hydrophobic control of regioselectivity in the reaction of juglone (3.1) and the unsymmetrical diene 3.2 [35] (see Scheme 1.2). When the reaction is performed in water, the ratio of 3.4 to 3.3 is 2.5/1, a considerable step down from the ratio of 19/1 reported for the reaction in diethyl ether [36]. This is consistent with water disrupting the intramolecular hydrogen bond that activates the C(4) carbonyl in 3.1b

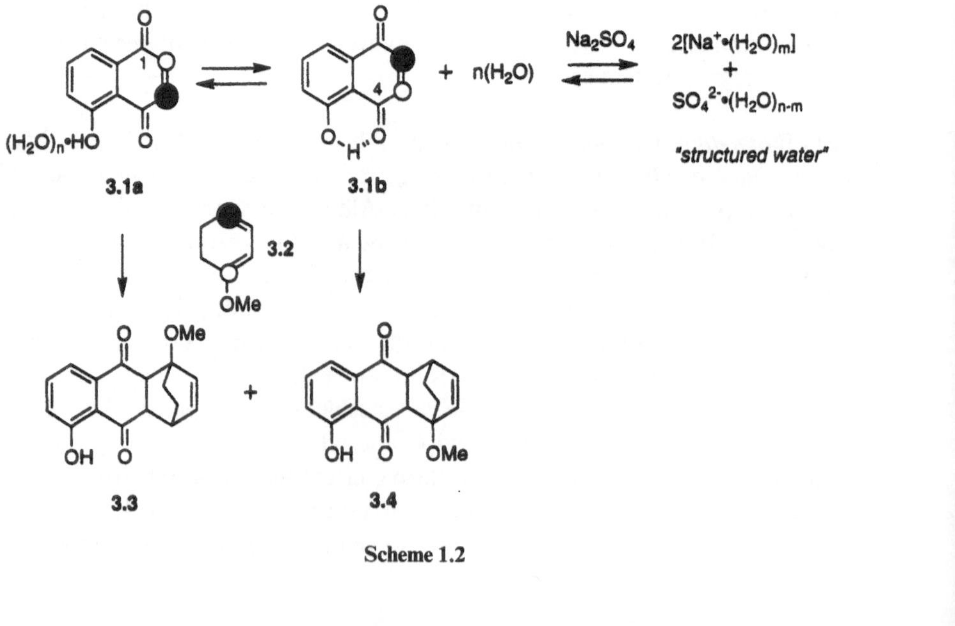

Scheme 1.2

leading to a hydrated species **3.1a**, in which the C(4) carbonyl is now deactivated (it is a vinylogous acid) relative to the C(1) ketone. In the presence of prohydrophobic ('structure-making') salts, the equilibrium is shifted back toward **3.1b**, resulting in a 7.4/1 ratio of regioisomers **3.4** and **3.3**.

A successful approach to regiocontrol in the aqueous Diels–Alder reaction using cyclodextrins was reported by Chung and Wang [37]. They found that, while the cycloaddition of 2-methyl-1,4-benzoquinone (**3.5**) and (*E*)-1,3-pentadiene (**3.6**) proceeded about an order of magnitude faster in water than in acetone, essentially the same modest level (ca. 2/1) of regioselection was observed in both:

conditions	rxn time	3.7/3.8	yield
acetone	48 h	64/36	52%
water	6 h	66/34	73%
β-CD, water	6 h	83/17	82%

Significantly, when this same aqueous Diels–Alder reaction was run in the presence of β-cyclodextrin, the regioselectivity favoring the '*ortho*' adduct **3.7** increased to ca. 5/1. An even more dramatic effect is seen with **3.5** and 2-methyl-1,3-butadiene (**3.9**). With this less reactive diene, the cycloadditions in both acetone and water are inefficient (low yielding) and relatively nonselective ('*para*'/'*meta*' ca. 1/1):

conditions	rxn time	3.10/3.11	yield
acetone	11 d	53/47	23%
water	24 h	56/44	30%
β-CD, water	12 h	14/86	86%

However, a reversal of regioselectivity, now favoring the '*meta*' adduct

3.11, was observed when the aqueous reaction was conducted in the presence of β-cyclodextrin. These examples of cyclodextrin-mediated regiocontrol are likely related to a favored transition-state geometry (a preorganizational effect) within the hydrophobic binding cavity.

Jaeger and co-workers have reported the results of experiments which demonstrate that expected orientational effects in micelles or related aggregates cannot always be counted upon to override the intrinsic FMO regiocontrol of the Diels–Alder reaction [38]. Thus, the surfactant diene **3.12** reacted with a four-fold excess of (non-orientational) dienophile **3.13a** (R = H) in water to give a mixture of *ortho* adducts **3.14a/3.15a** in 93% combined yield:

R	yield	ortho/meta
H	93%	100/0
C₆H₁₃	56%	91/−9

None of the regioisomeric *meta* adduct corresponding to **3.16a** was detected in the crude reaction mixture. When the reaction was repeated with the surfactant dienophile **3.13b** (R = C₆H₁₃), less than 10% of the product was (tentatively) identified as a mixture of the regioisomeric *meta* adducts **3.16b**. Since alignment of the hydrophobic 'tails' of **3.12** and **3.13b** in a micelle or aggregate would be expected to preorganize the reactants for *meta*-cycloaddition, it was concluded that such orientational effects were not strong enough to overcome FMO-dictated regiocontrol.

1.3.2   Endo- versus exo-addition

As with regioselectivity, the issue of endo- versus exo-selectivity in aqueous Diels–Alder reactions (simple diastereoselectivity) is also under FMO control. However, the unique solvent properties of water lead to an enhancement of endo-selectivity that goes beyond well-known solvent polarity effects. For example, Breslow's group looked at the endo/exo product ratios for the reaction of cyclopentadiene and several dienophiles (Table 1.5) [39,

40]. A dramatic increase in the endo/exo ratios (as well as reaction rates – see section 1.2) was observed for all four dienophiles studied on going from organic to aqueous reaction media (entries 1–3 in Table 1.5). The selectivity enhancements in entry 3 must be ascribed to the water, since these two-phase reactions did not exhibit the same selectivity (and rate) profiles as the reactions in neat cyclopentadiene (entry 1). It was suggested that the preference for endo attack in water was the result of a polar medium effect plus a hydrophobic effect that favored the more compact endo transition state.

Evidence implicating the hydrophobic effect came from a series of experiments with 2-butenone and cyclopentadiene at concentrations that resulted in true solutions (entries 4–7 in Table 1.5). That the endo-selectivity was essentially independent of diene concentration argued against micellar effects and/or aggregation, while the increased selectivity in the presence of the prohydrophobic additive LiCl and decreased selectivity in the presence of the antihydrophobic agent guanidinium chloride (GnCl) suggested that the hydrophobic effect was responsible for the selectivity of these aqueous Diels–Alder reactions.

Breslow also noted that the presence of detergents such as sodium dodecylsulfate and cetyltrimethylammonium bromide (data not shown) had little effect on the product ratios of these aqueous Diels–Alder reactions. These results suggest that the increased endo/exo ratios and rates measured for a series of similar Diels–Alder reactions in aqueous detergent solutions by Sauer and co-workers may not be due to micellar effects as proposed [41]. Further support for Breslow's 'hydrophobic solvent effect' hypothesis came from a study of Diels–Alder reactions of cyclopentadiene and a series of maleate and fumarate derivatives by Schneider and Sangwan [42]. They found a better correlation between the endo/exo product ratios for these reactions and solvophobicity parameters for various solvent systems as opposed to polarity effects alone. Subsequently, Cativiela et al. noted that a dual model (solvophobicity and polarity effects) can satisfactorily account

Table 1.5 Endo/exo product ratios in organic and aqueous media

Entry	Reaction medium	Reactant concentrations	Butenone	Methyl acrylate	Dimethyl maleate	Methyl methacrylate
1	cyclopentadiene	excess diene	3.85	2.9	2.8	0.43
2	ethanol	0.15 M each	8.5	5.2	4.5	0.6
3	water	0.15 M each	21.4	9.3	13.7	1.4
4	water	0.010 M each	24.0 ± 0.3			
5	water	0.001 M in diene 0.010 M in dienophile	25.0 ± 0.3			
6	water + lithium chloride	0.001 M in diene 0.010 M in dienophile	28.0 ± 0.4			
7	water + guanidium chloride	0.001 M in diene 0.010 M in dienophile	22.0 ± 0.8			

Adapted from references [39] and [40].

for the endo/exo selectivity (and rate) of intermolecular Diels–Alder reactions [43].

Grieco's group reported the effects of water on the reaction of diene **3.18** and α-formylenone **3.19** in their first publication on aqueous Diels–Alder chemistry (Table 1.6) [44]:

When 'conventional' Diels–Alder reaction conditions were employed (entries 1 and 2), a chromatographically inseparable mixture of stereoisomers **3.20/3.21** (R = Et) was obtained. Unfortunately, only **3.20** could be used for their proposed synthesis of glaucarubinone. This posed an interesting stereo-control problem since the formation of both products involved secondary orbital (endo) overlap of the diene with an activating group – the former with the ketone and the latter with the aldehyde.

Aqueous Diels–Alder chemistry provided the solution to this dilemma. In the end, reaction of **3.18** with diene carboxylate **3.19** (R = Na) in water (2.0 M in diene) was found to give the best ratio (diastereoselectivity ds = 3/1) of **3.20/3.21** in the shortest reaction time when compared to the analogous reaction with **3.19** (R = Et) in organic media. It was speculated that this

Table 1.6 Reaction of dienophile **3.18** with diene **3.19**[a]

Entry	R	Solvent	Diene concentration	Reaction time (h)	Combined yield (%)	3.20/3.21
1	Et	benzene	1.0 M	288	52[b]	0.85
2	Et	neat	—	144	69	1.3
3	Et	water	1.0 M	168	82	1.3
4	Na	water	0.1 M	120	46[b]	0.9
5	Na	water	1.0 M	8	83	2.0
6	Na	water	2.0 M	5	100	3.0

[a]All reactions were run at room temperature employing a five-fold excess of diene over dienophile.
[b]Starting dienophile was recovered: entry 1 (29%) and entry 4 (14%).
Adapted from reference [20].

increase in the product ratio **3.20/3.21** was due to differences in transition-state (TS) volumes. Thus, **3.20** would have to be formed via a TS with a smaller surface area, minimizing unfavorable hydrophobic interactions relative to the competing transition state leading to **3.21**. While this explanation is reasonable, the dependence of the stereoselectivity on the concentration of diene carboxylate (entries 3-6 in Table 1.6) suggests that aggregation may also be involved (compare Breslow's results above).

While enhanced endo-selectivity is the norm for aqueous Diels–Alder reactions, there are notable exceptions. One involves the reaction of dimethylpentafulvene (**3.22**) and 1,4-benzoquinone (**3.23**) to give a mixture of the endo and exo adducts **3.24** and **3.25**:

concentration of reactants	reaction time	3.24/3.25
0.001 M	24 h	14/86
0.019 M	16 h	23/77
0.090 M	12 h	40/60
0.472 M	8 h	76/24
1.600 M	8 h	88/12

This reaction exhibits the usual rate acceleration in water. Griesbeck reported that the endo/exo ratio of this reaction varies as a function of reactant concentration [45], with formation of the exo product **3.25** being favored at low concentrations of diene/dienophile and the endo product **3.26** being favored by high concentrations. Enhanced exo-selectivities have also been reported for aqueous Diels–Alder reactions run in the presence of baker's yeast [46].

1.4   Diastereofacial selectivity

In light of the pivotal role that asymmetric cycloadditions have played in many notable synthetic endeavors over the years, it is not surprising that this important aspect of aqueous Diels–Alder chemistry is of interest. Diastereofacial selectivity can be achieved by incorporating stereochemistry into either the diene **4.1** (chiral R^1) or dienophile **4.2** (chiral R^2) [47].

The degree of (kinetic) stereocontrol or asymmetric induction will then depend on the differential energies of the diastereomeric transition states corresponding to **4.3**. We will restrict our discussion here to intermolecular Diels–Alder reactions. The stereochemical aspects of intramolecular Diels–Alder reactions will be discussed separately. Even though asymmetric Diels–Alder reactions using chiral catalysts are known [48], an effective aqueous version of this reaction has not yet been developed.

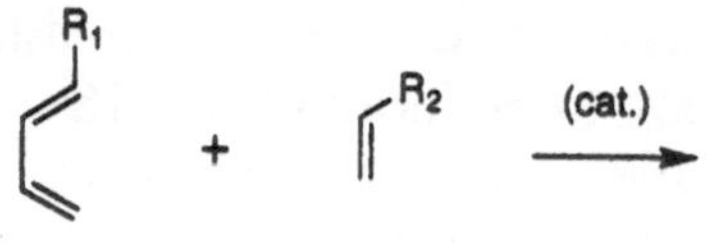

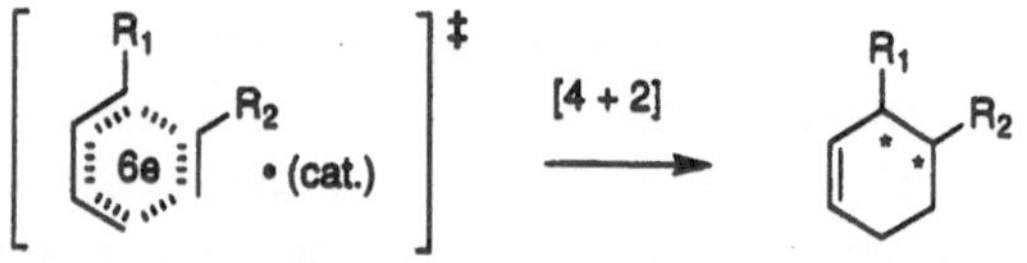

1.4.1 Chiral dienophiles

As noted already in the previous section, diene carboxylate **4.6** adds exclusively to the less-hindered α-face of the bicyclic dienophile **4.5** to give a mixture of cycloadducts **4.7**:

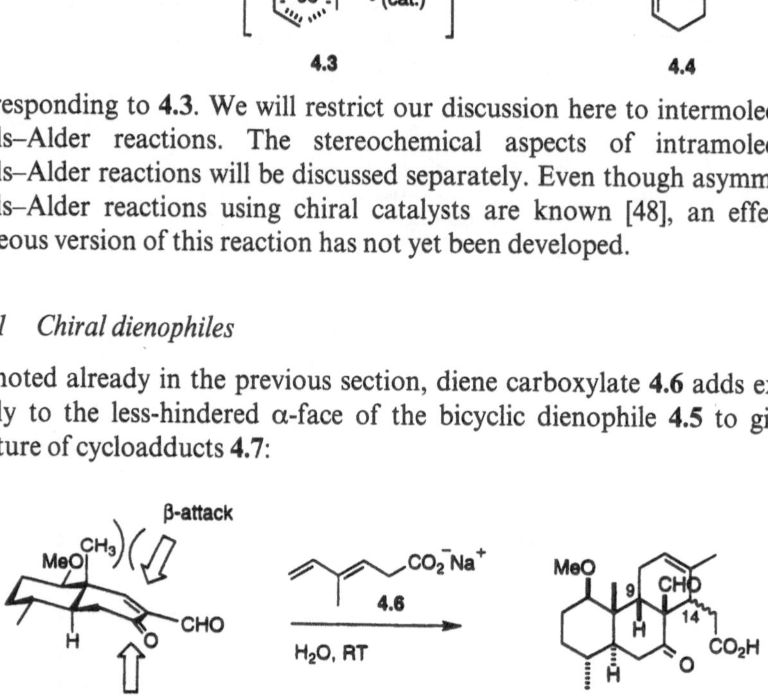

The axial methyl group at C(10) hinders the approach of the diene to the β-face of the rigid bicyclic dienophile **4.5**. While both the rate and endo/exo (14α/14β) ratio are increased when this reaction was performed in water rather than organic media, the observed diastereofacial selectivity appears to derive solely from the dienophile structure since the same α-selectivity was observed when the reaction was performed in benzene.

More recently, Zwanenburg and co-workers have reported the use of 4-hydroxycyclopent-2-en-1-one and its derivatives (cf. **4.9**) as chiral synthetic equivalents of cyclopentadienone in asymmetric Diels–Alder reactions [49].

The need for a synthetic equivalent in this instance arises from: (i) the fact that the parent cyclopentadienone is achiral; and (ii) the well-known tendency of cyclopentadienone to dimerize rather than undergo 'crossed' Diels–Alder reactions with other dienes. Thus, it was found that **4.8** adds to

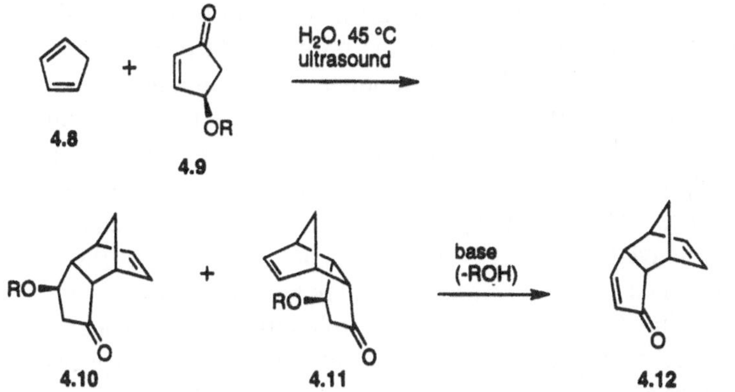

R = H, 82% ee
R = Ac, 63% ee
R = Bz, 67% ee

the cyclopentenone **4.9** to give a mixture of diastereomeric endo adducts **4.10** and **4.11** in good combined yield. Treatment of this mixture with base results in elimination (revealing the synthetic equivalence of **4.9** and cyclopentadienone) to give compound **4.12** as a mixture of enantiomers whose ratio parallels that of **4.10/4.11**. It is of interest to note that the use of water as a solvent favors the formation of the *anti*-diastereomer **4.10**, with the highest ratio (*ds* = 10/1) occurring with the parent 4-hydroxycyclopent-2-en-1-one (**4.9a**). Apparently, solvation of the hydroxyl group (and to some extent the esters in **4.9b** and **4.9c**) increases the effective steric bulk of this group and overrides the competing Cieplak stereoelectronic effect [50] which favors the *syn*-diastereomer **4.11** [51].

Of course, the use of water as a solvent need not have a beneficial effect on the diastereoselectivity of a Diels–Alder reaction whatsoever. This was, in fact, the case with the addition of cyclopentadiene to *cis*-3-phenylsulfonyl-prop-2-enoic acid (**4.13**) [52]:

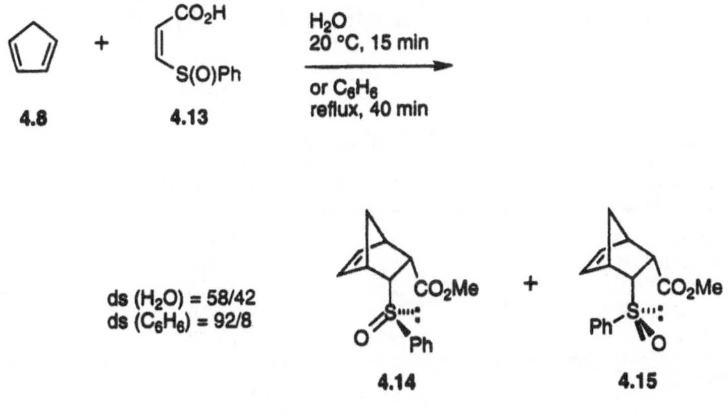

ds (H₂O) = 58/42
ds (C₆H₆) = 92/8

While the diastereoselectivity of this reaction in benzene was quite good in favor of adduct **4.14** (*ds* = 98/2), the ratio of **4.14** to **4.15** dropped off to 58/42 in water. A plausible explanation for this outcome may involve the effect of sulfoxide hydration on the accessibility of the dienophile faces. If one assumes that the favored sulfoxide rotamer would place the lone pair (the 'smallest' sulfur substituent) in the same plane as the carboxylic acid, then cyclopentadiene would approach from the least-hindered *re* face *anti* to the large phenyl group. However, in water one would expect the sulfoxide to be extensively hydrated, effectively equalizing the steric encumbrance of each dienophile face. Support for this explanation comes from the observation that the analogous Diels–Alder reaction with *trans*-3-phenylsulfonyl-prop-2-enoic acid (**4.16**), which is not expected to have an overwhelming rotamer preference, does not exhibit good selectivity in either benzene or water:

ds (H$_2$O) = 9/32/20/30
ds (C$_6$H$_6$) = 12/36/15/37

Waldmann has looked at the possibility of using proline esters as chiral auxiliaries for aqueous Diels–Alder reactions (Table 1.7) [53]:

Table 1.7 Reaction of cyclopentadiene with chiral acrylamides

R	Solvent	T (°C)	Yield (%)	*ds* (**4.22/4.23**)	endo/exo
Bn	toluene	0	27	81/19	70/30
Bn	H$_2$O/EtOH 4/1	−10	52	81/19	60/40
All	toluene	0	22	82/18	57/43
All	H$_2$O/EtOH 4/1	−5	43	64/36	78/22
All	H$_2$O	25	95	68/32	68/37

From reference [53].

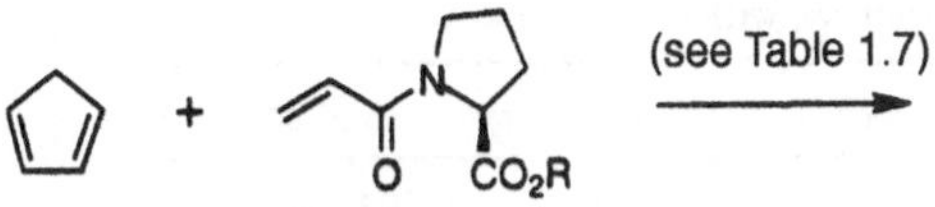

4.8   4.21   (see Table 1.7)

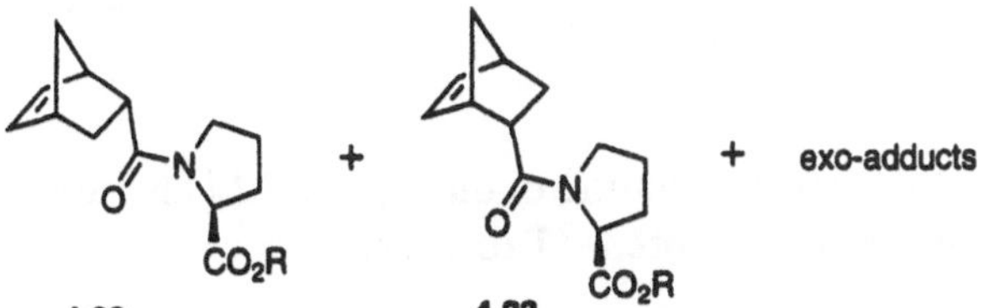

4.22   4.23   + exo-adducts

Although the use of water as a solvent for the reaction between cyclopenta-diene (**4.8**) and chiral acrylamide **4.21** led to the usual rate enhancement, it did not have a dramatic effect on either the endo/exo ratio or diastereoselectivity. In retrospect, the modest diastereoselectivity is not really surprising since the chiral center is quite remote from the olefin in the preferred *trans*-amide conformation.

The situation improves considerably when the bis-substituted fumaric acid derivative **4.24** is used as the chiral dienophile (Table 1.8):

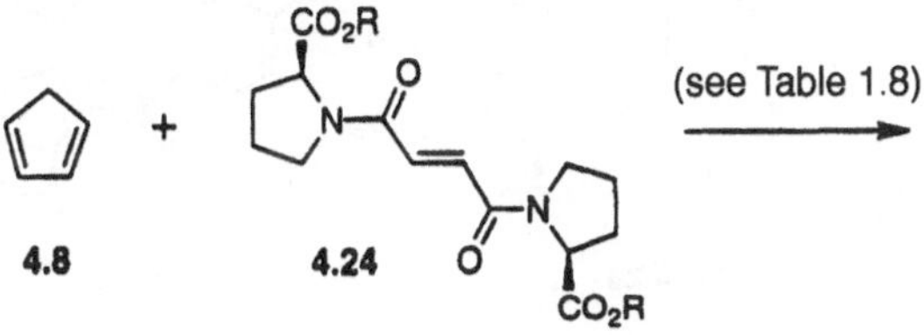

4.8   4.24   (see Table 1.8)

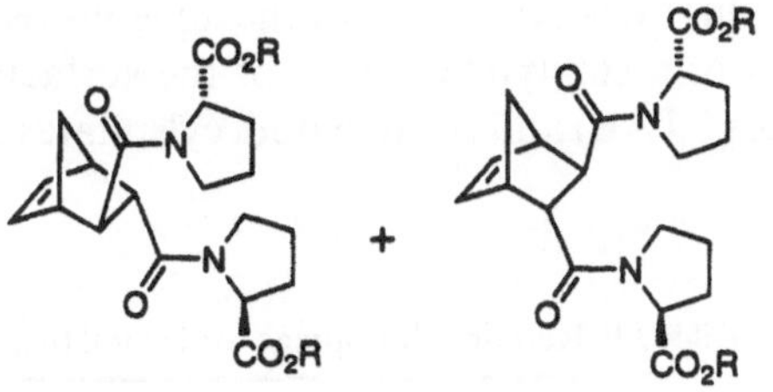

4.25   4.26

Here, there is no endo/exo problem and the (modest) diastereoselectivity associated with the proline auxiliary is amplified through the principle of double asymmetric induction [54]. This accounts fairly well for the diastereo-selectivity enhancements observed for both the organic and aqueous Diels–Alder additions to cyclopentadiene presented in Tables 1.7 and 1.8.

Table 1.8 Reaction of cyclopentadiene with chiral fumarates

R	Solvent	T (°C)	Yield (%)	ds (**4.25/4.26**)
Bn	toluene	0	100	93.5/6.5
Bn	H$_2$O/EtOH 4/1	−10	100	96/4
All	toluene	−30	91	97/3
All	H$_2$O/EtOH 4/1	−10	80	89/11

From reference [53].

The bis-activated nature of dienophile **4.24** also allows the use of less-reactive dienes such as isoprene (Table 1.9):

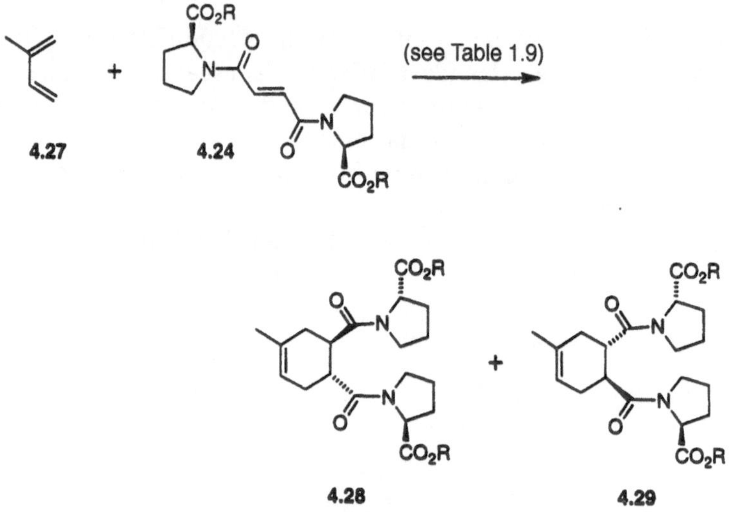

With this diene, the yields of the aqueous reactions benefited greatly by the addition of a catalytic amount of the surfactants cetyltrimethylammonium bromide (CTAB) and sodium dodecylsulfate (SDS).

Table 1.9 Reaction of isoprene with chiral fumarates

R	Solvent	T (°C)	yield (%)	ds (**4.28/4.29**)
Bn	toluene	25	24	81/19
Bn	H$_2$O/EtOH 4/1	25	8	82/18
Bn	H$_2$O/EtOH 4/1 + cat. CTAB	25	48	80/20
Bn	H$_2$O/EtOH 4/1 + cat. SDS	25	46	82/18

From reference [53].

1.4.2 Chiral dienes

In the examples cited so far, the controlling chirality has been localized on the dienophile component. It is also possible to construct chiral dienes that are capable of inducing asymmetry in intermolecular aqueous Diels–Alder reactions. For example, the chiral cyclic diene **4.30** – readily obtained via enzymatic oxidation of 7-phenylnorbornadiene – has been shown to be a good partner for asymmetric aqueous Diels–Alder reactions [55]. Thus, diene **4.30** reacts with dimethylacetylene dicarboxylate (**4.31**), N-ethylmaleimide (**4.33**) and nitrosobenzene (**4.36**) to give the cycloadducts **4.32**, **4.34 + 4.35** and **4.37** respectively (Scheme 1.3). However, it is curious that dienophiles **4.31** and **4.36** add to **4.30** with complete diastereofacial control, while the reaction with **4.33** is reported to give a diastereoselectivity of 1.6/1 (a Cieplak effect on the diene highest occupied molecular orbital (HOMO) is not expected).

Asymmetric induction is also possible with chiral acyclic dienes. Thus, Grieco and co-workers have shown that methacrolein (**4.38**) reacts with the

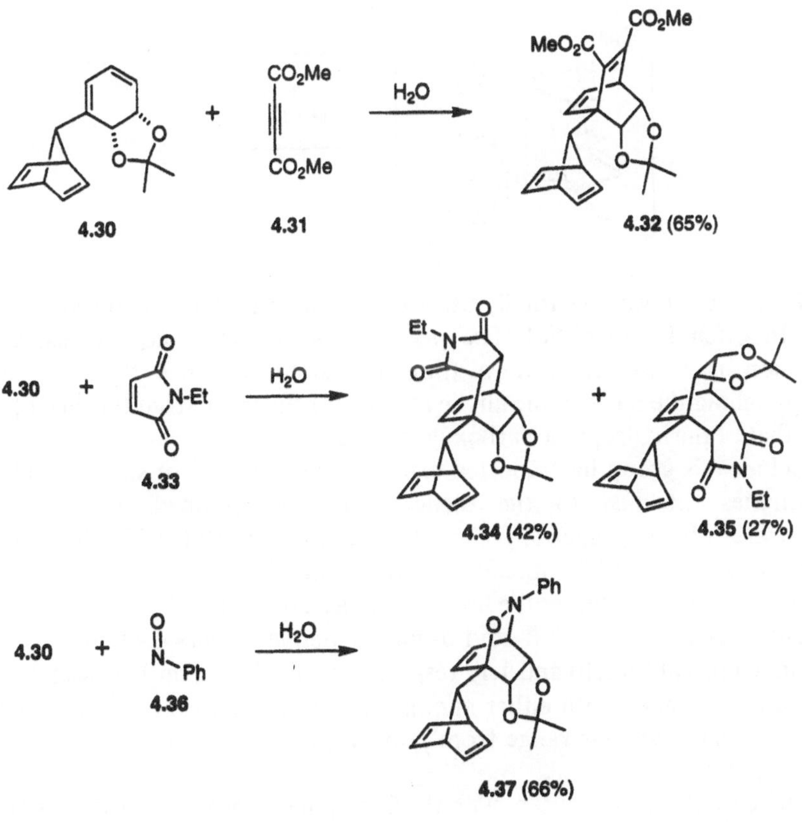

Scheme 1.3

chiral diene carboxylate **4.39** in water to give the endo-*si* adduct **4.40** and endo-*re* adduct **4.41** in a ratio of 4.7/1 [56]:

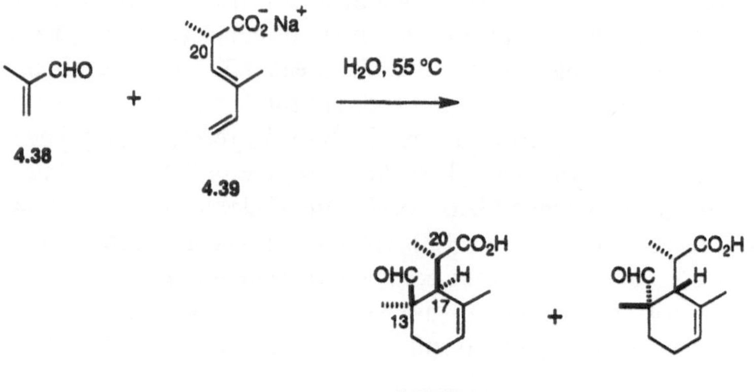

4.40 (70%)   **4.41** (15%)

The proposed transition-state conformation of [**4.38** + **4.39**] leading to cycloadduct **4.40** is shown below:

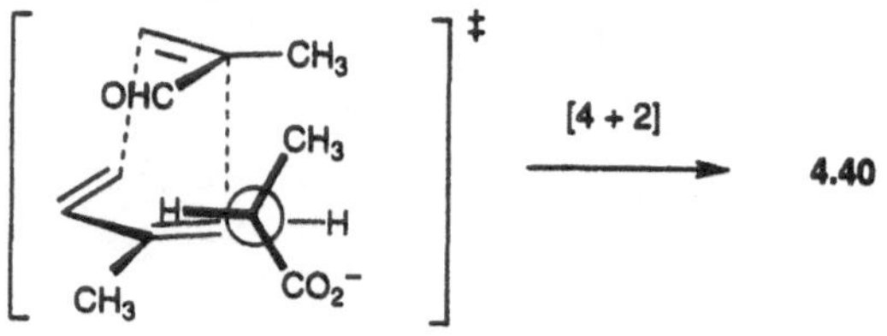

This reaction was eventually used to effect an efficient formal synthesis of the Inhoffen–Lythgoe diol. The *si*-selectivity of this aqueous Diels–Alder reaction is consistent with a transition state that has the dienophile approaching the most stable allylic rotamer of the diene from the face opposite that of the hydrophilic carboxylate group.

Lubineau's group has reported extensively on the use of sugars as chiral auxiliaries attached to the diene component (termed 'glyco-organic substrates') for asymmetric Diels–Alder reactions in water. The incorporation of such hydrophilic moieties confers some degree of amphiphilicity to the diene, which facilitates the use of water as a solvent.

Their results with both β- and α-glucose-derived dienes **4.42** and **4.45** are compiled in Tables 1.10 and 1.11 respectively [57]. While the endo-selectivity is reasonably good with either anomer, the diastereoselectivities associated with the endo adducts range from poor to good, depending on the substitution pattern.

These results are consistent with the dienophile approaching the *s-cis* diene from its least-hindered face, the latter being determined by the conformation

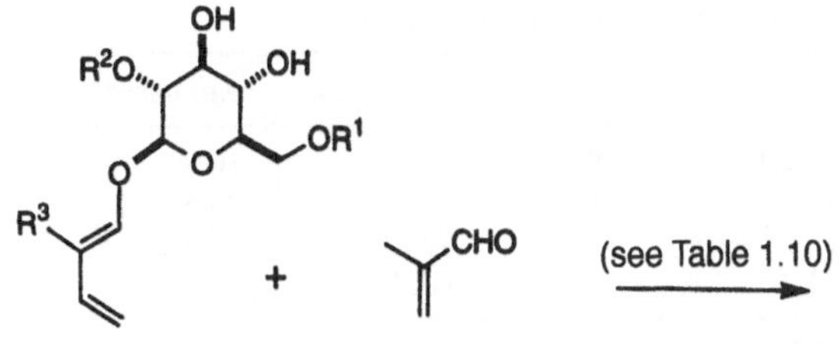

4.42   **4.38**   (see Table 1.10)

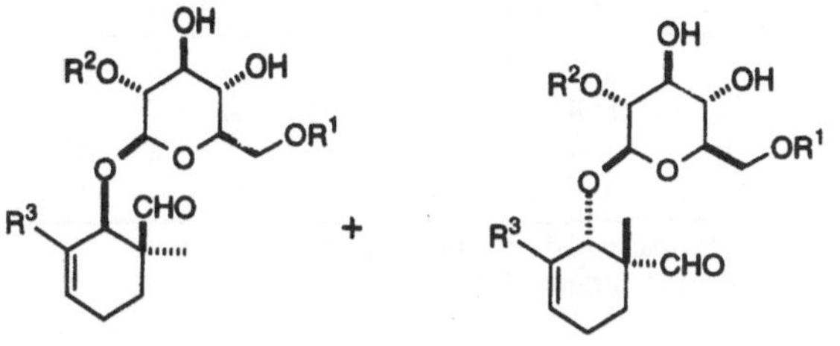

4.43   +   **4.44**

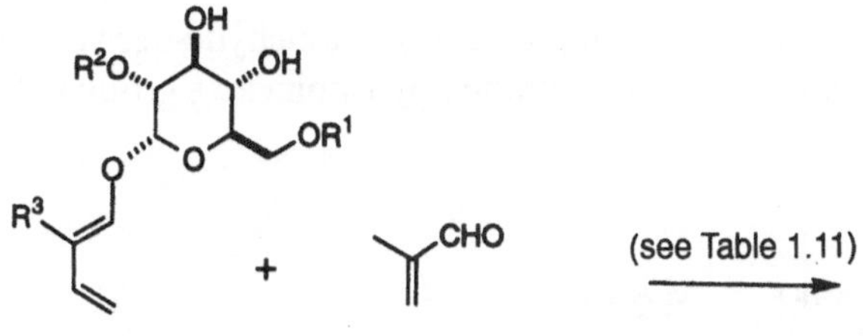

4.45   **4.38**   (see Table 1.11)

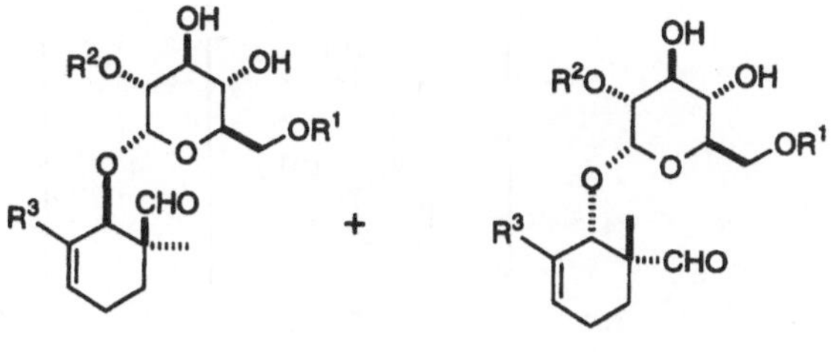

4.46   **4.47**

of the glycosidic linkage and sugar protection pattern. The glycoside conformation is governed by well-known anomeric and steric effects [58]. So while the parent (unprotected) auxiliary shows very little facial discrimination, 6-*O*-benzylation blocks the *si* face in the β series and the *re* face in the α series. Accordingly, 2-*O*-benzylation reverses the selectivity by hindering the dienophile approach to the opposite face of the diene. The effect of substi-

Table 1.10 Reaction of β-glycosyl dienes and methacrolein

Entry	Diene series	T (°C)	rel/si (endo)[a]	endo/exo[a]
1	a: $R^1 = R^2 = R^3 = H$	20	60/40	100/0
2	b: $R^1 = Bn, R^2 = R^3 = H$	40	69/31	95/5
3	c: $R^1 = R^3 = H, R^2 = Bn$	40	34/66	97/3
4	d: $R^1 = R^2 = H, R^3 = Me$	20	69/31	96/14

[a]Ratios were determined after peracetylation.
From reference [57].

Table 1.11 Reaction of α-glycosyl dienes and methacrolein

Entry	Diene series	T (°C)	rel/si (endo)[a]	endo/exo[a]
1	a: $R^1 = R^2 = R^3 = H$	20	36/64	93/7
2	b: $R^1 = Bn, R^2 = R^3 = H$	40	28/72	95/5
3	c: $R^1 = R^3 = H, R^2 = Bn$	40	57/43	93/7
4	d: $R^1 = R^2 = H, R^3 = Me$	20	18/82	95/5

[a]Ratios were determined after peracetylation.
From reference [57].

tuting a methyl group at the 2-position of the diene may simply favor the extended conformation shown.

The reaction of the structurally simpler D-glyceraldehyde-derived diene **4.48** and acrolein (**4.49**) has also been examined by Lubineau's group [59]:

The proposed transition-state conformation of [**4.48** + **4.49**] leading to cycloadduct **4.50** is shown below:

Here, the initially formed endo adducts **4.50** and **4.51** spontaneously cyclize to the lactols **4.52** and **4.53**. No exo adducts were obtained in this Diels–Alder reaction. While the observed diastereoselection is only modest in this case, it is better than that obtained in organic solvents and is consistent with a Felkin–Anh type of transition-state conformation.

1.5 Synthetic applications

The true measure of a new synthetic method or technique is its applicability to specific problems and the extent of its adoption by the organic synthesis community at large. Accordingly, not long after the first report of hydrophobic acceleration of the Diels–Alder reaction by the Breslow group, applications of this reaction to synthesis began. This is not so surprising when one considers how important a place the Diels–Alder reaction occupies in the chemist's arsenal of synthetic transformations. In this section, applications of the aqueous Diels–Alder reaction to specific synthetic targets will be highlighted. The focus will be on the aqueous Diels–Alder reaction itself and how it relates to and benefits the overall synthetic strategy. The reader should consult the original references for specific details concerning the other steps in each synthesis.

1.5.1 Terpenes

In an early application, the aqueous Diels–Alder reaction of diene carboxylates was used to effect a very concise synthesis of the vernolepin precursor **5.2** [60]. This compound had previously been prepared by Schlessinger via another route in 11 steps from ethyl crotonate [61]. Reaction of the substituted methacrolein **5.3** with sodium (*E*)-3,5-hexadienoate (**5.4**) in water proceeded at 50°C to give the Diels–Alder endo adduct **5.5** along with a minor amount of the exo adduct (not shown) quantitatively in a ratio of 10/1 (Scheme 1.4). Sodium borohydride was added directly to this reaction mixture to produce the δ-hydroxycarboxylate **5.6**, which lactonized spontaneously upon acidic workup. This one-pot operation resulted in the isolation of vernolepin AB-ring

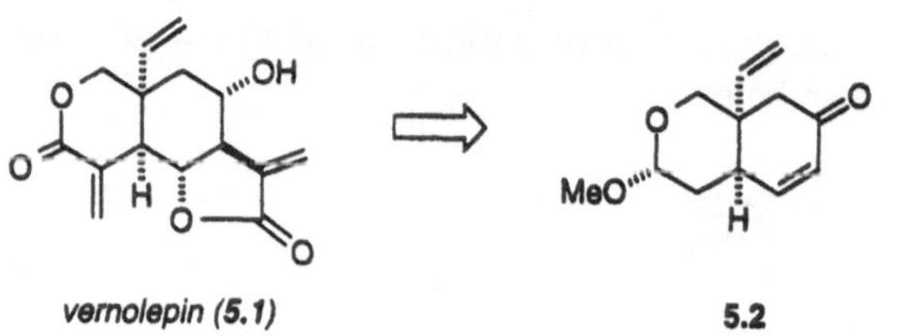

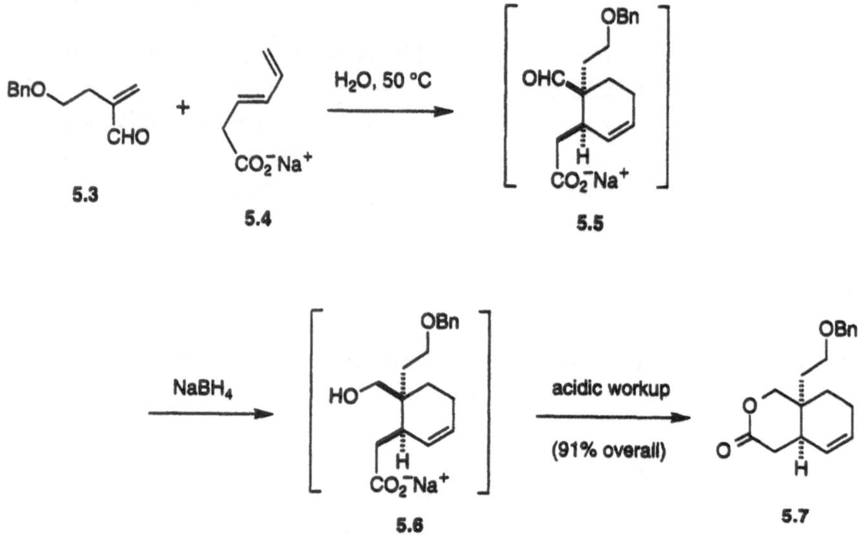

vernolepin (5.1)

5.2

5.3 + 5.4 → H₂O, 50 °C → 5.5

5.6 → acidic workup (91% overall) → 5.7

Scheme 1.4

precursor **5.7** (91% overall yield), which was then converted to the Schlessinger intermediate **5.2** using a standard five-step sequence. The aqueous Diels–Alder reaction of diene carboxylate **5.4** was about an order of magnitude faster than using its organic counterpart methyl (*E*)-3,5-hexadienoate.

The aqueous Diels–Alder reaction was also used to effect a short synthesis of *dl*-pyroangolensolide (**5.9**) and *dl*-epi-pyroangolensolide (**5.15**) [62]. Sesquiterpene **5.9** had been obtained from pyrolysis of methyl angolensate (**5.8**) as well as hydroiodic acid treatment of calodendrolide (a C(15)-degraded limonoid). This synthesis began with the reaction of diene carboxylate **5.10** with methacrolein (**5.11**) in water to give, after treatment of the initially formed cycloadduct with diazomethane, compound **5.12** in very good overall yield (Scheme 1.5). Low-temperature addition of 3-furyllithium to aldehyde **5.12** led to a mixture of diastereomers favoring the 'Cram product' **5.13** over the required diastereomer **5.14**. Each of these compounds was subjected to a standard α-selenation/oxidation/elimination protocol to afford *dl*-epi-pyroangolensolide (**5.15**) and *dl*-pyroangolensolide (**5.9**) respectively. The structure of the former compound was confirmed by X-ray crystallography.

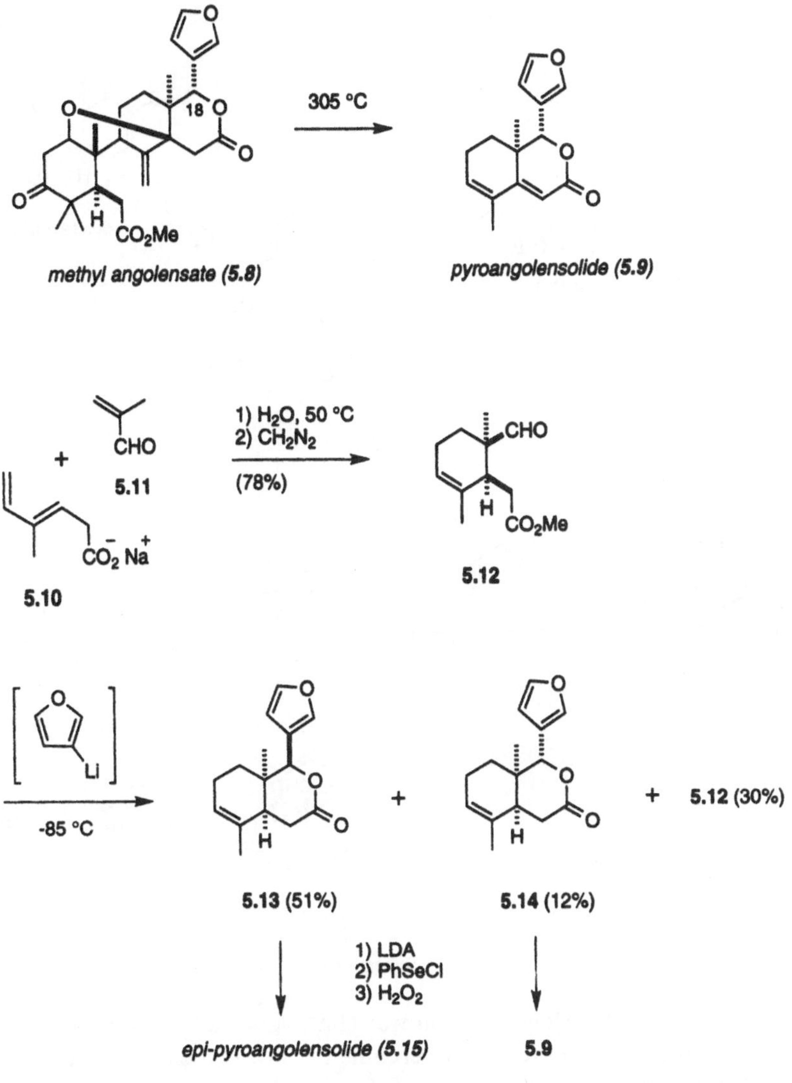

Scheme 1.5

The successful total synthesis of the highly oxygenated quassinoids chapparinone (5.16), glaucarubolone (5.17), and glaucarubinone (5.18) was very much dependent on the aqueous Diels–Alder reaction of diene carboxylates (Scheme 1.6). Retrosynthetic analysis of the pentacyclic structure common to these targets suggested that a Diels–Alder transform could be used to append the C-ring to an intact AB-unit with direct control of the newly developing stereocenters at C(8) and C(14). (The C(9) stereocenter would necessarily be introduced with the wrong configuration but is correctable later on in the synthesis.) This plan was predicated on Grieco's successful application of this same Diels–Alder strategy to the total synthesis of quassin (5.19). The proto-

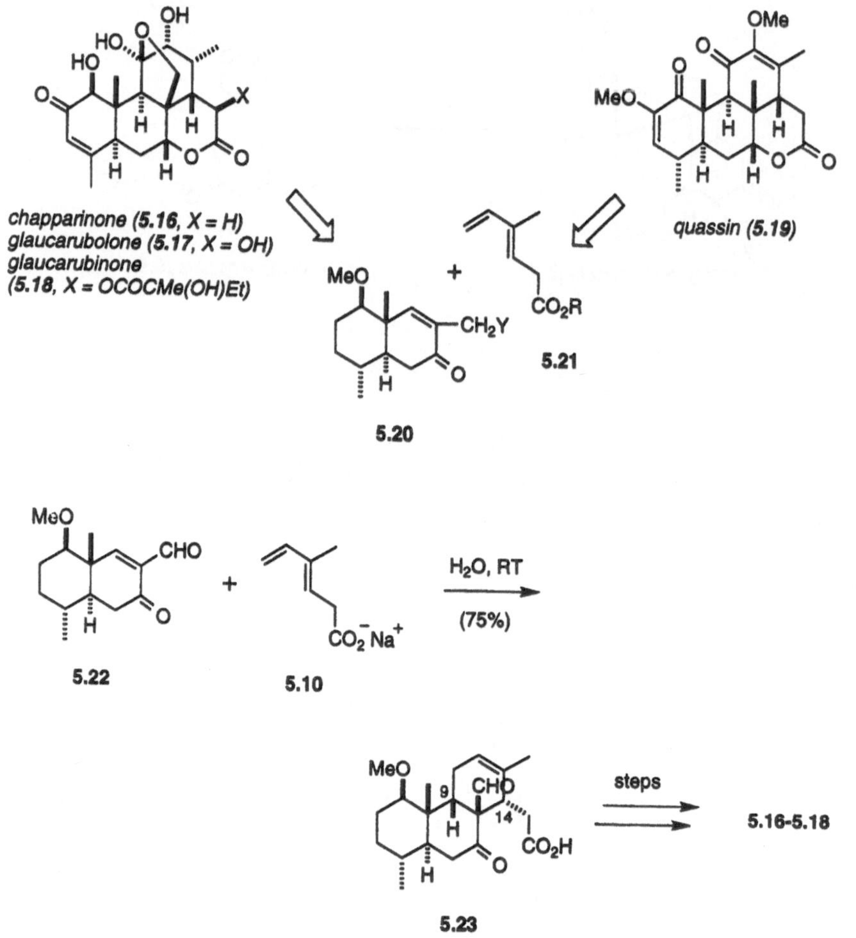

chapparinone (5.16, X = H)
glaucarubolone (5.17, X = OH)
glaucarubinone
(5.18, X = OCOCMe(OH)Et)
quassin (5.19)
MeO
CH2Y
CO2R
5.21
5.20
MeO
CHO
5.22
CO2⁻Na⁺
5.10
H2O, RT
(75%)
MeO
CHO
CO2H
5.23
steps
5.16-5.18

Scheme 1.6

type quassin Diels–Alder reaction was characterized by: (i) the need to use a strong Lewis acid ($EtAlCl_2$) to overcome the low reactivity of the enone **5.20** (Y = H); (ii) steric shielding of the β face of the enone by the C(10) methyl group; (iii) formation of the regioisomer that avoids substitution at the sterically congested C(10) position; and (iv) complete endo-selectivity, which leads to the desired stereochemistry at C(14).

However, in contrast to quassin, the synthesis of glaucarubinone and its congeners required the installation of a hydroxymethyl group (or its surrogate) at C(8) (cf. **5.20**, Y = OH). It was decided that a formyl group would not only serve this purpose but also increase the reactivity of the dienophile. Unfortunately, the incorporation of an additional carbonyl onto the dienophile also compromised the stereochemistry at C(14) by providing an alternative mode for endo-cycloaddition. As noted already, the breakthrough came when **5.22** was combined with the diene carboxylate **5.10** in water at ambient temperature [63]. These aqueous conditions resulted in the isolation of pure ABC-tricycle **5.23** in 75% yield and paved the way for the successful synthesis of **5.16**–**5.18** [64].

1.5.2   Steroids

Early studies on the use of chiral diene carboxylates for asymmetric aqueous Diels–Alder reactions resulted in a novel approach to the C/D *trans*-fused hydrindan ring and side-chain found in many steroid systems. The strategy was exemplified by a stereocontrolled synthesis of the *trans*-hydrindan **5.26** [65]. Since Trost had shown that **5.26** could be converted to the Inhoffen–Lythgoe diol (**5.25**) [66], this corresponded to a formal synthetic entry to vitamin D_3 and related metabolites [67]. Retrosynthetic disconnection of **5.26** led back to **5.30** through an aldol cyclization to form ring C via **5.27**, Claisen rearrangement of **5.28** to install a two-carbon unit at C(14) with the desired stereochemistry, and an aldol condensation to form ring D via **5.29** (Scheme 1.7). This densely functionalized system was to arise from the oxidative cleavage of cyclohexene **5.30** in which the Diels–Alder retron can now be recognized.

In the event, the reaction of methacrolein (**5.11**) with the chiral diene carboxylate **5.31** proceeded at 55°C in water to give a 4.6/1 mixture of endo adducts **5.32** and **5.33** in 85% combined yield. These products were then separately reduced with lithium aluminum hydride to give the diastereomeric crystalline diols **5.34** and **5.35**. Unambiguous stereochemical assignments rested on an X-ray structure determination of diol **5.35**. Diol **5.34** was then sequentially O-protected to give compound **5.30**, which was processed as described above. This approach to the C/D *trans*-fused hydrindan ring system possessing side-chain stereochemistry differs from most in that the intact C(20) stereocenter is used to control the introduction of the developing stereochemistry at C(13) and C(17) directly and C(14) indirectly.

On a related note, the Claisen rearrangement used in this synthesis was later modified to take advantage also of the rate acceleration of this reaction in water. Thus, the water-soluble carboxylate salt **5.36** rearranged very cleanly at 95°C to give desired Claisen product **5.37** in 82% yield [68]. For comparison, the rearrangement of **5.28** was performed at 190°C (refluxing decalin). The aqueous Claisen rearrangement leading to Inhoffen–Lythgoe diol intermediate **5.37** is shown below:

5.36   **5.37**

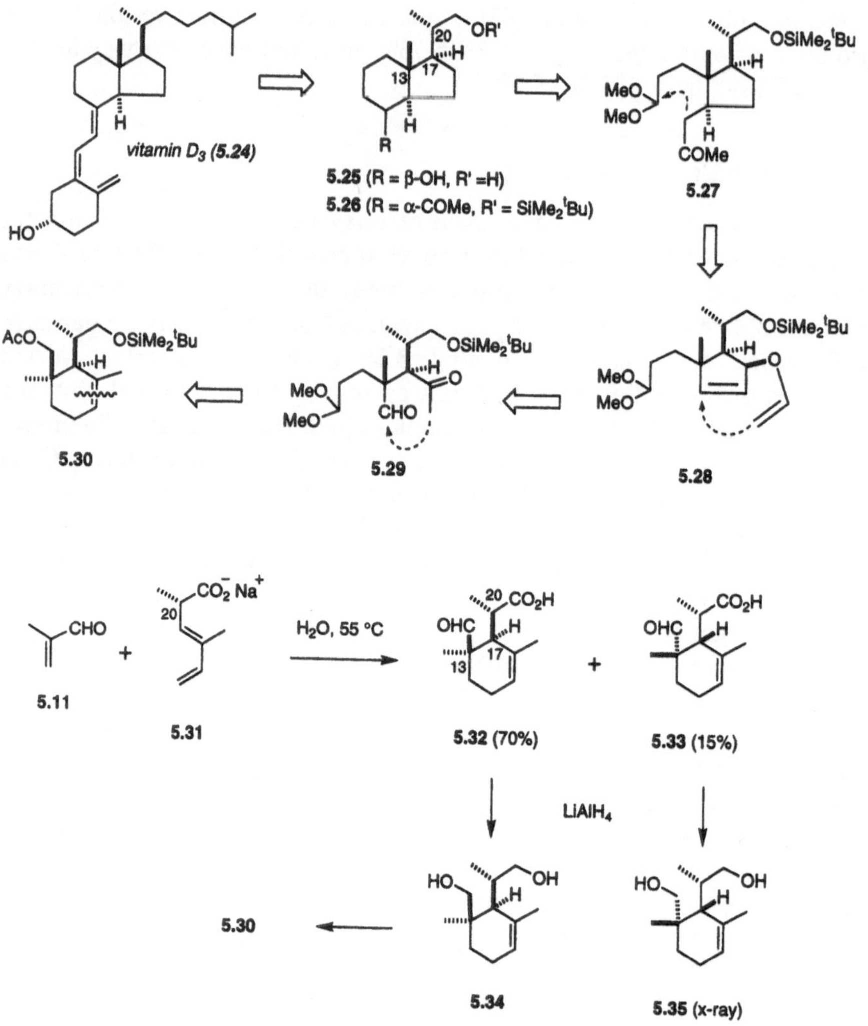

Scheme 1.7

1.5.3 Alkaloids

A synthetic approach to the A/E/F-ring substructure of the neurotransmission inhibitor methyllycaconitine (**5.38**) starts out with an aqueous Diels–Alder reaction between diene carboxylate **5.43** and methacrolein (**5.44**, X = H) to give the endo adduct **5.45** (corresponding to ring A) in excellent yield [69] (Scheme 1.8). This was followed by a series of standard functional-group transformations to aldehyde **5.42**, which now serves as a template for an intramolecular nitrone cycloaddition resulting in isoxazolidine **5.41** (and ring E). Finally, the N–O bond is cleaved and ring F formed through reductive amination of the aminoaldehyde **5.40**. It should be relatively straight-

methyllycaconitine (5.38)

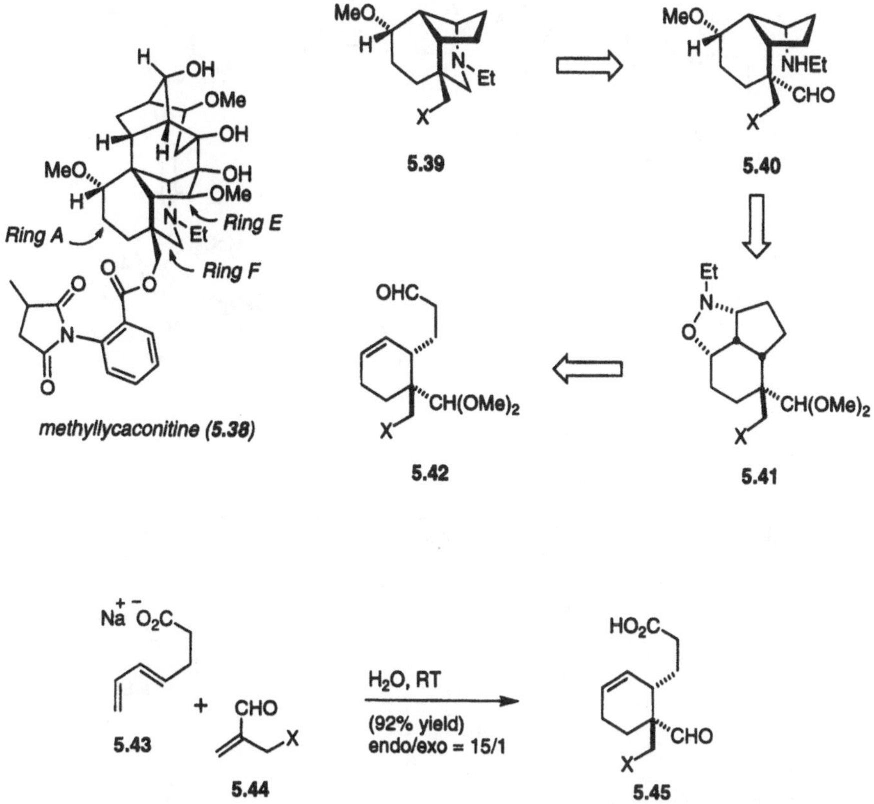

Scheme 1.8

forward to repeat this synthesis with a hydroxymethyl acrolein derivative **5.44** (X = OR), so that esters related to the natural product can be made.

1.5.4 Miscellaneous systems

Because of the practical implications of using water as a reaction solvent, the aqueous Diels–Alder reaction should find particular acceptance within the pharmaceutical industry. In this context, Saksena and co-workers at Schering-Plough developed a stereoselective synthesis of antifungals based on an aqueous Diels–Alder reaction [70]. The target structures (**5.46**) may be accessed via regioselective manipulation of the (bis)hydroxymethyltetra-hydrofuran derivative **5.47**, which is obtained via oxidative degradation of the 7-oxonorbornene derivative **5.48** (Scheme 1.9). This intermediate was obtained by chemoselective hydrogenation of the Diels–Alder adduct **5.51**. The use of water as a solvent for the Diels–Alder reaction of **5.49** and **5.50** led to a notable increase in yield when compared to the conventional thermal reaction, which is compromised by the retrograde Diels–Alder process.

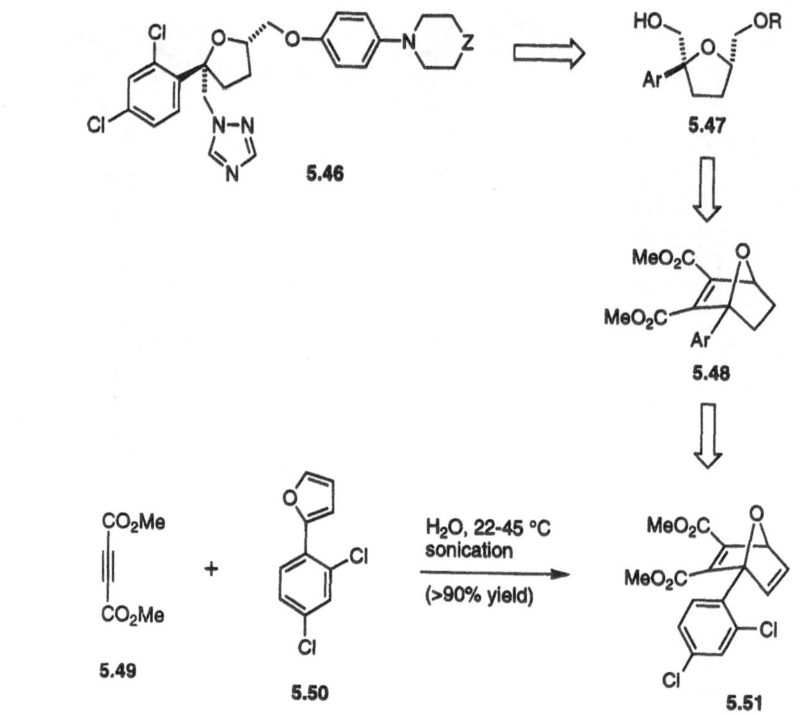

Scheme 1.9

1.6 Intramolecular cycloadditions

1.6.1 Hydrophobic effects

Intramolecular versions of the aqueous Diels–Alder reactions have also been investigated, though not to the same extent as with their bimolecular counterparts. However, since intramolecular reactions often exhibit considerable rate enhancements due to lowering of the entropy of activation, and the unique conformational aspects of the (necessarily cyclic) transition state can result in enhanced selectivities, it is not always clear what effect water has on a particular intramolecular Diels–Alder process. The situation is often complicated by the routine addition of β-cyclodextrin to the reaction medium. Blokzijl and Engberts have reported quantitative data on the intramolecular cycloadditions of substrates **6.1** in water and various solvents as a function of substituent R [71].

The observation that the rate enhancement in water decreases as the substrate attains more hydrophobic character is consistent with the concept of an enforced hydrophobic effect. Keay reported that the extent of (reversible) intramolecular Diels–Alder reactions of substituted 6-(2-furyl)hex-1-en-3-ones can be favorably affected by the prohydrophobic salt $CaCl_2$ [72].

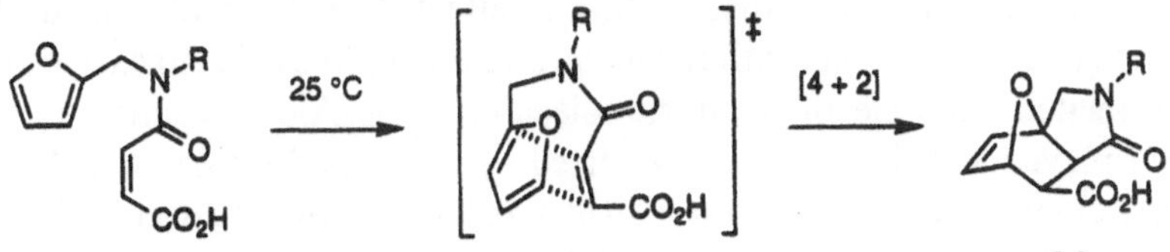

R =	k(H$_2$O)/k(n-PrOH)
CH$_3$	86.2
C$_2$H$_5$	80.5
n-C$_3$H$_7$	37.7
n-C$_8$H$_{17}$	32.1

1.6.2 Synthetic applications

The first report of an intramolecular Diels–Alder reaction that benefited from being performed in water came from Sternbach's laboratory [73]. While these workers did find that heating an aqueous solution of substrate **6.4** resulted in [4 + 2] cycloaddition, the best yield (91%) of diastereomeric cycloadducts **6.5** and **6.6** was obtained when a full equivalent of β-cyclodextrin was included in the reaction medium:

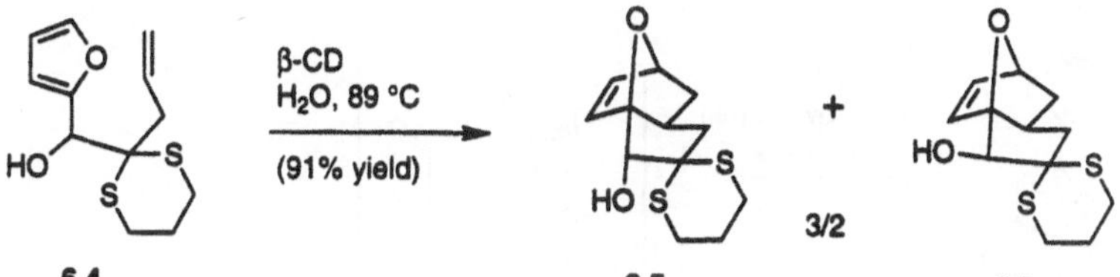

Unfortunately, the diastereoselectivity of this reaction was not very good (*ds* = 3/2). Two models were offered to explain the rate-accelerating effect of β-cyclodextrin on this Diels–Alder reaction. These two models illustrating possible binding of β-cyclodextrin to intramolecular cycloaddition substrate **6.4** are shown below:

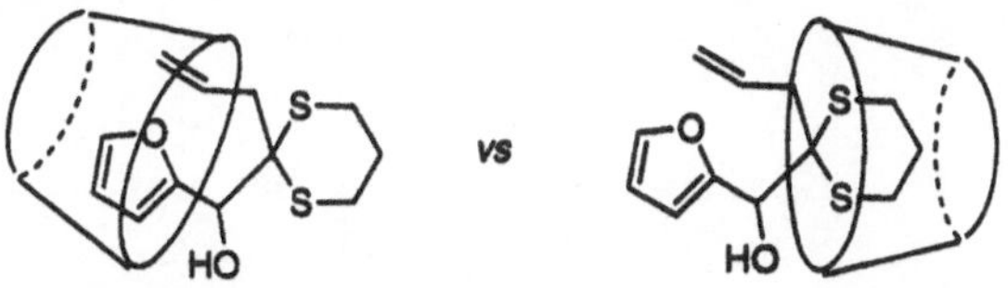

One involves the simultaneous encapsulation of the diene and dienophile components within the cyclodextrin cavity, overcoming the intrinsic entropic barrier associated with bringing these two reactive centers together. This was, in fact, the same argument put forth by Breslow to explain the effect of cyclodextrins on analogous intermolecular Diels–Alder reactions. Alternatively, one could also imagine encapsulation of the dithiane moiety by the cyclodextrin. Such complexation might also accelerate the reaction by accentuating a geminal disubstitution effect, which would again favor substrate conformations with the diene and dienophile close to each other.

A related example of an intramolecular Diels–Alder reaction in which water played a critical role was reported by Wang and Roskamp [74]. While initial attempts to perform Diels–Alder reactions with **6.7** were unsatisfactory, adsorption of the hemiketal substrate onto silica gel followed by saturation of the mixture with water and microwave irradiation resulted in the formation of a 1/1 mixture of cycloadducts **6.9** and **6.10** in 64% combined yield (Scheme 1.10). These rather unusual reaction conditions were based on the observation that **6.7** was partially converted to **6.9** and **6.10** during flash chromatography on silica gel. It was noted that water was critical for the success of this reaction. Possible roles for the water include: (i) the efficient generation of heat during the microwave process; (ii) hydrophobic acceleration of the Diels–Alder reaction; and/or (iii) facilitation of the equilibrium between **6.7** and the ketone **6.8**. The adduct **6.9** was exploited successfully in a Grob fragmentation/homoallylic elimination sequence (**6.11 → 6.12**), providing a route to the bicyclo[6.2.1] ring system commonly found in the germacranolide sesquiterpenes.

Hudlicky and co-workers have followed Sternbach's lead and used water

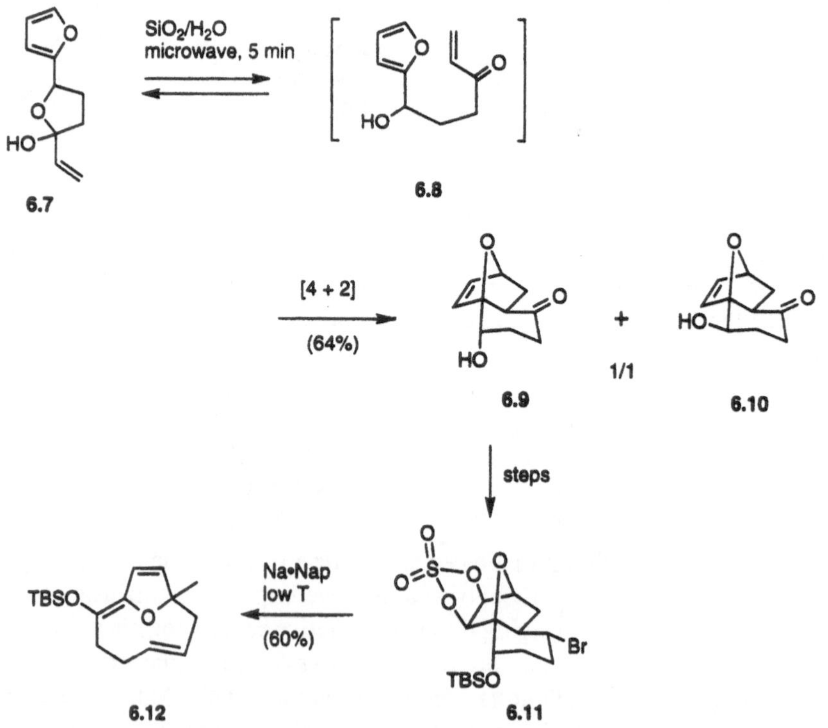

Scheme 1.10

plus β-cyclodextrin to facilitate the intramolecular Diels–Alder reaction of substituted furan **6.13** [75]. Thus, heating an equimolar quantity of **6.13** and β-CD in water resulted in the isolation of cycloadduct **6.14** in 84% yield:

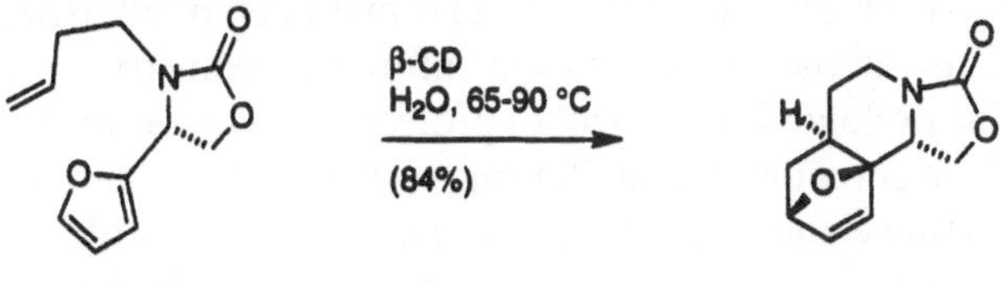

6.13   **6.14**

This stereoselective reaction proceeds through an exo transition state with a *trans*-oxazohydrindan bridge conformation. The stereocontrolled assembly of this compound in one step represents a direct route to functionalized isoquinoline derivatives, which may prove useful for alkaloid synthesis.

An aqueous intramolecular Diels–Alder reaction formed the cornerstone of De Clercq's synthesis of (±)-11-ketotestosterone (**6.15**) and (±)-adrenosterone (**6.16**) [76]. Thus, a suspension of substrate **6.19** in water was shaken vigorously or, more efficiently, subjected to ultrasonic agitation to effect intramolecular cycloaddition to give a single product **6.21** via the exo transition state **6.20** (Scheme 1.11). Since the cycloadduct **6.21** undergoes a very

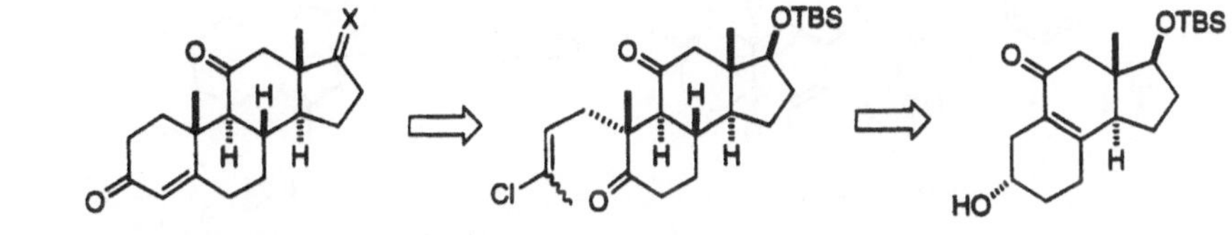

11-ketotestosterone (**6.15**, X = H, β-OH)   **6.17**   **6.18**
adrenosterone (**6.16**, X = O)

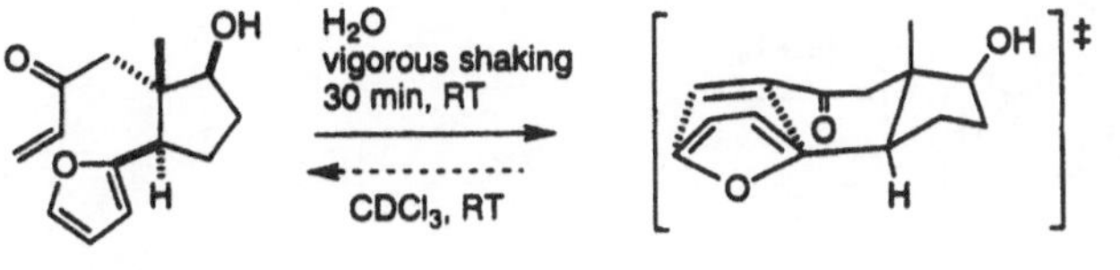

6.19   **6.20**

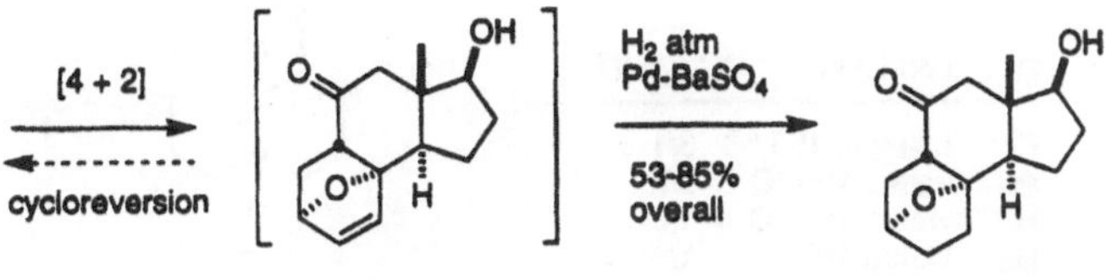

6.21   **6.22**

Scheme 1.11

facile cycloreversion back to **6.19** in organic media, it was found best to hydrogenate **6.21** directly to give the stable compound **6.22**. It was proposed that the forward reaction (**6.19**→**6.21**) is not favored in non-hydrogen-bonding solvents because an intramolecular hydrogen bond between the free hydroxyl group and the enone carbonyl enforces a geometry not suitable for intramolecular cycloaddition. Alcohol protection followed by β-elimination gave **6.18**, which embodies the B/C/D-ring system of the target structures and is ready for A-ring elaboration via **6.17** using Stork's protocol.

Williams and co-workers have reported a remarkable reversal of regio-selectivity in the intramolecular Diels–Alder reaction of bis-diene substrate **6.25** in connection with efforts toward the total synthesis of ilicicolin (**6.23**) [77]. They found that heating a toluene solution of **6.25** (R = Et) at 165°C in a sealed tube resulted in the production of regioisomeric cycloadducts **6.26** and **6.27** in the ratio of 3/1 (Scheme 1.12). However, when this Diels–Alder reaction was conducted in boiling water, the product ratio was reversed in favor of **6.27**, the highest selectivity (**6.27/6.26** = 8/1) being observed when the free carboxylic acid (R = H) or its sodium salt (R = Na) were used. The forma-tion of both cycloadducts is consistent with exo-bridged Diels–Alter transi-

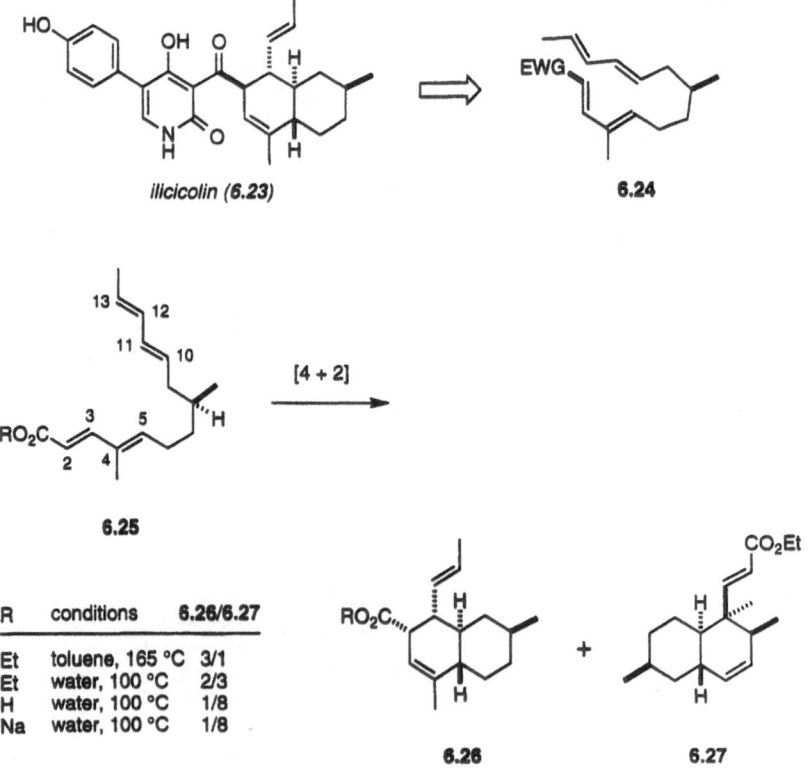

R	conditions	6.26/6.27
Et	toluene, 165 °C	3/1
Et	water, 100 °C	2/3
H	water, 100 °C	1/8
Na	water, 100 °C	1/8

Scheme 1.12

tion states in which the C(2)–C(5) diene reacts with the C(10)–C(11) olefin (to give **6.26**) or the C(4)–C(5) olefin reacts with the C(10)–C(13) diene (to give **6.27**). The two competing intramolecular cycloaddition transition states for bis-diene substrate **6.25** are shown below:

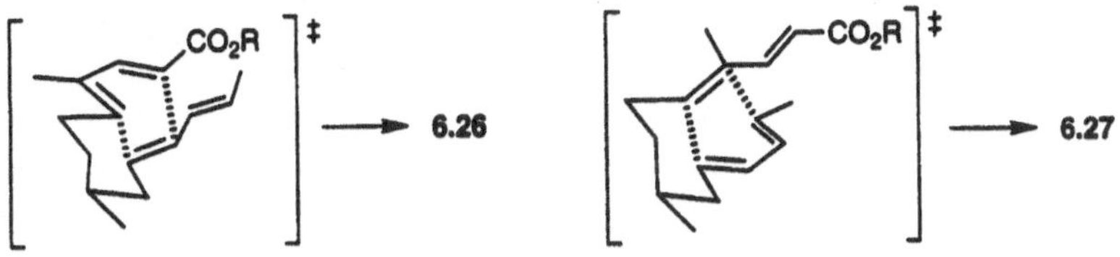

The high stereoselectivity of these reactions simply reflects the fact that the C(8) methyl group prefers to occupy a pseudoequatorial position in each TS independent of the medium. Note that the 'classical' Diels–Alder reaction between the enoate C(2)–C(3) dienophile and the remaining C(10)–C(13) diene (which is not observed) would result in the formation of an eight-membered ring incorporating a *trans*-olefin.

Finally, Kibayashi used aqueous versions of the intramolecular acyl-nitroso Diels–Alder reaction to good effect in his stereocontrolled syntheses of (–)-swainsonine (**6.28**) [78] and (–)-pumiliotoxin C (**6.34**) [79] (Schemes 1.13 and 1.14). In each case, oxazinolactam cycloadducts (cf. **6.32** and **6.38**) served as key functional chiral building blocks for the remaining synthetic operations. For example, retrosynthetic analysis of the trihydroxy-indolizidine system of swainsonine (**6.28**) reveals that the pyrrolidine ring of this target can be formed by amine alkylation (cf. **6.29**) while the vicinal diol can be installed via substrate-controlled *cis*-hydroxylation of a (Z)-allyl

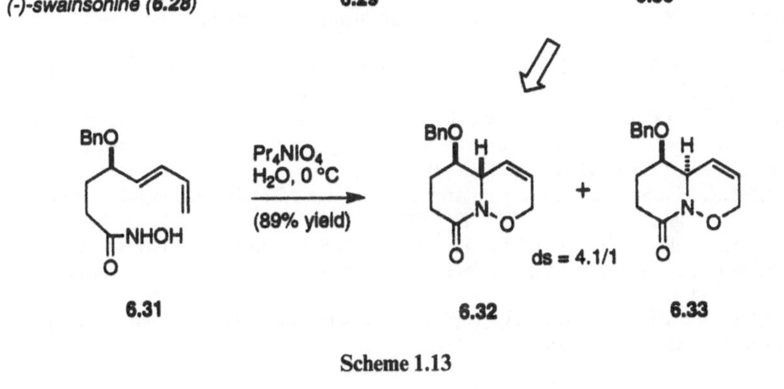

Scheme 1.13

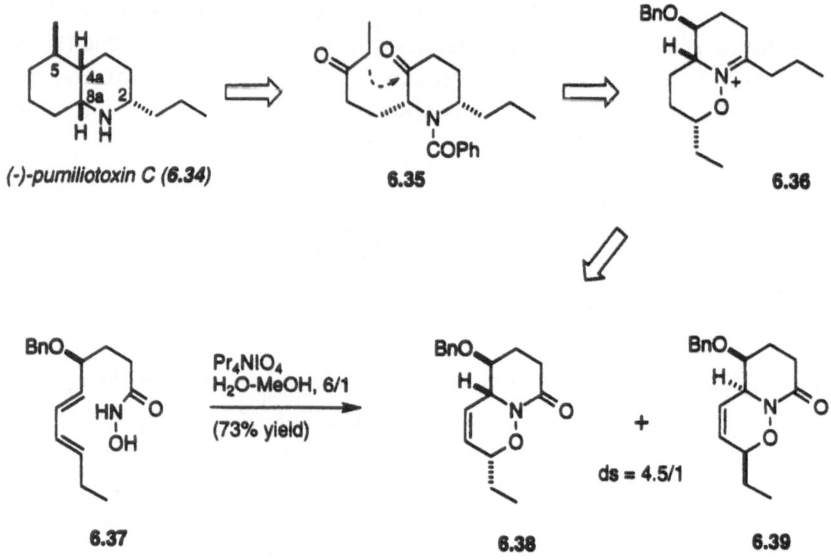

Scheme 1.14

alcohol derivative **6.30** obtained by reductive cleavage of the oxazinolactam N–O bond. His synthesis of (–)-pumiliotoxin C also emanates from an oxazinolactam but utilizes the stereocontrolled reduction of iminium ion **6.36** to give the α-propyl substituent at C(2). An aldol condensation is used to assemble the carbocyclic ring (cf. **6.35**) and the resulting enone subjected to a hydrogenation/epimerization sequence to install the remaining stereocenters at C(4a) and C(5). The increased diastereoselectivity of these hetero Diels–Alder reactions in water was the key development that made both of these syntheses possible.

References

[1] Diels, O.; Alder, K., Synthesen in der hydroaromatische Reihe, *Ann. Chem.*, 1931, **490**, 243–57.

[2] Woodward, R.B.; Baer, H., The reaction of furan with maleic anhydride, *J. Am. Chem. Soc.*, 1948, **70**, 1161.

[3] Hopff, H.; Rautenstrauch, C.W., Production of addition products from dienes and compounds having an unsaturated carbon linkage activated by substituents, US Patent 2262002, 1939; *Chem. Abstr.*, 1942, **36**, 10469.

[4] Lane, L.C.; Parker, C.H., Jr., Fumaric acid-conjugated hydrocarbon adducts, US Patent 2444263, 1948; *Chem. Abstr.*, 1948, **42**, 7102.

[5] Bandaranayake, W.M.; Banfield, J.E.; Black, D.S.C., Postulated electrocyclic reactions leading to endriandric acid and related natural products, *J. Chem. Soc., Chem. Commun.*, 1980, 902–3.

[6] Breslow, R.; Rideout, D.C., Hydrophobic acceleration of Diels–Alder reactions, *J. Am. Chem. Soc.*, 1980, **102**, 7816–17.

[7] Schneider, H.-J.; Sangwan, N.K., Diels–Alder reactions in hydrophobic cavities: a quantitative correlation with solvophobicity and rate enhancements by macrocycles, *J. Chem. Soc., Chem. Commun.*, 1986, 1787–9.

[8] Sangwan, N.K.; Schneider, H.-J., The kinetic effects of water and of cyclodextrins on Diels–Alder reactions. Host–guest chemistry. Part 18, *J. Chem. Soc., Perkin Trans. 2*, 1989, 1223–7.

[9] Breslow, R.; Rizzo, C.L., Chaotropic salt effects in a hydrophobically accelerated Diels–Alder reaction, *J. Am. Chem. Soc.*, 1991, **113**, 4340–1.

[10] Breslow, R., Hydrophobic effects on simple organic reactions in water, *Acc. Chem. Res.*, 1991, **24**, 159–64.

[11] Breslow, R., Hydrophobic and antihydrophobic effects on organic reactions in aqueous solution, in *Structure and Reactivity in Aqueous Solution*, Cramer, C.J.; Truhlar, D.G., Eds.; ACS Symposium Series 568; American Chemical Society, Washington, DC, 1994, pp. 291–302.

[12] Blokzijl, W.; Blandamer, M.J.; Engberts, J.B.F.N., Diels–Alder reactions in aqueous solutions. Enforced hydrophobic interactions between diene and dienophile, *J. Am. Chem. Soc.*, 1991, **113**, 4241–6.

[13] Blokzijl, W.; Engberts, J.B.F.N., Initial-state and transition-state effects on Diels–Alder reactions in water and mixed aqueous solvents, *J. Am. Chem. Soc.*, 1992, **114**, 5440–2.

[14] Blokzijl, W.; Engberts, J.B.F.N., Enforced hydrophobic interactions and hydrogen bonding in the acceleration of Diels–Alder reactions in water, in *Structure and Reactivity in Aqueous Solution*, Cramer, C.J.; Truhlar, D.G., Eds.; ACS Symposium Series 568; American Chemical Society, Washington, DC, 1994, pp. 303–17.

[15] Breslow, R.; Zhu, Z., Quantitative antihydrophobic effects as probes for transition rate structures. 2. Diels–Alder reactions, *J. Am. Chem. Soc.*, 1995, **117**, 9923–4.

[16] Lubineau, A.; Bienaymé, H.; Queneau, Y.; Scherrmann, M.-C., Aqueous cycloadditions using glyco-organic substrates. Thermodynamics of the reaction, *New J. Chem.*, 1994, **18**, 279–85.

[17] Blake, J.F.; Lim, D.; Jorgensen, W.L. Enhanced hydrogen bonding of water to Diels–Alder transition states. *Ab initio* evidence, *J. Org. Chem.*, 1994, **59**, 803–5.

[18] Jorgensen, W.L.; Blake, J.F.; Lim, D.; Severence, D.L., Investigation of solvophobic effects on pericyclic reactions by computer simulations, *J. Chem. Soc., Faraday Trans.*, 1994, **90**, 1727–32.

[19] Otto, S.; Blokzijl, W.; Engberts, J.B.F.N., Diels–Alder reactions in water. Effects of hydrophobicity and hydrogen bonding, *J. Org. Chem.*, 1994, **59**, 5372–6.

[20] Grieco, P.A.; Garner P.; He, Z.-M., 'Micellar' catalysis in the aqueous intermolecular Diels–Alder reaction: rate acceleration and enhanced selectivity, *Tetrahedron Lett.*, 1983, **24**, 1897–900.

[21] Grieco, P.A.; Yoshida, K.; Garner P., Aqueous intermolecular Diels–Alder chemistry: reactions of diene carboxylates with dienophiles in water at ambient temperature, *J. Org. Chem.*, 1983, **26**, 3137–9.

[22] Yoshida, K.; Grieco, P.A., Synthesis and reactivity of (*E*)-4,6,7-octatrienoic acid sodium salt in the aqueous Diels–Alder reaction, *Chem. Lett.*, 1985, 155–8.

[23] Grieco, P.A.; Galatsis, P.; Spohn, R.F., Aqueous intermolecular Diels–Alder chemistry. Reaction of (*E*)-2,4-pentadienyl ammonium chloride and related ammonium salts with dienophiles in water, *Tetrahedron*, 1986, **42**, 2847–53.

[24] Unpublished results of Mr. Zen-Min He, Indiana University.

[25] Keana, J.F.W.; Guzikowski, A.P.; Morat, C.; Volwerk, J.J., Detergents containing a 1,3-diene group in the hydrophobic segment. Facile chemical modification by a Diels–Alder reaction with hydrophobic dienophiles in aqueous solution, *J. Org. Chem.*, 1983, **48**, 2661–6.

[26] See references [7] and [8].

[27] Braun, R.; Schuster, F.; Sauer, J., (4+2)-Cycloadditionen in Micellen: Ein Vergleich des Produktspectrums und der Reaktionsgeschwindigkeit mit Reaktionen in Lösung, *Tetrahedron Lett.*, 1986, **27**, 1285–8.

[28] Singh, V.; Raju, B.N.S, p4S + p2S cycloaddition of spiro[4,*n*] cyclic 1,3-dienes with quinones in homogeneous and aqueous micellar media: synthesis of novel polycyclic cage diones, *Ind. J. Chem.*, 1996, **35B**, 303–11.

[29] Otto, S.; Engberts, J.B.F.N., Lewis-acid catalysis of a Diels–Alder reaction in water, *Tetrahedron Lett.*, 1995, **36**, 2645–8

[30] Kobayashi, S.; Hachiya, I.; Araki, M.; Ishitani, H., Scandium trifluoromethanesulfonate (Sc(OTf)₃). A novel reusable catalyst in the Diels Alder reaction, *Tetrahedron Lett.*, 1993, **34**, 3755–8.

[31] Breslow, R.; Guo, T., Diels–Alder reactions in nonaqueous polar solvents. Kinetic effects of chaotropic and antichaotropic agents and of β–cyclodextrin, *J. Am. Chem. Soc.*, 1988, **110**, 5613–17.

[32] Dunams, T.; Hoekstra, W.; Pentaleri, M.; Liotta, D, Molecular aggregation and its applicability to synthesis. The Diels–Alder reaction, *Tetrahedron Lett.*, 1988, **29**, 3745–8.

[33] Jaeger, D.A.; Tucker, C.E., Diels–Alder reactions in ethylammonium nitrate, a low-melting fused salt, *Tetrahedron Lett.*, 1989, **30**, 1785–8.

[34] Pai, C.K.; Smith, M.B., Rate enhancement in dilute salt solutions of aqueous ethanol: the Diels–Alder reaction, *J. Org. Chem.*, 1995, **60**, 3731–5.

[35] Casetta, M.; Colona, S.; Manfredi, A., Hydrophobic control of organic stereochemistry. Changes of regioselectivity in Diels–Alder reactions by salt effects, *Gazz. Chim. Ital.*, 1989, **119**, 533–5.

[36] Kelley, T.R.; Montury, M., Lewis acid-catalyzed Diels–Alder reactions of peri-hydroxylated naphthoquinones: a regiochemical divergence, *Tetrahedron Lett.*, 1978, 4311–14.

[37] Chung, W.-S.; Wang, J.-Y., Control of regioselectivity in the Diels–Alder reactions of alkyl-substituted 1,4-benzoquinones by β-cyclodextrin and its derivatives, *J. Chem. Soc., Chem. Commun.*, 1995, 971–2.

[38] Jaeger, D.A.; Shinozaki, H.; Goodson, P.A., Diels–Alder reactions of a surfactant 1,3-diene, *J. Org. Chem.*, 1991, **56**, 2482–9.

[39] Breslow, R.; Maitra, U.; Rideout, D., Selective Diels–Alder reactions in aqueous solutions and suspensions, *Tetrahedron Lett.*, 1983, **24**, 1901–4.

[40] Breslow, R.; Maitra, U., On the origin of product selectivity in aqueous Diels–Alder reactions, *Tetrahedron Lett.*, 1984, **25**, 1239–40.

[41] Braun, R.; Schuster, F.; Sauer, J., (4+2)-cycloadditionen in Micellen: Ein Vergleich des Produktspektrums und der Reaktiongeschwendigkeit mit Reaktionen in Lösung, *Tetrahedron Lett.*, 1986, **27**, 1285–8.

[42] Schneider, H.-J.; Sangwan, N.K., Changes of stereoselectivity in Diels–Alder reactions by hydrophobic solvent effects and by β-cyclodextrin, *Angew. Chem. Int. Ed. Engl.*, 1987, **26**, 896–7.

[43] Cativiela, C.; Garcia, J.I.; Mayoral, J.A.; Avenoza, A.; Peregrina, J.M.; Roy, M.A., Development of a model to explain the influence of the solvent on the rate and selectivity of Diels–Alder reactions, *J. Phys. Org. Chem.*, 1991, **4**, 48–52.

[44] See reference [20].

[45] Griesbeck, A.G., Fulvene cycloaddition reactions in water: influence on rate and selectivity, *Tetrahedron Lett.*, 1988, **29**, 3477–80.

[46] Rao, K.R.; Srinivasan, T.N.; Bhanumathi, N., A stereoselective biocatalytic Diels–Alder reaction, *Tetrahedron Lett.*, 1990, **31**, 5959–60.

[47] Paquette, L.A., Asymmetric cycloaddition reactions, in *Asymmetric Synthesis*, Morrison, J.D., Ed.; Academic Press, Orlando, 1984, vol. 3, Ch. 7.

[48] Ghosh, A.K.; Mathivanan, P.; Cappiello, J., Conformationally contrained bis(oxazoline) derived chiral catalyst: A highly effective enantioselective Diels–Alder reaction, *Tetrahedron Lett.*, 1996, **37**, 3515–18, and references therein.

[49] Dols, P.P.M.A.; Klunder, A.J.H.; Zwanenburg B., 4-Hydroxycyclopent-2-en-1-one and derivatives as chiral synthetic equivalents of cyclopentadienone in asymmetric Diels–Alder reactions, *Tetrahedron*, 1994, **28**, 8515–38.

[50] Cieplak, A.S., Stereochemistry of nucleophilic addition to cyclohexanone. The importance of two-electron stabilizing interactions, *J. Am. Chem. Soc.*, 1981, **103**, 4540–52.

[51] Jeroncic, L.O.; Cabal, M.P.; Danishefsky, S.J.; Shulte, G.M., On the diastereofacial selectivity of Lewis acid catalyzed carbon–carbon bond forming reactions of conjugated cyclic enones bearing electron-withdrawing substituents at the γ-position, *J. Org. Chem.*, 1991, **56**, 387–95.

[52] Proust, S.M.; Ridley, D.D., Effect of catalyst and solvent on the stereochemistry of

Diels–Alder reactions between cyclopentadiene and 3-phenylsulfinylprop-2-enoic acids and methyl esters, *Aust. J. Chem.*, 1984, **37**, 1677–88.

[53] Waldmann, H.; Dräger, M., Thermische Diels–Alder-Reactionen mit *N*-(2-Alkenoyl)-(*S*)-prolinestern als chiralen Dienophilen in organischen und wässrigen Reaktionsmedien, *Ann. Chem.*, 1990, 681–5.

[54] Masamune, S.; Choy, W.; Petersen, J.S.; Sita, L.R., Double asymmetric synthesis and a new strategy for stereochemical control in organic synthesis, *Angew. Chem. Int. Ed. Engl.*, 1985, **24**, 1–30.

[55] Geary, P.J.; Pryce, R.L.; Roberts, S.M.; Ryback, G.; Winders, J.A., The oxidations of norbornadiene and some derivatives using *Pseudomonas* sp., *J. Chem. Soc., Chem. Commun.*, 1990, 204–5.

[56] Brandes, E.; Grieco, P.A.; Garner P., Diastereoselection in an aqueous Diels–Alder reaction: A formal total synthesis of the Inhoffen–Lythgoe diol, *J. Chem. Soc., Chem. Commun.*, 1988, 500–2.

[57] Lubineau, A.; Queneau, Y., Stereochemical variations in aqueous cycloadditions using glyco-organic substrates as a consequence of chemical manipulations on the sugar moiety, *Tetrahedron*, 1989, **45**, 6697–712, and references therein.

[58] Goeljian, P.G.; Wu, T.-C.; Kishi, Y., Preferred conformation of *C*-glycosides. 6. Conformational similarity of glycosides and corresponding *C*-glycosides, *J. Org. Chem.*, 1991, **56**, 6412–22.

[59] Lubineau, A.; Augé, J.; Lubin, N., Aqueous cycloaddition using glyco-organic substrates. Facial selectivity in Diels–Alder reactions of a chiral diene derived from D-glyceraldehyde, *J. Chem. Soc., Perkin Trans. 1*, 1990, 3011–15.

[60] Yoshida, K.; Grieco, P.A., Aqueous Diels–Alder chemistry: vernolepin revisited, *J. Org. Chem.*, 1984, **49**, 5257–60.

[61] Kieczykowski, G.R.; Queseda, M.L.; Schiessinger, R.H., Total synthesis of *dl*-bisnorvernolepin, *J. Am. Chem. Soc.*, 1980, **102**, 782–90.

[62] Drewes, S.E.; Grieco, P.A.; Huffman, J.C., A short synthesis of *dl*-epi-pyroangolensolide and *dl*-pyroangolensolide: confirmation of the structures of pyroangolensolide and calodendrolide, *J. Org. Chem.*, 1985, **50**, 1309–11.

[63] See reference [20].

[64] Grieco, P.A.; Collins, J.L.; Moher, E.D.; Fleck, T.J.; Gross, R.S., Synthetic studies on quassinoids: total synthesis of (–)-chapparinone, (–)-glucarubolone, and (+)-glaucarubinone, *J. Am. Chem. Soc.*, 1993, **115**, 6078–93, and references therein.

[65] See reference [57].

[66] Trost, B.M.; Bernstein, P.R.; Funfschilling, P.C., A stereocontrolled approach toward vitamin D metabolites. A synthesis of the Inhoffen–Lythgoe diol, *J. Am. Chem. Soc.*, 1979, **101**, 4378–80.

[67] Lythgoe, B.; Moran, T.A.; Nambudiry, M.E.N.; Tideswell, J.; Wright, P.W., *J. Chem. Soc., Perkins Trans. 1*, 1978, 590.

[68] Brandes, E.; Grieco, P.A.; Gajewski, J.J., Effect of polar solvents on the rates of Claisen rearrangements: asessment of ionic character, *J. Org. Chem.*, 1989, **54**, 515–16.

[69] Baillie, L.C.; Bearder, J.R.; Whiting, D.A., Synthesis of the A/E/F tricyclic section of the norditerpenoid alkaloid methyllycaconitine, a potent inhibitor of neurotransmission, *J. Chem. Soc., Chem. Commun.*, 1994, 2487–8.

[70] Saksena, A.K.; Girijavallabhan, V.M.; Chen, Y.-T.; Jao, E.; Pike, R.E.; Desai, J.A.; Rane, D.; Ganguly, A.K., Aqueous Diels–Alder reactions of electron deficient 2-arylfurans: a highly stereoselective route to 2,2,5-trisubstituted tetrahydrofurans towards a novel class of orally active azole antifungals, *Heterocycles*, 1993, **35**, 129–34.

[71] See reference [14].

[72] Keay, B.A., Intramolecular Diels–Alder reaction of the diene unit of furan in 2.0 M CaCl$_2$, *J. Chem. Soc., Chem. Commun.*, 1987, 419–21.

[73] Sternbach, D.D.; Rossana, D.M., Cyclodextrin catalysis in the intramolecular Diels–Alder reaction with the furan diene, *J. Am. Chem. Soc.*, 1982, **104**, 5853–4.

[74] Wang, W.-B.; Roskamp, E.J., New technology for the construction of bicyclo[6.2.1] ring systems, *Tetrahedron Lett.*, 1992, **33**, 7631–4.

[75] Hudlicky, T.; Butora, G.; Fearnley, S.P.; Gum, A.G.; Persichini, P.J., III; Stabile, M.R.; Merola, J.S., Intramolecular Diels–Alder reactions of the furan diene (IMDAF); rapid

construction of highly functionalised isoquinoline skeletons, *J. Chem. Soc., Perkin Trans. 1*, 1995, 2393–8.

[76] Van Royen, L.A.; Mijngheer, R.; De Clercq, P.J., Intramolecular Diels–Alder reaction with furan-diene. Total synthesis of (±)-11-ketotestosterone and (±)-adrenosterone, *Tetrahedron*, 1985, **41**, 4667–80.

[77] Williams D.R.; Gaston, R.D.; Horton, I.B., III, Intramolecular Diels–Alder cycloadditions of bis-diene substrates, *Tetrahedron Lett.*, 1985, **26**, 1391–4.

[78] Naruse, M.; Aoyagi, S.; Kibayashi, C., New chiral route to (–)-swainsonine via an aqueous acylnitroso cycloaddition approach, *J. Org. Chem.*, 1994, **59**, 1358–64.

[79] Naruse, M.; Aoyagi, S.; Kibayashi, C., Total synthesis of (–)-pumiliotoxin C by aqueous intramolecular acylnitroso Diels–Alder approach, *Tetrahedron Lett.*, 1994, **35**, 9213–16.

2 Hetero Diels–Alder reactions
D.T. PARKER

2.1 Introduction

Hetero Diels–Alder reactions have generally required activation of the requisite dienophile through substitution with electron-withdrawing substituents, Lewis-acid catalysis and/or the use of highly reactive dienes. These reactions, therefore, have traditionally been performed in aprotic organic solvents for the solubility and compatibility of the required reagents and catalysts. As a consequence, there are inherent limitations with respect to the scope and application of the hetero Diels–Alder reaction. Recently, the use of water as a solvent in hetero [4+2] cycloadditions has served both to complement existing methodology and to open up new reaction avenues for further synthetic exploitation. This chapter will highlight the developments over the past decade with respect to the use of water in heterocycloaddition [1] and related cycloreversion processes as well as applications in heterocyclic synthesis.

2.2 Aza Diels–Alder reactions in aqueous media

2.2.1 Simple protonated iminium ions as heterodienophiles

Simple imines are generally unreactive toward dienes without activation. In 1985, Grieco and Larson [1] first reported that simple unactivated iminium ions, generated *in situ* under Mannich-like conditions, react with a variety of dienes in water to effect an exceptionally mild and convenient cyclocondensation reaction. Water as a solvent is particularly suited to using commercially available 37% aqueous formaldehyde in combination with an amine hydrochloride to form the incipient protonated imine heterodienophile.

Preliminary studies focused on reacting a variety of dienes with benzyliminium ion 1 derived from formaldehyde and benzylamine hydrochloride. For example, addition of neat cyclopentadiene (2.0 equiv) to a 2.5 M aqueous solution of benzylamine hydrochloride (1.0 equiv) and 37% aqueous formaldehyde solution (1.4 equiv) gives rise to a near-quantitative yield of *N*-benzyl-2-azanorbornene after vigorous stirring of the heterogeneous reaction for 3 h at room temperature:

Use of alcoholic solvents leads to a decreased reaction rate; however, tetrahydrofuran as a cosolvent with water can be employed without any noticeable effect on reaction rate or yield. Substituting the less-reactive cyclohexadiene for cyclopentadiene in the above equation requires heating (55°C) and results in a substantially lower yield (35%) of the corresponding fused bicyclic [2.2.2] product.

Attempts to use other aldehydes in this reaction met with limited success. Substituting acetaldehyde in place of formaldehyde results in a 47% yield of the corresponding exo and endo azanorbornenes **2** and **3** in a 1.5/1 ratio:

The reaction rate is notably slower than the formaldehyde case (16 h vs. 3 h) with concomitant formation of more by-products. No Diels–Alder products are detected when simple ketones (e.g. acetone) are employed in the above equation. Substituting benzylamine with methylamine or ammonia in this reaction provides the corresponding Diels–Alder products, albeit in somewhat lower yields.

The overall cyclocondensation of iminium ions proceeds with the regio- and stereoselectivity expected for a true Diels–Alder process. The formation of a single diastereomer in the reaction of (*E,E*)-2,4-hexadiene with iminium ion **1** is consistent with a concerted [2+4] cycloaddition rather than an ionic, stepwise process:

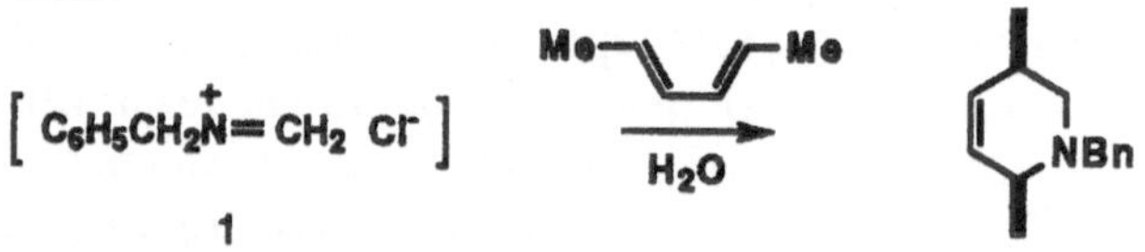

In addition to demonstrating the feasibility of intermolecular iminium-ion-based [4+2] cycloadditions in aqueous media for entry into the monocyclic and fused bicyclic piperidine backbones, a few intramolecular variants are also detailed, which allow for the construction of azabicyclo [4.3.0] and [4.4.0] systems. For example, cyclization of dienylamine hydrochlorides, **4** and **5**, in the presence of aqueous formaldehyde at 50°C for 48 h gives rise to the bicyclic 5,6 (indolizidine) and 6,6 (quinolizidine) ring systems in 95% and 65% yields, respectively:

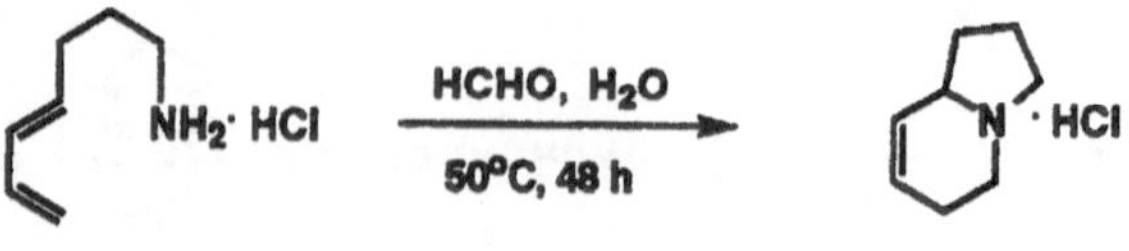

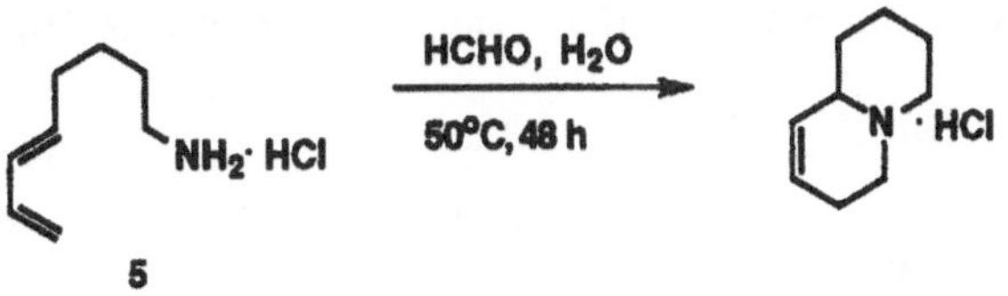

5

A representative octahydroquinoline backbone can be constructed via cyclocondensation of dienyl aldehyde **6** with benzylamine hydrochloride. Slow addition of **6** over 20 h to a 1.0 M solution of the amine hydrochloride (5.0 equiv) in 50% aqueous ethanol at 70°C affords a 63% yield of adducts **7** and **8** in a 2.5/1 ratio:

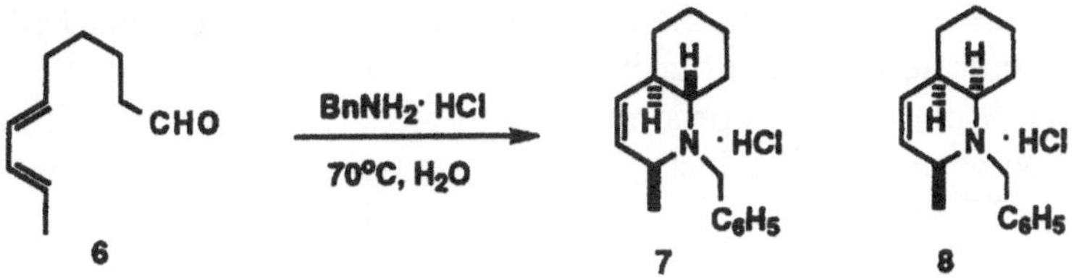

6   **7**   **8**

This seminal work set the stage for further improvements and extensions of the aqueous aza Diels–Alder reaction as well as providing practical methodological alternatives for the synthesis of heterocycles and alkaloids (section 2.3). The following reactions highlight additional straightforward examples of this methodology.

It was demonstrated that [²H]-diethoxymethane could be used as a deuteroformaldehyde equivalent with benzylamine hydrochloride and cyclohexadiene in the aqueous aza Diels–Alder reaction. The monolabelled 3-[²H]-2-benzyl-2-azabicyclo[2.2.2]oct-5-ene is obtained in 32% yield [2]:

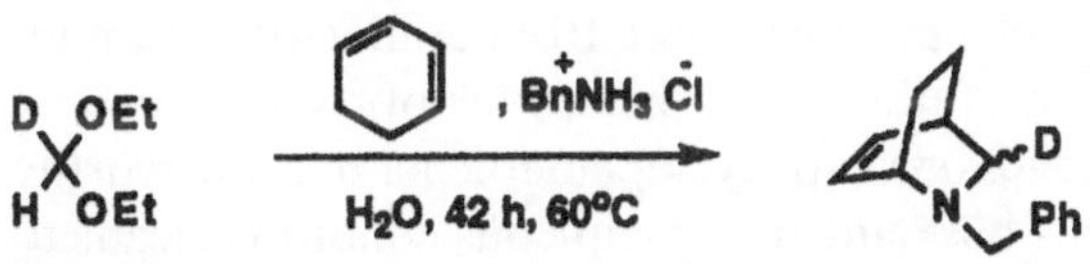

In another straightforward example of this methodology, investigators reacted the dihydrochloride salt of *N,N*-dimethylpropane-1,3-diamine with aqueous formaldehyde and cyclopentadiene to provide the expected 2-azanorbornene **9** in 60% yield [3]:

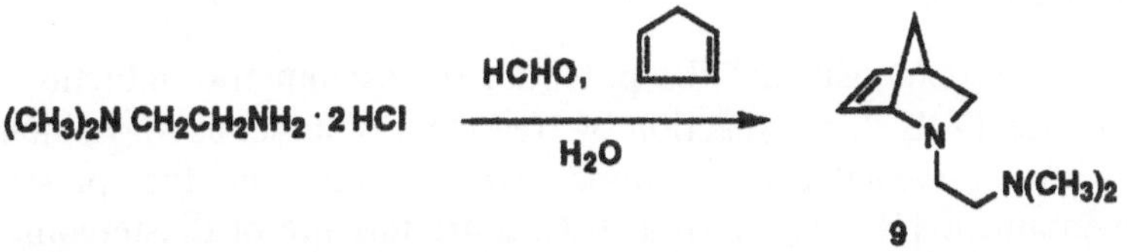

9

The diene, 11,12-dimethylene-9,10-ethanoanthracene (**10**), reacts with methyleneimine hydrochloride, transiently produced from ammonium chloride and formalin in 50% water–ethanol, yielding 2-azatriptycene (96%) [4]:

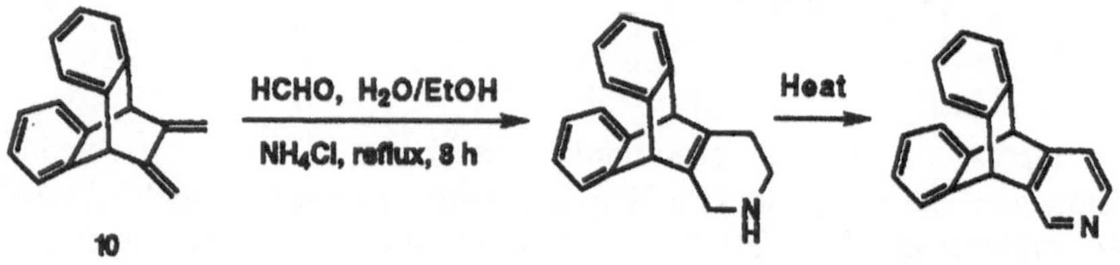

An interesting role reversal is observed in the cyclocondensation of cyclopentadiene with iminium ions derived from aryl amines and aldehydes [5]. The use of aniline as the amine component with formaldehyde and cyclopentadiene in aqueous acetonitrile fails to produce any of the corresponding *N*-phenyl-2-azanorbornene **11**. Instead, the novel tetrahydroquinoline-based pentacyclic products **12** and **13** are isolated in a near-quantitative yield as a 3.7/1 isomeric ratio:

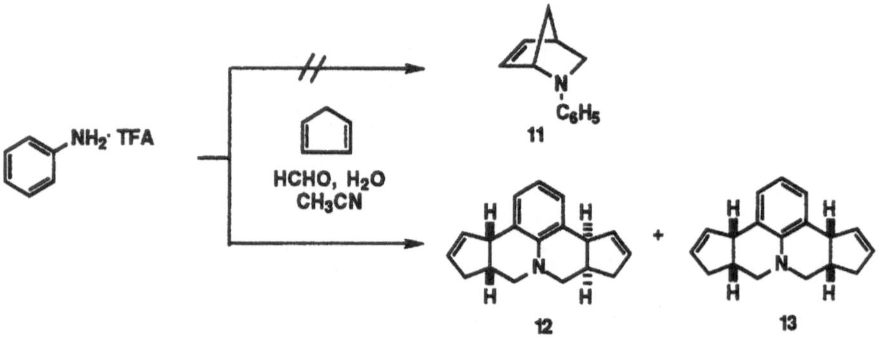

These pentacyclic products arise from aniline-derived iminium ions functioning as dienes rather than heterodienophiles in two sequential [4+2] cycloaddition reactions with cyclopentadiene. It is noteworthy that use of the preformed Schiff base under non-aqueous conditions leads to the formation of tricyclic tetrahydroquinoline derivatives wherein one cyclopentene unit is incorporated [5].

2.2.2   Asymmetric aza Diels–Alder reactions with simple protonated iminium ions in aqueous media

Grieco *et al.* demonstrated the potential for asymmetric induction in the aqueous aza Diels–Alder reaction by treatment of aqueous formaldehyde at 0°C with (–)-α-methylbenzylamine hydrochloride in the presence of cyclopentadiene [1]. After 20 h at 0°C, a 4/1 mixture of diastereomers was obtained in 86% yield. The absolute configurations of the diastereomers were later established to correspond to adducts **14** and **15** [6]:

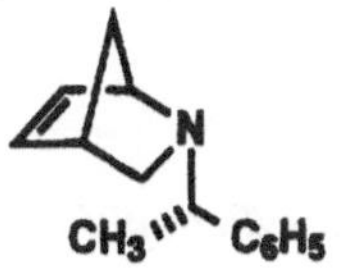

14 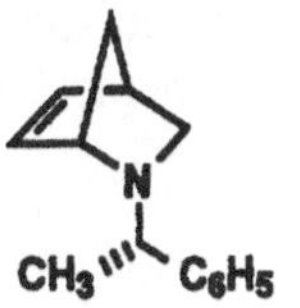   **15**

Waldman *et al.* [7] extended and improved the asymmetric methodology by utilizing (*R*)- and (*S*)-amino acid ester hydrochlorides as chiral auxiliaries in analogous aza Diels–Alder processes. A variety of chiral amino acids were evaluated, with isoleucine providing the best asymmetric induction. A typical protocol involves reacting (*S*)-isoleucine methyl ester hydrochloride with aqueous formaldehyde in the presence of the diene in water–tetrahydrofuran (11/1) at 0°C. Use of cyclopentadiene provides the corresponding azanorbornene products, **16** and **17**, in excellent stereoisomeric ratio (93/7) in 57% yield (Scheme 2.1). In the case of the less-reactive cyclohexadiene, the corresponding diastereomers are formed at 25°C in a 80/20 ratio in 35% yield.

The (S)-amino acid methyl ester auxiliary results in the (1*R*,4*S*) configuration of the fused bicyclic products; the (*R*)-configured auxiliary produces the (1*S*,4*R*) bridgehead stereochemistry. The observed stereochemistries are explained by assuming a compact dienophile conformation where the ester carbonyl is proximal to the imine group. The subsequent approach of the diene is governed by secondary orbital interactions with the ester carbonyl and steric interactions from the amino acid side-chain (see Scheme 2.1).

The reaction also proceeds with open-chain 1,3-dienes, giving the expected regiochemical control, although diastereoselectivities and yields are moderate. Use of isoprene and *trans*-2-methyl-1,3-pentadiene under the standard reaction conditions results in the formation of one regioisomer, as has previously been observed with simple achiral amines [1]. The reaction with *trans,trans*-2,4-hexadiene provided one pair of diastereomers. The absence of

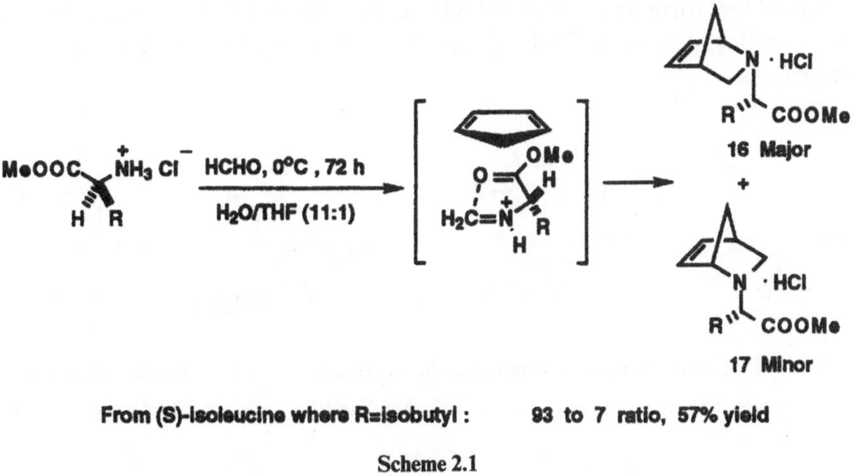

From (S)-Isoleucine where R=isobutyl :   93 to 7 ratio, 57% yield

Scheme 2.1

the other possible stereoisomers in this reaction is also indicative of a truly concerted process (Scheme 2.2).

2.2.3 Protonated C-acyl iminium ions as heterodienophiles

Grieco et al. have also reported the use of C-acyl iminium ions as hetero-dienophiles in the aqueous Diels–Alder reaction with cyclopentadiene [8]. This work details the cyclocondensation of C-acyl N-alkyl iminium ions **18** and C-acyl N-unsubstituted iminium ions **19** with cyclopentadiene in water:

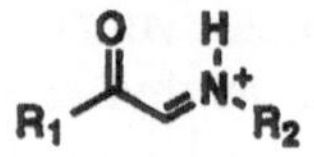

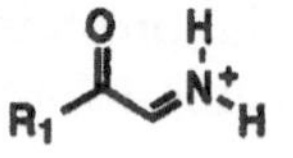

In a typical protocol, the C-phenacyl N-benzyl iminium ion **20** is generated from a 2.0 M aqueous solution of benzylamine hydrochloride (1.0 equiv) containing phenylglyoxal (1.2 equiv). Addition of neat cyclopentadiene (2.0 equiv) affords a heterogeneous mixture, which after 22 h at ambient temperature gives rise to an 88% yield of adducts **21** and **22** in a 3/2 ratio:

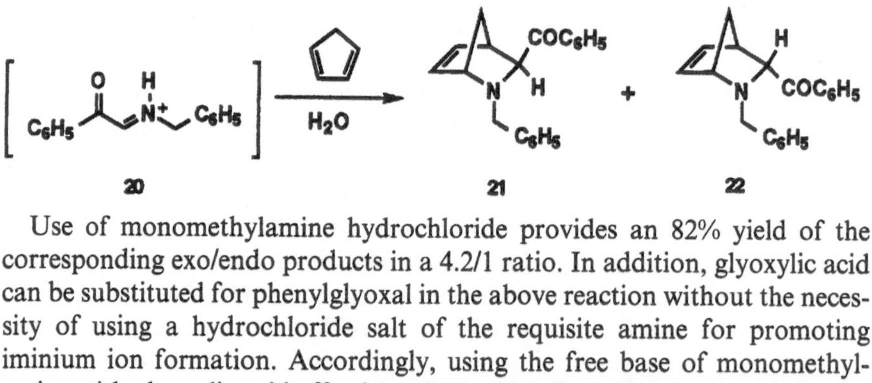

Use of monomethylamine hydrochloride provides an 82% yield of the corresponding exo/endo products in a 4.2/1 ratio. In addition, glyoxylic acid can be substituted for phenylglyoxal in the above reaction without the necessity of using a hydrochloride salt of the requisite amine for promoting iminium ion formation. Accordingly, using the free base of monomethyl-amine with glyoxylic acid affords an 86% yield of a 1.9/1 mixture of exo and endo acids.

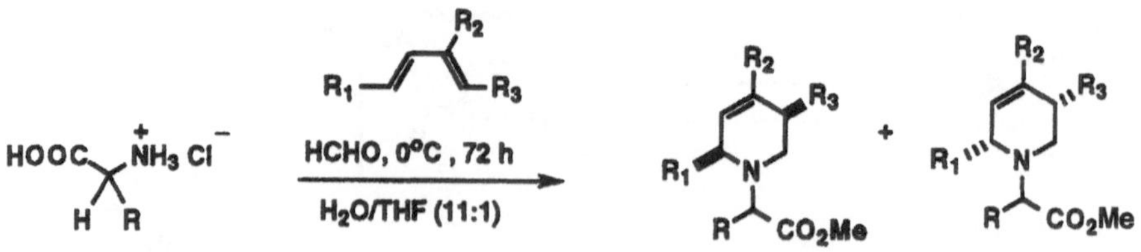

From (S)-isoleucine where $R_1 = R_2 = Me$, $R_2 = H$:   71 to 29 ratio, 45% yield
From (R)-phenylglycine where $R_1 = R_3 = Me$, $R_3 = H$:   37 to 63 ratio, 22% yield

Scheme 2.2

The *N*-unsubstituted iminium ion **23** generated *in situ* from an aqueous saturated ammonium chloride solution and phenylglyoxal reacts with cyclopentadiene providing an 89% yield of the exo/endo adducts **24** and **25** (1/2 ratio):

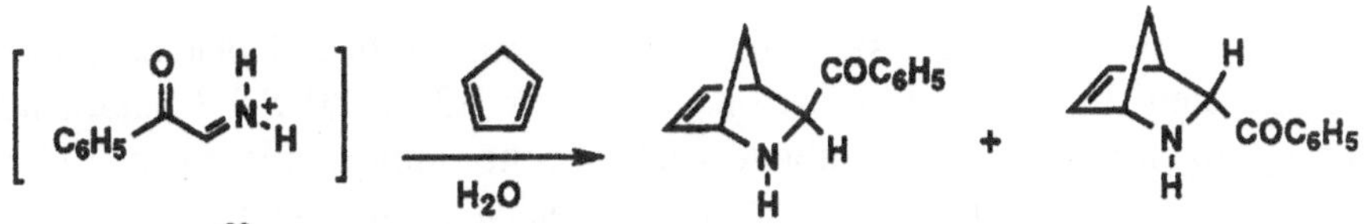

Use of pyruvic aldehyde in place of phenylglyoxal resulted in the corresponding Diels–Alder adducts in similar proportion and yield.

2.2.4 Lanthanide(III) triflate catalysis of the imino Diels–Alder reaction in aqueous media

Kobayashi *et al.* were the first to report the use of lanthanide triflates as catalysts in imino Diels–Alder reactions [9]. The successful three-component cyclocondensation of an aldehyde, amine and diene in acetonitrile was realized in the presence of lanthanide triflates as the Lewis-acid catalyst. This process is reminiscent of the protonated iminium-ion-based methodologies whereby the coordinated imine formation and [4+2] cycloaddition occur in one pot.

The same researchers found that lanthanide triflates could catalyze aldol reactions and allylations in aqueous media [10]. However, Wang *et al.* exploited the potential of lanthanide triflates to act as stable Lewis-acid catalysts in the aqueous imino Diels–Alder reaction [11]. This variant of the aqueous heterocycloaddition protocol also expands the scope of such reactions. The use of higher aldehydes in such reactions generally met with limited success under the conventional protocol. For example, under the standard conditions, the reaction of hexanal and benzylamine hydrochloride with cyclopentadiene in water is sluggish and affords only 4% of the Diels–Alder adducts **26** and **27** in a 2.7/1 ratio. In sharp contrast, the addition of various lanthanide(III) triflates (0.25 M) to this reaction results in substantial increases in both the rate and yield of Diels–Alder adduct formation. In particular, use of praseodymium(III) triflate results in a 68% yield of adducts **26** and **27**:

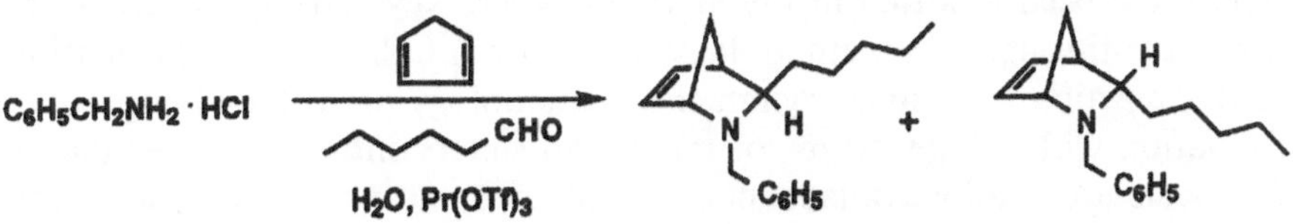

In general, lanthanide catalysis does not change the endo/exo product ratio as compared to the uncatalyzed reaction. Neutralization of the reaction mixture followed by extractive workup allows for recycling of the lanthanide-containing aqueous phase in subsequent reactions with no diminished catalytic effect.

Several other aldehydes were utilized, with similar rate and yield enhancements noted. In substituting cyclohexadiene or 2,3-dimethyl-1,3-butadiene for cyclopentadiene in the above example, no appreciable Diels–Alder adducts are detected. However, when formaldehyde is utilized with the same unreactive dienes, excellent yields of the corresponding heterocycloaddition products are observed. Again, this is in contrast to the reactions run without lanthanide mediation, which are substantially slower and lower yielding.

Noteworthy is the neodymium(III)-triflate-catalyzed reaction of cyclohexadiene with formalin and L-phenylalanine methyl ester, which provides a 3/1 ratio of diastereomers **28** and **29** in an 84% yield. The uncatalyzed reaction results in only a 27% yield of **28** and **29**:

Among the seven lanthanides screened, praseodymium(III), ytterbium(III) and neodymium(*III*) triflates were shown to be most effective. Magnesium chloride and lithium chloride showed no substantial influence on these reactions, thereby ruling out the salt effect for explaining the rate and yield enhancements of lanthanides. The details of the catalytic mechanism of this intriguing lanthanide-mediated effect are under investigation.

2.3   Synthetic exploitation of the aqueous aza Diels–Alder reaction

The availability of practical and convenient aqueous imino Diels–Alder methodology allows for the synthesis of multigram quantities of the corresponding nitrogen heterocycles. It is not surprising that synthetic applications have been reported in the literature where key starting materials are generated through this methodology. Furthermore, the successful extension of this reaction to include asymmetric methodology has resulted in product formation with a high degree of regio- and stereocontrol. The asymmetric protocols use readily available chiral auxiliaries that provide easy access to enantiopure heterocycles for exploitation in alkaloid syntheses. The following is a brief account of the synthetic applications of the aqueous aza Diels–Alder reaction.

2.3.1 Synthetic applications of the aqueous aza Diels–Alder reaction involving simple protonated iminium ions

Ready availability [12] of azabicyclo[2.2.1]heptene and azabicyclo-[2.2.2]octene derivatives has facilitated their use as starting points for accessing other bicyclic systems. For example, the bicyclic piperidones **32** and **33** are prepared in 61% and 53% yield by reacting substituted 2-azabicyclo[2.2.1]hept-5-enes **30** and **31** respectively with diphenylketene [13]:

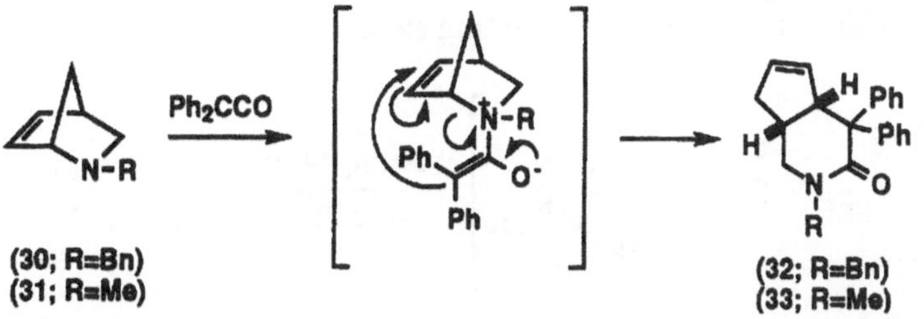

(30; R=Bn)
(31; R=Me)

(32; R=Bn)
(33; R=Me)

The mechanistic rationale put forth for product formation involved ketene attack on the amine moiety followed by an amino Claisen rearrangement of the zwitterionic enolate. An analogous reaction is observed with methyl propiolate and fused azabicyclic systems [14].

Interestingly, reaction of diphenylketene with spirocyclic azanorbornene **34**, prepared by the method of Grieco et al. [1] from spirocyclopropyl-cyclopentadiene and benzyliminium ion in 62% yield, results in the formation of an unstable tricyclic intermediate **35**, which upon workup gives allylic alcohols **36** in 51% yield (Scheme 2.3) [13]. A single cyclopentenone derivative

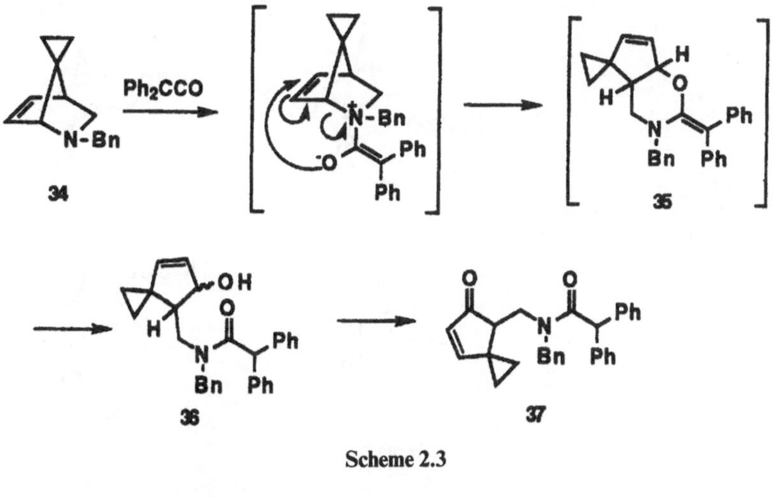

34

35

36

37

Scheme 2.3

37 is obtained on oxidation. Product formation is explained by initial attack of diphenylketene, as before, except with the oxyanion participating in the subsequent rearrangement. No explanation was given for the different reaction course as compared to the previous equation.

2-Methylazabicyclo[2.2.1]octene and 2-methylazabicyclo[2.2.2]heptene are also convenient precursors to the heretofore-unknown 3,6-disubstituted 3,6-diazabicyclo[3.2.2]nonane and 3,6-disubstituted 3,6-diazabicyclo[3.2.1]-octane skeletons respectively [15]. The methodology entails subjecting the hydrochloride salt of the azanorbornene starting material to an ozonolysis–reductive amination sequence. For example, the tandem transformations result in a 40% yield of the corresponding 3,6-disubstituted 3,6-diazabicyclo[3.2.1]octane derivative using benzylamine/sodium cyano-borohydride in the reductive amination step:

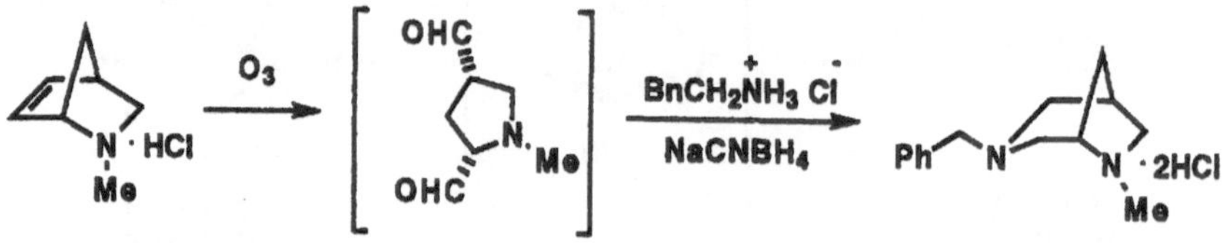

As part of a medicinal chemistry program [16] directed toward discovering new muscarinic receptor ligands, a series of endo- and exo-2-alkyl-2-aza-bicyclo[2.2.1]heptan-5-ol (**38** and **39**) and endo- and exo-2-alkyl-2-aza-bicyclo[2.2.1]heptan-6-ol (**40** and **41**) esters have been synthesized from 2-alkyl-2-azabicyclo[2.2.1]heptan-5-enes (Scheme 2.4). Oxidation of the olefin and stereochemical manipulation of the secondary alcohols followed by esterification of the azanorbornene products provided the different isomers required for systematic investigation of the structure–activity relationships of this structural series.

The feasibility of employing the intramolecular iminium-ion-based Diels–Alder reaction for the construction of nitrogen-containing ring systems

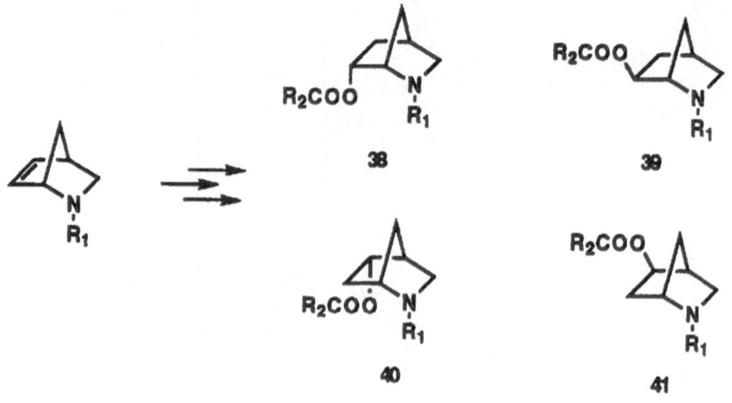

Scheme 2.4

was established in the preliminary studies of Grieco *et al.* [1]. Several applications of this methodology from this laboratory have been applied toward the synthesis of natural alkaloids as a structural framework to explore and extend the scope of the aqueous aza Diels–Alder protocol.

An application toward the synthesis of alloyohimbane required reacting the protected tryptamine derivative **42** with cyclohexadiene and formalin to obtain the requisite [2.2.2] bicyclic adduct **43** [17]. Under the standard aqueous aza Diels–Alder conditions, a modest 22% yield of **43** is obtained, which upon further elaboration involving Mariano's [18] aza Claisen rearrangement protocol yields alloyohimbane **44** (Scheme 2.5).

Grieco *et al.* have also probed an intramolecular variant of the iminium reaction with cyclohexadiene [19]. The intramolecular cyclization of cyclohexadienyl aldehyde **45** provides the precursor to dihydrocannivonine **46**. Initial attempts to execute the intramolecular iminium-ion Diels–Alder reaction under standard conditions gave rise to poor yields (<10%), probably due to the inherent instability of dienyl aldehyde **45**. Yields are dramatically improved by slow addition of a solution of **45** in ethanol over 20–30 h to monomethylamine hydrochloride in aqueous ethanol at 70°C. This modified protocol results in a 66% yield of tricyclic amine **46**. Subsequent hydrogenation of **46** affords racemic dihydrocannivonine **47**:

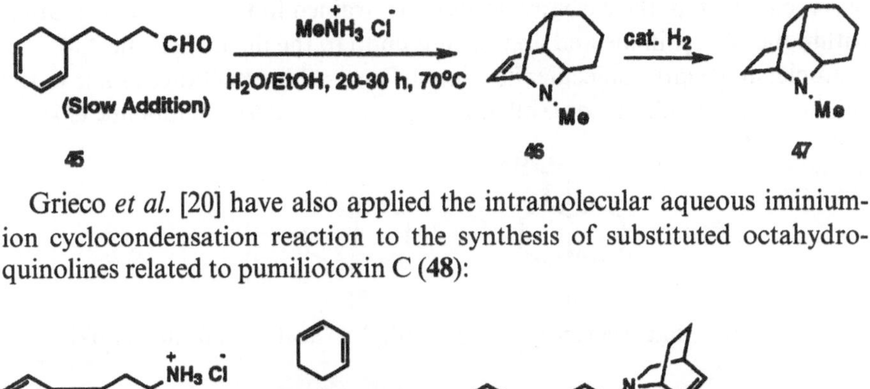

Grieco *et al.* [20] have also applied the intramolecular aqueous iminium-ion cyclocondensation reaction to the synthesis of substituted octahydroquinolines related to pumiliotoxin C (**48**):

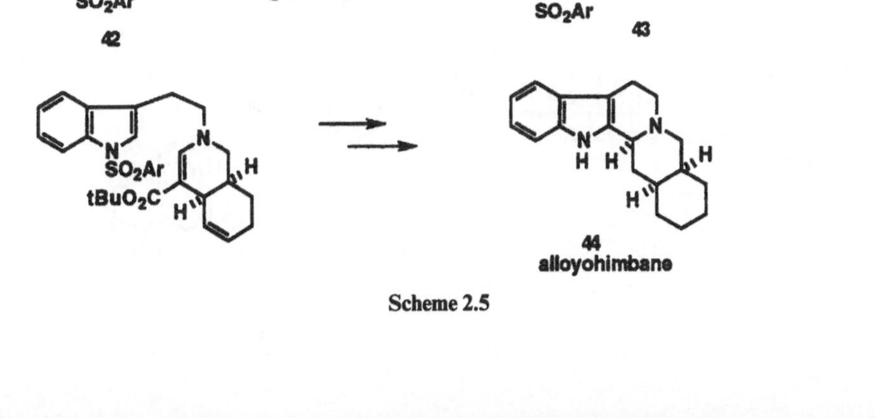

Scheme 2.5

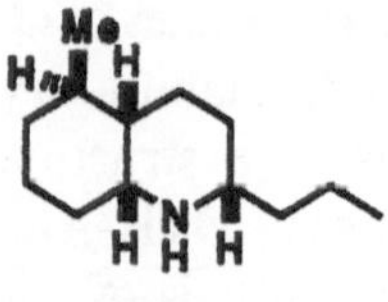

48

Accordingly, a dilute ethanol solution of chiral dienyl aldehyde **49** is treated with an equal volume of a saturated aqueous ammonium chloride solution at 75°C for 48 h. Workup provides a 55% yield of two octahydroquinolines, **50** and **51**, in a 2.2/1 ratio (Scheme 2.6); no other stereoisomers were detected.

The stereochemical rationale put forth for the observed stereoisomers involves examination of four transition states (**I–IV**). Adducts **52** and **53** would arise from the chair-like conformations **I** and **II** respectively (Scheme 2.7). The formation of **52** as the major product was anticipated since the related transition-state conformation **II** leading to **53** is destabilized by an eclipsing interaction between H_a and H_b. Not surprisingly, the octahydro-quinoline products, **54** and **55**, derived from the two boatlike conformations, **III** and **IV**, are not detected.

It is noteworthy that, in the corollary intramolecular imino Diels–Alder reaction reported by Weinreb *et al.* where *N*-acyl imines are employed as heterodienophiles [21], exclusive preference for boat-like transition states is observed, wherein the carbonyl group is restrained in the *s-cis* conformation within the *N*-acyl imine and approach is endo to the diene (Scheme 2.8). For example, in the intramolecular acyl imine Diels–Alder reaction *en route* to the synthesis of the quinolizidine alkaloids epi-lupinine (**56**) and lupinine (**57**):

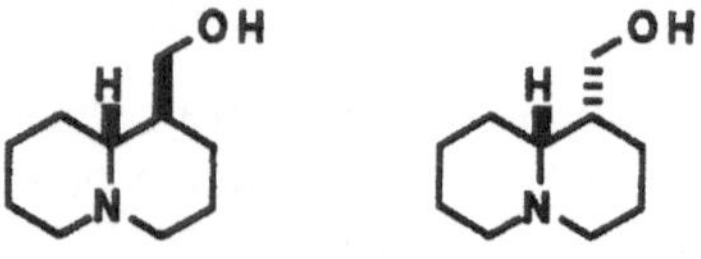

56   **57**

it was found that thermolysis of methylol acetate **58** in refluxing *o-*

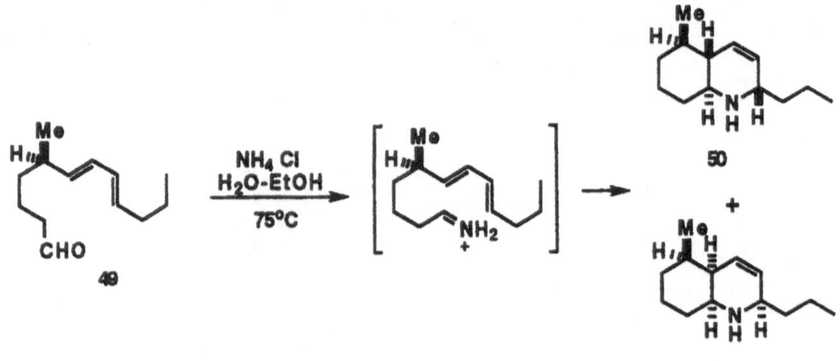

Scheme 2.6

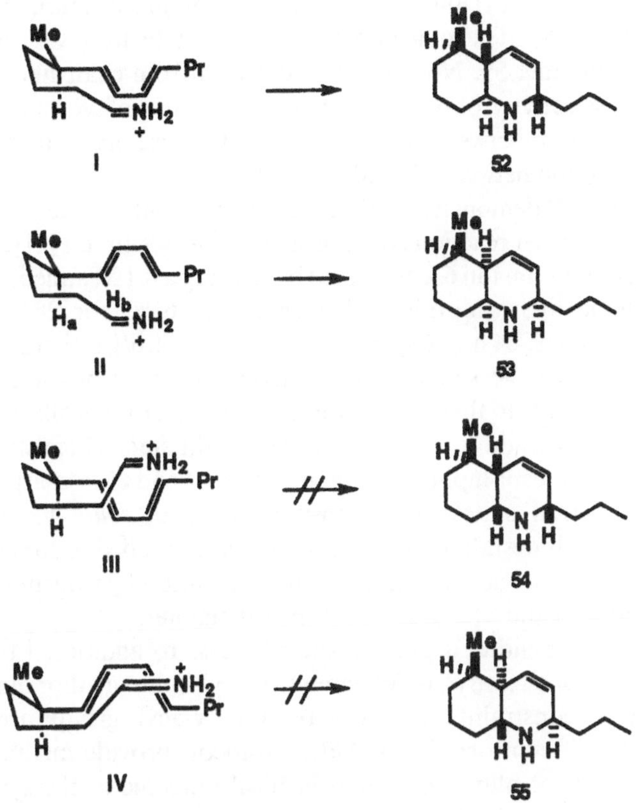

Scheme 2.7

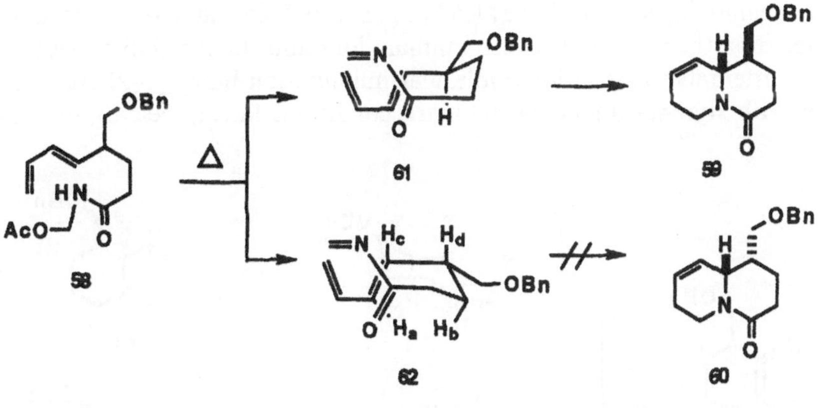

Scheme 2.8

dichlorobenzene provides solely bicyclic lactam **59**, the precursor to epi-lupi-nine (Scheme 2.8). None of the epimeric lactam **60** with the relative stereo-chemistry embodied in lupinine is detected. These results are rationalized by consideration of the two transition states **61** and **62** that emerge from the

requirement of the *N*-acyl imine to adopt the *s-cis* endo transition-state orientation (Scheme 2.8). The quasi-boat transition state **61** gives rise to the
observed cycloadduct **59**. No product is detected from chair-like transition
state **62**. The selectivity observed is attributed to an unfavorable, 1,4-non-
bonded interaction between H_a and H_b and, more importantly, to an
eclipsing interaction between H_c and H_d in **62**.

Grieco *et al.* [22] demonstrated that the intramolecular aqueous iminium-
ion-based Diels–Alder reaction is complementary to Weinreb's [21] acyl imine
methodology. Iminium ion **63**, generated by treating a 0.1 M aqueous solution
of dienyl amine hydrochloride with formalin at 65°C for 28 h, provides
Diels–Alder adducts, **64** and **65**, in an 82% yield in a 1.6/1 ratio (Scheme 2.9)
[22]. In contrast to the stereospecific reaction pathway of the acyl imine
methodology *en route* to the precursor to epi-lupinine, the corollary iminium-
ion-based [4+2] cycloaddition provides access to the stereochemistry *en route*
to both lupinine and epi-lupinine. Both products, **64** and **65**, can arise through
chair-like transition states. The transition state *en route* to the minor
stereoisomer (**65**) is destabilized by the interactions alluded to for chair transition state **62** in Scheme 2.8. Catalytic hydrogenation/hydrogenolysis of **64**
and **65** provides **56** and **57** in a straightforward manner.

The iminium-ion-induced process allows access to additional transition-
state conformations relative to Wienreb's *N*-acyl imine methodology, since
transition-state constraints imposed by the *N*-acyl group are absent.
Therefore, the different aza Diels–Alder protocols provide methodological
alternatives for controlling the stereochemical outcome in the synthesis of
alkaloids embodying a bridgehead nitrogen.

Applications of the aqueous iminium-ion-based Diels–Alder reaction to
more demanding systems have also been explored. One area of investigation
centered on the reaction between iminium ions and stabilized diene systems.
Initial attempts to effect intermolecular iminium-ion-based [4+2] cycloadditions with styrenes under the standard conditions failed, yielding only by-

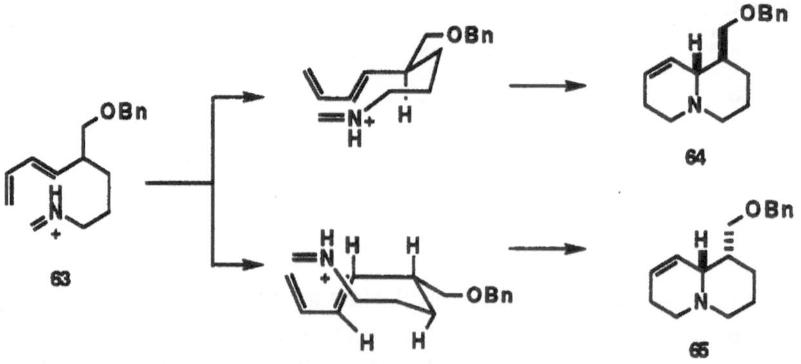

Scheme 2.9

products as shown in Scheme 2.10 [23]. However, subsequent studies success-fully applied the intramolecular protocol to the construction of the natural quinolizidines, julandine **68** and cryptopleurine **70** [22].

The key step in the synthesis of julandine (**68**) is executed by treatment of an ethanol–water (1/1) solution of the diaryl amine hydrochloride **66** with formalin at 110°C in a sealed tube for 23 h. A single diastereomer, **67**, is obtained in 60% yield and converted to julandine by an acid-catalyzed double-bond isomerization:

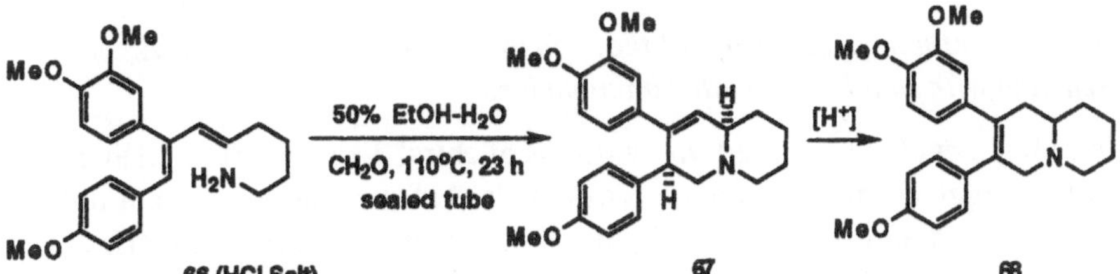

Similarly, the hydrochloride of phenanthryl amine **69** yields cryptopleurine **70** (84%) in an ethanol–water (1/1) formalin solution at 180°C for 10 h.

The higher reaction temperature is required in the imino Diels–Alder reaction leading to cryptopleurine, relative to julandine, since lower reaction temperatures give rise to numerous uncharacterized by-products. This is not surprising since effecting the [4+2] cycloaddition to produce cryptopleurine involves disruption of the aromatic phenanthrene unit followed by isomerization *in situ* to regain aromaticity.

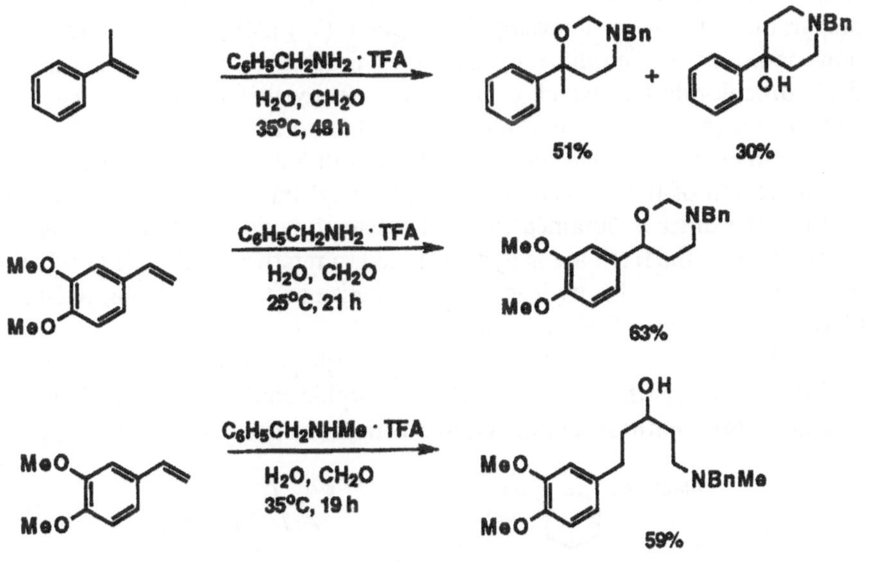

Scheme 2.10

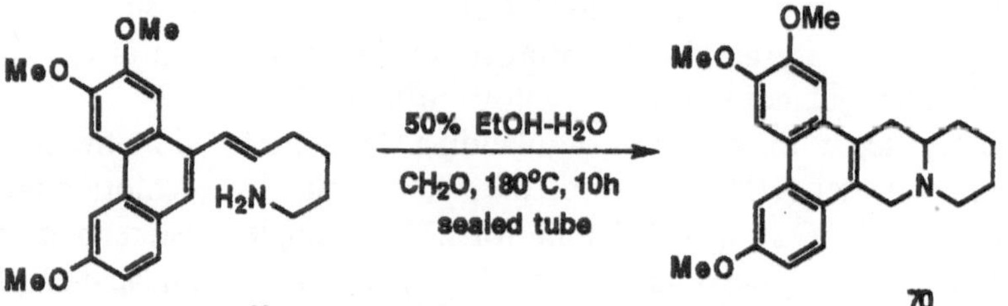

2.3.2   Synthetic applications of the asymmetric aqueous aza Diels–Alder reaction with simple protonated iminium ions

Waldman *et al.* [7] reported the synthesis of chiral 2-azabicyclo[2.2.1]heptane (**72**) through asymmetric aqueous aza Diels–Alder methodology that utilizes (*R*)- and (*S*)-amino acid ester hydrochlorides as the chiral auxiliaries. The chiral azanorbornene **71** derived from (*S*)-phenylglycine is deprotected by hydrogenolysis followed by hydrogenation of the remaining olefin. The resulting (–)-2-azabicyclo[2.2.1]heptane (**72**) is isolated as the hydrochloride salt:

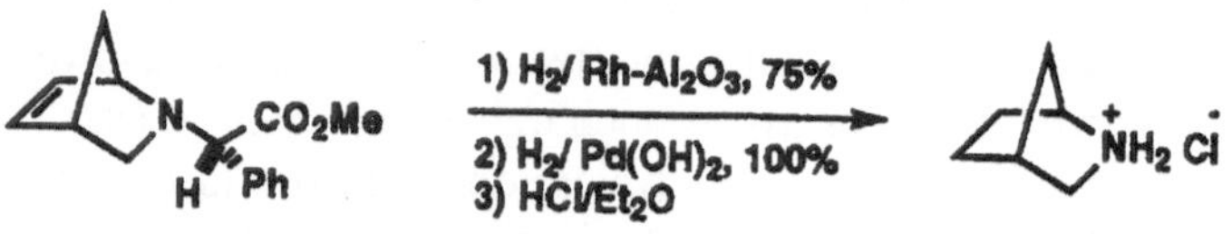

Recently, Grieco's asymmetric protocol [1], utilizing chiral methylbenzylamine hydrochloride as the chiral auxiliary, has been extended to large-scale preparation of (–)-2-azabicyclo[2.2.1]heptane (**73**) [24]. The standard reaction conditions were modified by using the chiral amine acetate salt instead of the hydrochloride salt, which results in identical yields (80%) and diastereomeric ratios (80:20) of Diels–Alder adducts **14** and **15**. The mixture of amine diastereomers is separated on a large scale with high recovery by fractional crystallization of the corresponding L-dibenzoyl tartaric acid (DBTA) salt. The major isomer is obtained in good diastereomeric purity (>95%). After neutralization, the free base is hydrogenated to remove the double bond and then deprotected under hydrogenolysis conditions to provide chiral 2-azabicyclo[2.2.1]heptane (**73**) (Scheme 2.11).

Rapid access to the chiral azabicyclo[2.2.1]heptane backbone has been exploited in the synthesis of rigid enantiomorphic conformations (e.g. **75**) of arecoline (**74**), a natural achiral alkaloid with muscarinic agonist activity [25]:

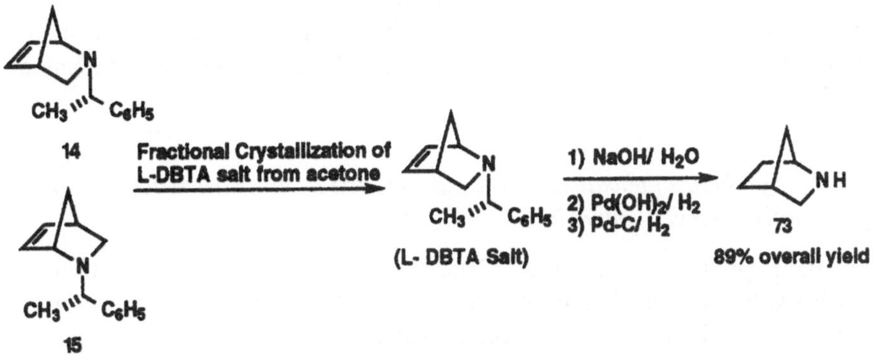

Scheme 2.11

The two-step process effecting the rearrangement of the chiral azanor-bornene backbone commences through treatment of **76** with bromine, providing aziridinium **77**, followed by exposure to tetraethylammonium cyanide (Scheme 2.12). Stereospecific S_N2 ring opening of the aziridine results in an overall clean inversion to the substituted azanorbornene **78**, which is appropriately functionalized for straightforward elaboration to **75**. The enantiomeric series of compounds is prepared by using (S)-phenylethylamine as the starting chiral auxiliary.

The (S)-phenylethylamine-derived norbornene **14** also serves as the starting point for the synthesis of the natural alkaloid, (–)-γ-N-normethylsky-tanthine (**79**) [26]. The key transformation of the 2-azabicyclo[2.2.1]heptane skeleton into the 3-azabicyclo[4.3.0]nonane backbone, embodied in the natural alkaloid, involves the ketene amino Claisen rearrangement as reported by Roberts *et al.* [27] (see Scheme 2.3 and the prior equation in section 2.3.1). Treatment of **14** with dichloroketene, generated *in situ* from dichloroacetyl chloride and Hunig's base, at 2°C for 16 h, gives lactam **80** in

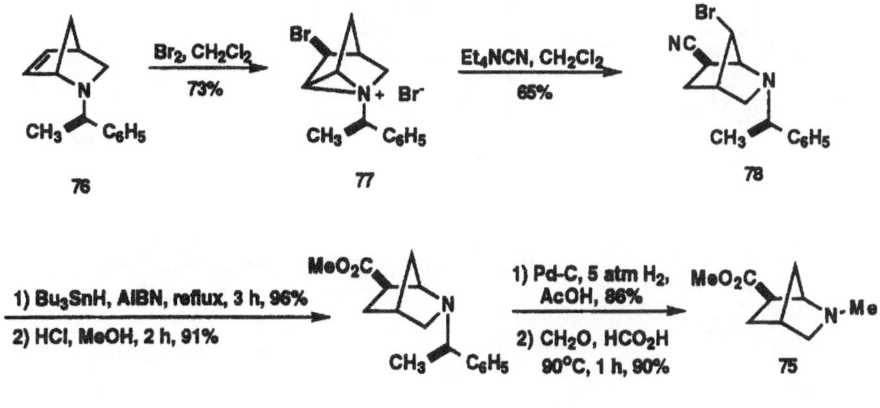

Scheme 2.12

61% yield (Scheme 2.13). Reductive dehalogenation with excess zinc in methanol and ammonium chloride provides intermediate **81**, which is further elaborated to (–)-γ-*N*-normethylskytanthine (**79**) and other related members of this family [28].

2.3.3   Synthetic applications of the aqueous aza Diels–Alder reaction with protonated C-acyl iminium ions

In their preliminary studies, Grieco *et al.* [8] demonstrated that the Diels–Alder adducts formed with cyclopentadiene and *C*-acyl iminium ions can be used in the construction of substituted cyclopentene derivatives. Diels–Alder adducts **24** and **25** (see section 2.2.3) are readily converted into cyclopentene derivative **82** in 76% yield upon exposure to zinc in glacial acetic acid. Cyclopentene derivatives such as **82** can serve as precursors to carbocyclic analogs of purine ribo- and deoxyribonucleosides [29]:

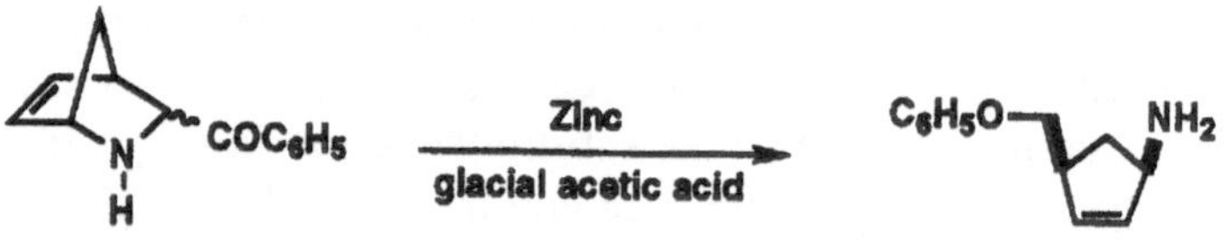

A novel acid-induced rearrangement of *N*-acylated 2-azabicyclo[2.2.1]-hept-5-ene-3-carboxylic acid into the 2-oxabicyclo[3.3.0]oct-7-en-3-one skeleton has been reported by Kobayashi *et al.* [30]. The requisite starting azanorbornene was formed in 84% yield via the aqueous aza Diels–Alder reaction of iminium ion **83**, generated from saturated ammonium chloride and ethyl glyoxylate, and cyclopentadiene. Acylation of the nitrogen with benzoyl- or *p*-nitrobenzoyl chloride followed by alkaline hydrolysis of the

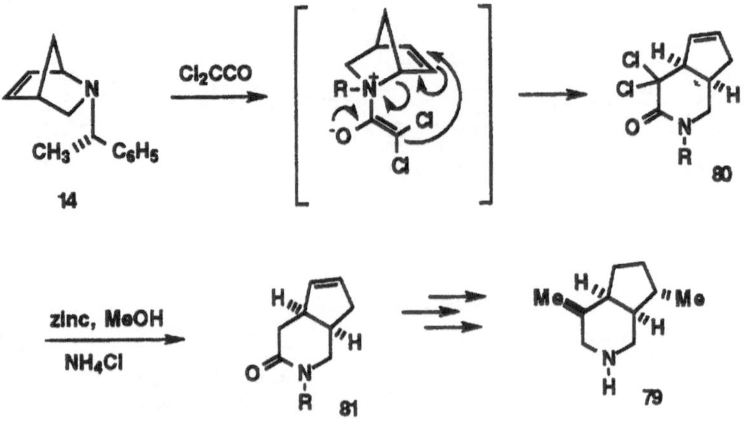

Scheme 2.13

isolated exo/endo isomers, without epimerization, provides the 3-exo- and 3-endo-carboxylates **84** and **85**, respectively. Treatment of the separate exo/endo isomers with trifluoroacetic acid mediates a stereospecific azanor-bornene backbone rearrangement to produce bicyclic lactones **86** and **87** in good yield (Scheme 2.14).

The mechanistic rationale for this transformation involves the acid-induced cleavage of the allylic C–N bond and carboxyl trapping of the resulting stabilized cation from the same face. A stereospecific stepwise process or a concerted hetero [3,3] sigmatropic rearrangement (in the endo isomer case) can explain the observed relative stereochemical outcome.

Another reported [31] transformation starting with Diels–Alder adducts **88** and **89**, formed from cyclopentadiene and *C*-acyl iminium ion **90**, exploits the intrinsically weak bond between the allylic nitrogen and the bridgehead carbon to effect a palladium-catalyzed hydride reduction, forming cyclopent-enyl glycine products. Under an optimized set of conditions, treatment of exo adduct **88** with sodium cyanoborohydride (2 equiv) in the presence of 10% palladium tetrakistriphenylphosphine in a refluxing tetrahydrofuran solution for 16 h provides the cyclopentenyl glycine derivatives **91** and **92** (7/1) in 77% yield (Scheme 2.15). Applying the same set of conditions to the endo adduct **89** results in a lower yield (45%) of the corresponding cyclopentenyl glycine derivatives **93** and **94** with less regioselectivity (1.3/1 ratio).

Various fused rings have been incorporated onto the 2-azanorbornyl skeleton to provide novel heterocycles by Kobayashi *et al.* [32]. The aqueous cycloaddition adduct **95** is converted into hydantoin **96** and thiohydantoin **97**

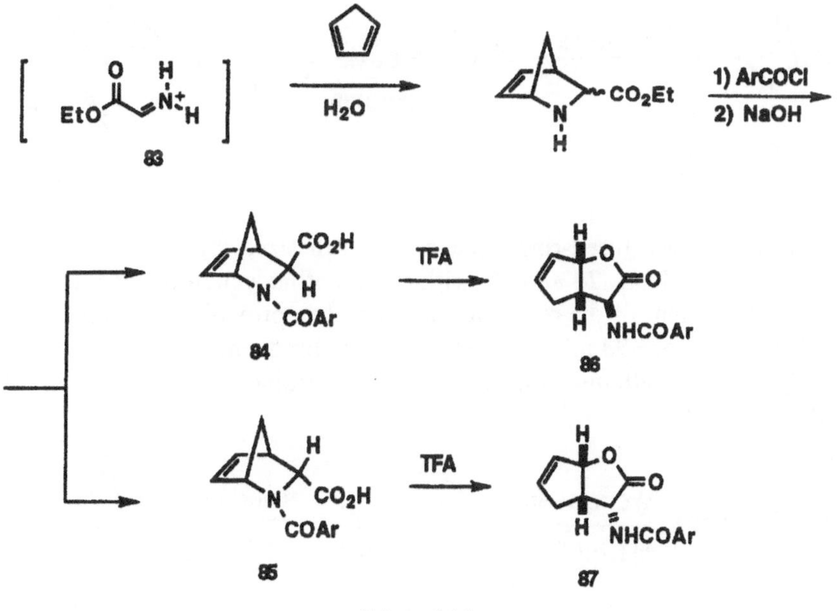

Scheme 2.14

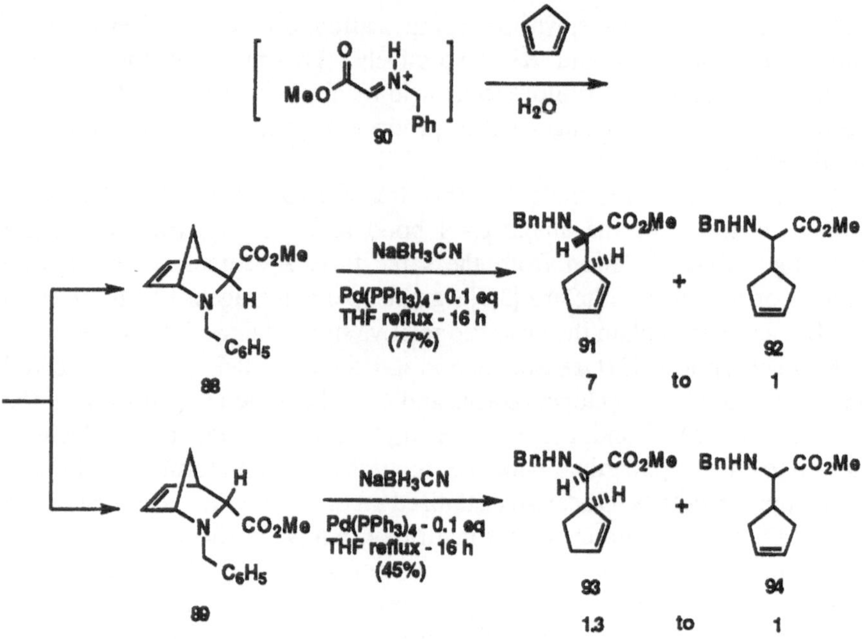

Scheme 2.15

by reacting with phenyl isocyanate or isothiocyanate followed by treatment of the resulting ureas with 1,8-diazabicyclo[5.4.0]undec-7-ene in refluxing benzene:

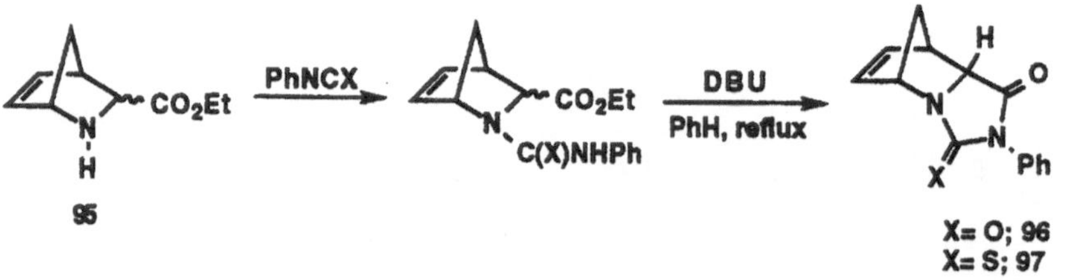

Another novel fused-ring system is synthesized by reacting the aza Diels–Alder adducts, **24** and **25**, with ethyl isothiocyanatoformate followed by base treatment of the resulting thiourea, thus providing 1,3-dihydro-2*H*-imidazole-2-thione **98** in 48% overall yield. Subsequent desulfurization with Raney nickel in ethanol provides the novel strained imidazole **99** in 29% yield:

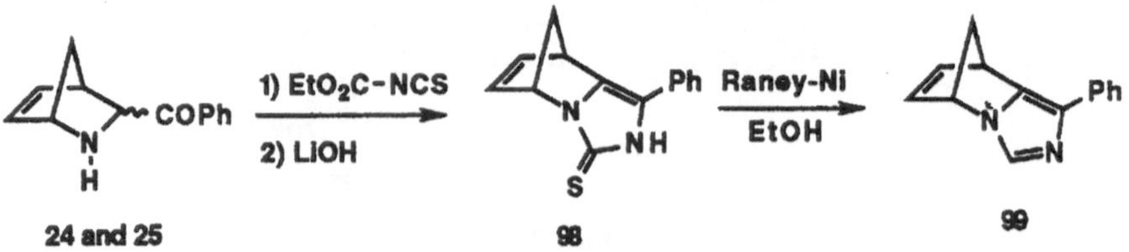

2.4 Retro aza Diels–Alder reactions in aqueous media

2.4.1 Acid-catalyzed heterocycloreversion of 2-azanorbornenes

The imino variant of the retro Diels–Alder process is known to require exceedingly high activation energies. Heterocycloreversion of 2-azanorbornene has traditionally been conducted at temperatures in the range of 400–600°C *in vacuo* [33]:

Until recently, these extreme conditions have, for the most part, precluded extensive investigation and exploitation of the imino variant of the retro Diels–Alder process. In 1987, Grieco *et al.* reported that 2-azanorbornene as well as its *N*-alkyl derivatives undergo smooth acid-catalyzed heterocycloreversion at room temperature in water [34]. The reverse [4+2] reaction generates the transient iminium ion, providing the unmasked amine component after hydrolysis. The resultant cyclopentadiene is trapped *in situ* with *N*-methylmaleimide, allowing smooth heterocycloreversion. For example, benzylamine is isolated in 81% yield after warming an aqueous solution of *N*-benzyl-2-azanorbornene hydrochloride to 50°C for 2 h in the presence of *N*-methylmaleimide (Scheme 2.16). The corresponding Diels–Alder adduct **100** is isolated in 91% yield (cf. the first equation in section 2.2.1).

The reaction in Scheme 2.16 proceeds, albeit more slowly, at room temperature in water; however, no detectable reaction occurs in benzene, acetonitrile, or tetrahydrofuran even at 50°C. At elevated temperatures (80°C) in organic solvents, products of the retro Diels–Alder process are observed in

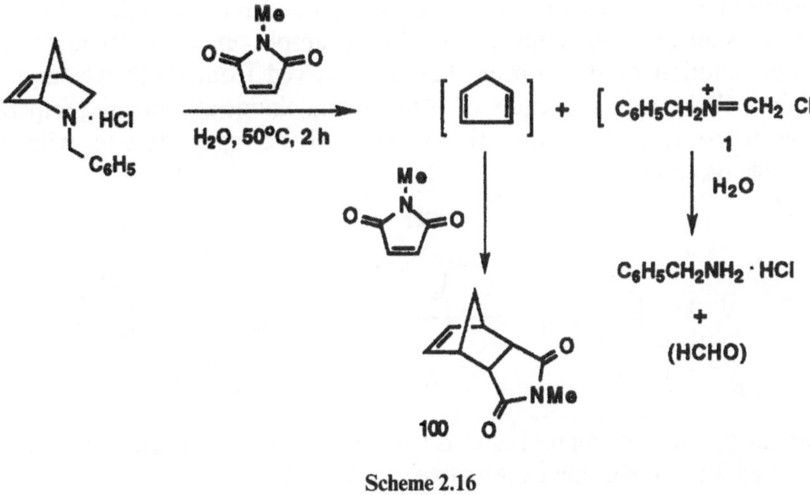

Scheme 2.16

substantially lower yields. In general, the optimal conditions for conducting the heterocycloreversion of 2-azanorbornenes involve heating the hydrochloride salt in water at 50°C in the presence of N-methylmaleimide. Under these conditions, the parent 2-azanorbornene undergoes the retro Diels–Alder reaction to provide an 80% yield of trapped cyclopentadiene adduct **100**:

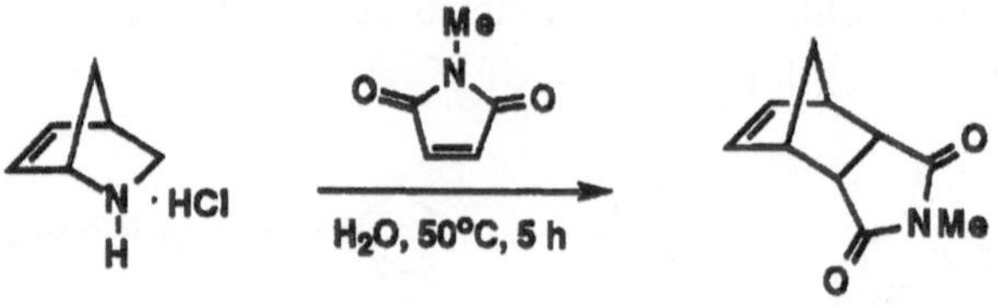

These relatively mild conditions stand in sharp contrast to those employed in the first equation in this section.

In the case of N-homoveratryl-2-azanorbonene, heterocycloreversion gives rise to a 61% yield of homoveratrylamine (**103**) along with 18% of the Pictet–Spengler cyclization product **102** derived from internal trapping of iminium ion **101**:

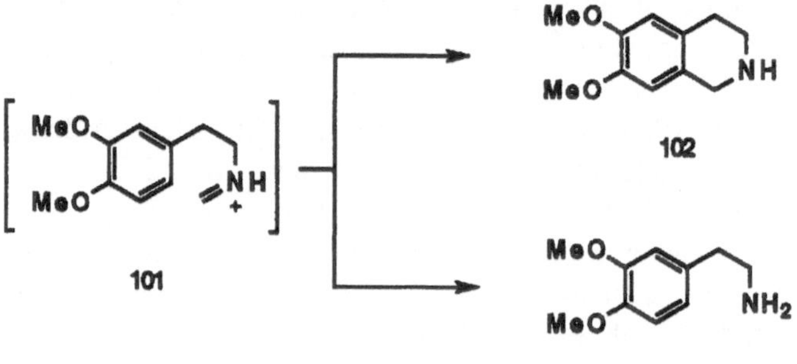

More hindered N-substituted 2-azanorbornenes smoothly undergo heterocycloreversion at room temperature. For example, exposure of a 0.84 M aqueous solution of the 2-azanorbornene derived from L-phenylalanyl-L-leucine methyl ester to 1.3 equiv of N-methylmaleimide at ambient temperature gives rise to a nearly quantitative yield of the dipeptide ester with no detectable racemization:

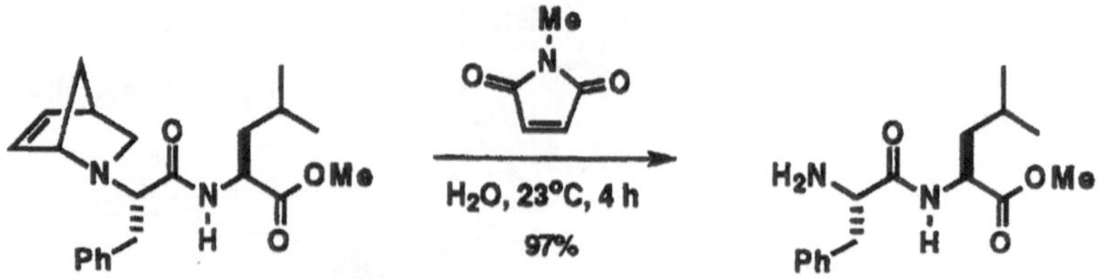

Similarly, L-leucine methyl ester is recovered stereochemically intact from its masked 2-azanorbornene derivative in 95% yield.

Grieco *et al.* [35] have also reported alternative methods for effecting the cycloreversion of 2-azanorbornenes that do not require the *in situ* trapping of cyclopentadiene with a reactive dienophile. The first method utilizes catalytic amounts of copper(II) to initiate the reverse [4+2] process on the parent 2-azanorbornene as the free amine. The general reaction conditions involve heating a 50% aqueous ethanol solution of the 2-azanorbornene in the presence of a catalytic amount of copper sulfate. The protocol has been applied to several substrates. One illustrative example details the protocol for *N*-homoveratryl-2-azanorbornene (**104**):

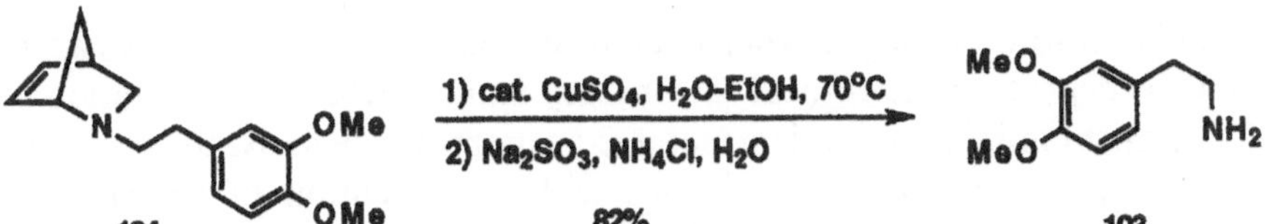

A 1.0 M solution of the parent 2-azanorbornene (**104**) in 50% aqueous ethanol at 70°C is treated with 0.05 equiv of copper sulfate for 2 h. The reaction mixture is worked up after addition of 0.05 equiv of sodium sulfite and 0.1 equiv of ammonium chloride at ambient temperature, thus providing an 82% yield of homoveratrylamine (**103**).

It is noteworthy that none of the cyclized product **102** is detected in the copper(II)-promoted process as compared to the conventional acid-induced reaction (see above) which produces 18% of **102**. This underscores a major difference associated with the copper(II)-catalyzed process, which obviates potentially reactive intermediate iminium ions such as **101**. However, while the copper(II)-catalyzed process is milder in some respects as compared to the original acid-induced protocol, some limitations do apply. For example, during attempts to unmask the terminal amine function of simple di- and tripeptide esters, considerable ester hydrolysis occurs under these alternative conditions. Another procedure was developed both to circumvent this problem and to eliminate the requirement of an internal cyclopentadiene trapping agent. The new protocol utilizes the sulfonic-acid-based ion-exchange resin with 2% crosslinking, Bio-Rad AG 50W-X2, 200–400 mesh. Reactions are conducted between 40 and 70°C, employing 2–3 weight equivalents of resin. An illustrative example involves unmasking the 2-azanorbornene derivative of L-phenylalanyl-L-leucine methyl ester (see above). A solution of **105** in water–ethanol (1/2) is exposed (2 h) to 2.0 wt equiv of resin at 40°C, providing after workup an 80% yield of the dipeptide ester **106**, stereochemically intact.

The ability to incorporate a primary amino function readily into the 2-azanorbornene framework, coupled with mild and efficient methodological options for reversion back to the starting amine, renders this overall process a viable new protection–deprotection scheme.

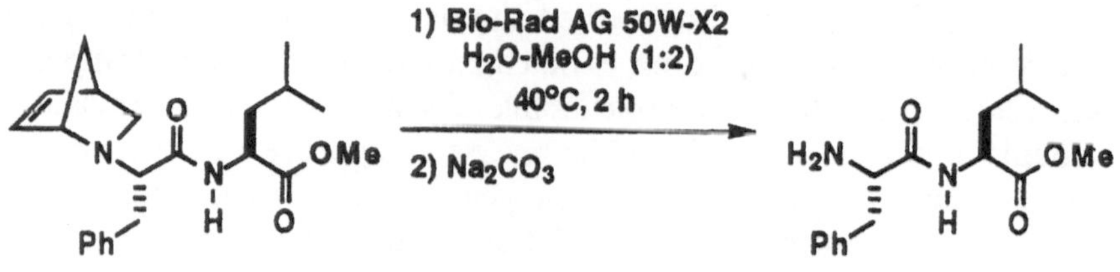

2.4.2 Synthetic applications involving the heterocycloreversion of 2-azanor-bornenes

In addition to utilizing the heterocycloreversion of 2-azanorbornenes as an aqueous deprotection protocol for primary amines, where the formation of the transient iminium ion is incidental, 2-azanorbornene derivatives can also serve as precursors to iminium ions with the intention of synthetically exploiting these unstable intermediates. Such heterocycloreversion applications do not involve the use of aqueous media to avoid undesired hydrolysis of the newly generated iminium ion. The next few examples illustrate the use of 2-azanorbornenes as iminium-ion synthons unmasked by the reverse Diels–Alder process (see section 2.4.1).

The first synthetic application of this type features a novel method for the N-methylation of amino acid and polypeptide derivatives [36]. This methodology involves trapping of the incipient iminium ions, formed from facile heterocycloreversion of N-substituted 2-azanorbornenes, with triethylsilane/trifluoroacetic acid. For example, treatment of a 0.1 M solution of the 2-azanorbornene derived from L-phenylalanyl-L-leucine methyl ester (105) in chloroform/trifluoroacetic acid (1/1) with 3.0 equiv of triethylsilane provides after 20 h at room temperature an 84% yield of N-methyl-L-phenylalanyl-L-leucine methyl ester, 107:

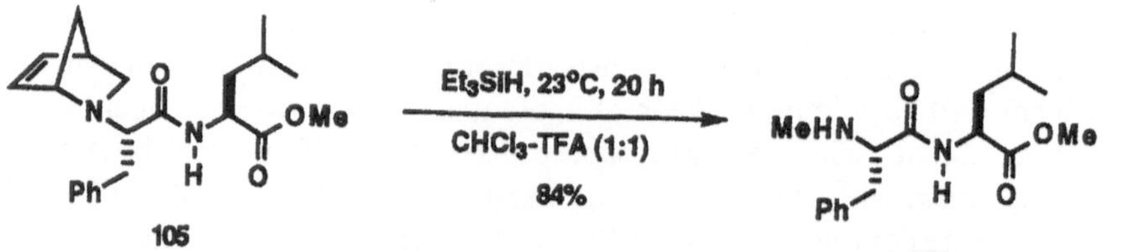

The process lacks any detectable racemization. Even racemization-prone phenylglycine can be monomethylated by this protocol without detectable epimerization.

This two-step methylation sequence, involving 2-azanorbornene synthesis and retro Diels–Alder initiated iminium-ion reduction, is applicable to a variety of amino acid derivatives. The protocol is compatible with unprotected phenols and hydroxyl groups. The 2-azanorbornenes derived from the methyl esters of unprotected L-tyrosine and L-serine both undergo the N-methylation procedure smoothly. Likewise, the unprotected amino acid, L-leucine, is monomethylated without incident in good yield.

Another application of the *N*-methylation sequence was disclosed in the synthesis of physostigmine [37]. The methodology features a unique protocol for the aminoethylation or the *N*-methyl aminoethylation of carbonyl compounds with a novel spiroaziridinium salt, 2-azanorbornene-2-spiro-1'-aziridinium triflate (**109**). The spiroaziridinium salt, **109**, is prepared in a two-step process:

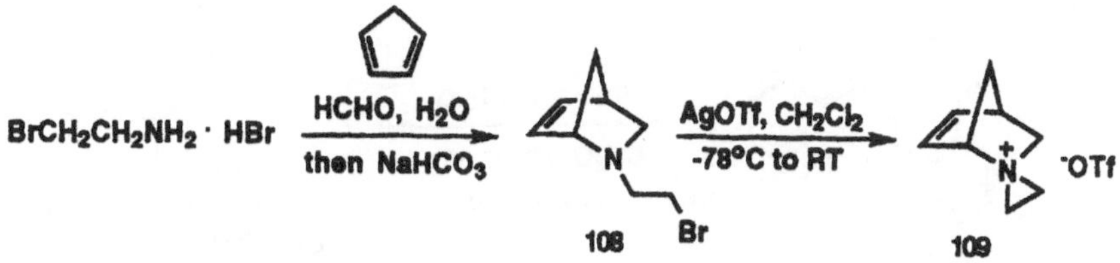

First, the 2-azanorbornene precursor, **108**, is obtained by the aqueous imino Diels–Alder reaction between cyclopentadiene and the iminium ion derived from formaldehyde and 2-bromoethylamine hydrobromide. Exposure of a solution of amino bromide **108** to silver triflate provides the spiroaziridinium salt, **109**, in good overall yield. The ability of **109** to function as an alkylating agent has been demonstrated with a variety of enolates [37].

The *N*-methyl aminoethylation procedure is illustrated in the synthesis of physostigmine, which entails the alkylation of 1,3-dimethyl-5-methoxyoxindole (**110**). The enolate of **110**, generated from lithium diisopropylamide (LDA) in THF, is added to a slurry of **109** (1.2 equiv) in THF at room temperature to provide, after 10 min, an 86% yield of azanorbornene **111**. Submitting **111** to the conditions for *N*-methylation as detailed above gives **112** in 92% yield. Compound **112** is elaborated to physostigmine in a straight-forward manner (Scheme 2.17).

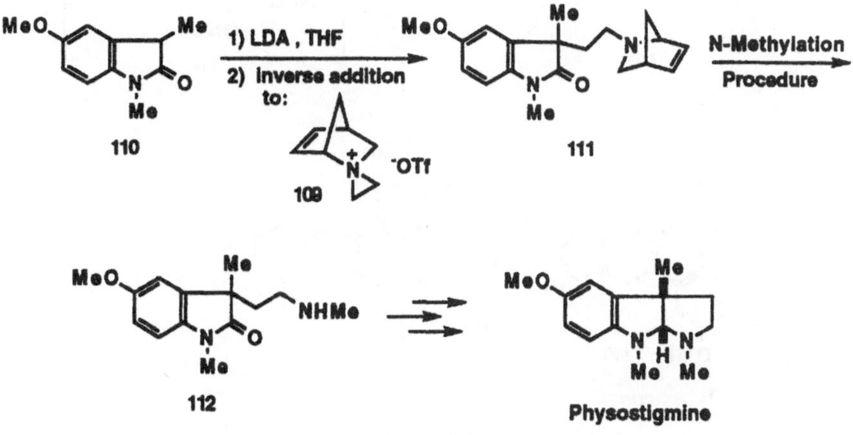

Scheme 2.17

Grieco's synthesis of cryptopleurine **70** and julandine **68** (see end of section 2.3.1), via an intramolecular imino Diels–Alder reaction for construction of the quinolizidine nucleus, failed in the case of the related indolizidine alkaloid, tylophorine [38]. Submitting phenanthryl amine **113** to the identical aza Diels–Alder conditions used for cryptopleurine only yields by-products (**114**), as a result of a Clarke–Eschweiler type of cyclization. No tylophorine is detected under these conditions:

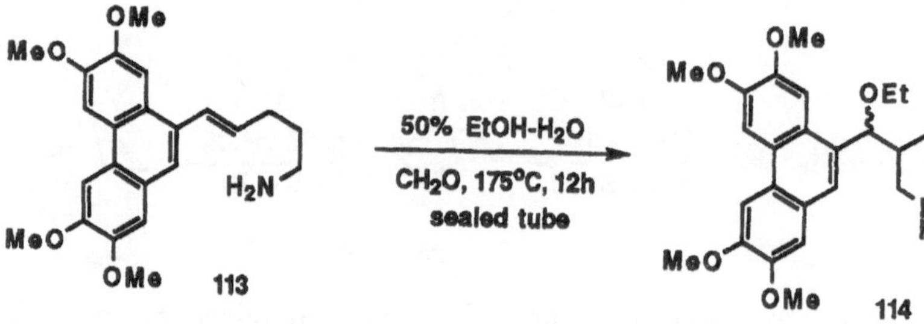

A unique tandem retro Diels–Alder/intramolecular aza Diels–Alder sequence was successfully applied to the synthesis of tylophorine **116** [38]. The requisite 2-azanorbornene derivative, **115**, of phenanthryl amine **113** is prepared under the conventional conditions with cyclopentadiene and formalin. Exposure of a solution of **115** in o-dichlorobenzene containing camphorsulfonic acid (1 equiv) at 155–160°C in a sealed tube for 5 h provides a 35% yield of tylophorine (**116**) (Scheme 2.18). In addition, the piperidine

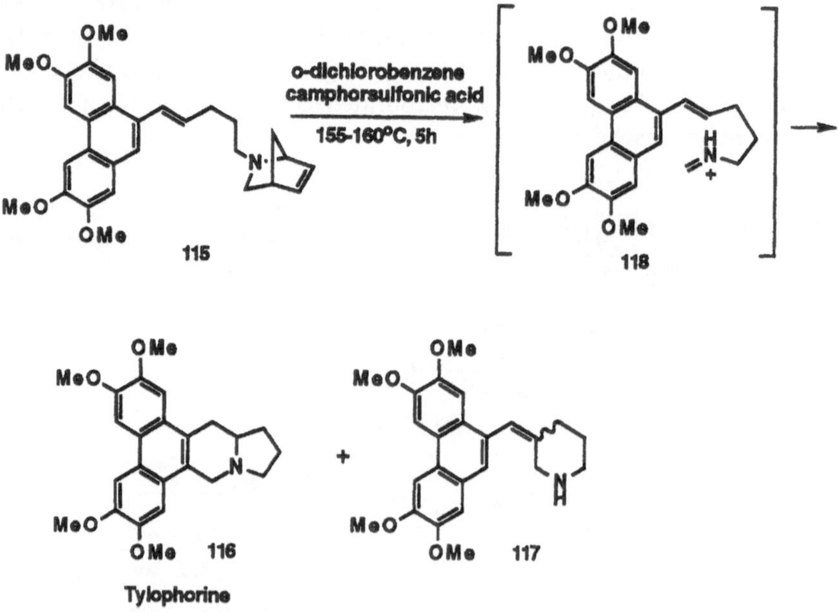

Scheme 2.18

by-products, **117**, are obtained in a 50% yield. The observed products are derived from heterocycloreversion of azanorbornene **115** to generate the highly reactive iminium ion, **118**, which reacts with the internal double bond. Tylophorine (**116**) results from a [4+2] cycloaddition reaction mode, while the piperidine by-products (**117**) arise through a Clarke–Eschweiler type of cyclization/elimination.

A total synthesis of pseudotabersonine (**119**) by Grieco *et al.* [39] utilizes both the aminoethylation (cf. Scheme 2.17) and tandem retro Diels–Alder/intramolecular aza Diels–Alder (cf. Scheme 2.18) protocols to construct spirofused indolizidine oxindoles, **122** and **123**, which serve as precursors to dehydrosecodine **125** (Scheme 2.19). The pentacyclic backbone of the *Aspidosperma* alkaloid, pseudotabersonine, is postulated to be biogenically [40] derived from a dehydrosecodine.

The synthesis commences with alkylation of oxindole **120** with spiroaziridinium triflate **109**, providing the 3,3-disubstituted **121** in 53% yield (cf. Scheme 2.17). Treatment of **121** with boron trifluoride etherate at 100°C in toluene initiates the tandem retro Diels–Alder/intramolecular aza Diels–Alder process, leading to spiro-tetracyclic oxindoles **122** and **123** (1.5/1) in 61% yield. Addition of 2-lithio-1,1-diethoxy-2-propene to oxindole **122** provides carbinolamine **124** (95%). Exposure of **124** to *p*-toluenesulfonic acid in acetone–water followed by treatment with excess triethylamine in acetonitrile at 80°C effects the biomimetic transformation to adduct **126**, which possesses the pentacyclic carbon framework of pseudotabersonine. This unique two-step one-pot transformation generates the inherently unstable dihydropyridine portion of dehydrosecodine **125**, which participates in an intramolecular reverse electron-demand Diels–Alder reaction, providing **126** in 50% yield. The total synthesis is completed by transformation of the formyl group into the requisite carbomethoxy unit followed by *N*-benzyl deprotection (Scheme 2.19).

2.5 Oxo Diels–Alder reactions in aqueous media

2.5 1 *Carbonyl groups as heterodienophiles in aqueous media*

The use of carbonyl groups acting as heterodienophilic components in [4+2] cycloaddition reactions has been well established for the construction of the dihydropyran backbone. The oxo Diels–Alder process, like the imino variant (see section 2.2), has generally required reaction activation through Lewis-acid catalysis, elevated temperature/pressure, and/or the use of electron-deficient carbonyl compounds. In 1991, Lubineau *et al.* [41] reported that glyoxylic acid in aqueous solution reacts with cyclic dienes to produce α-hydroxy-γ-lactones through rearrangement of the intermediary [4+2] cycloadducts. The reaction of aqueous glyoxylic acid with cyclopentadiene

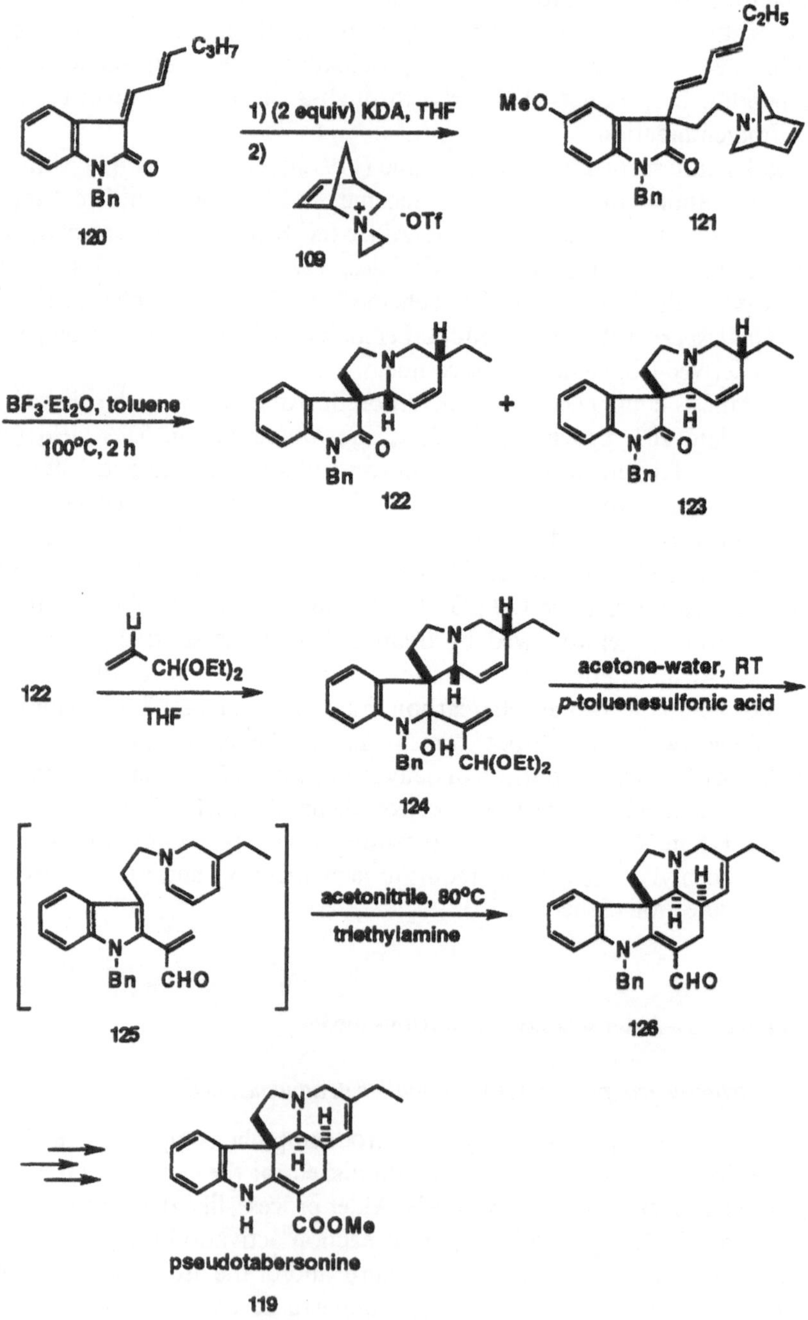
C3H7
1) (2 equiv) KDA, THF
2)
-OTf
MeO
C2H5
N
Bn
O
120
109
121
BF3·Et2O, toluene
100°C, 2 h
H
122
123
Li
CH(OEt)2
THF
acetone-water, RT
p-toluenesulfonic acid
OH
124
acetonitrile, 80°C
triethylamine
CHO
125
126
COOMe
pseudotabersonine
119

Scheme 2.19

was also independently reported by Grieco *et al.* [42] in the total synthesis of racemic sesbanimide A and B (see section 2.5.2). Water as solvent is advantageous by allowing direct use of inexpensive commercial aqueous glyoxylic acid solutions. In addition, the relative rate of the hetero Diels–Alder reaction is enhanced over the competing cyclopentadiene dimerization process in aqueous media.

The initial studies of Lubineau *et al.* [41] focused on reacting cyclopentadiene with aqueous glyoxylic acid at different pH levels. The reaction rate increased substantially at lower pH. At pH 0.9, the reaction of cyclopentadiene in aqueous glyoxylic acid provides α-hydroxy-γ-lactones **128** and **129** in 83% yield (73/27 ratio) after stirring the heterogeneous mixture for 1.5 h at 40°C:

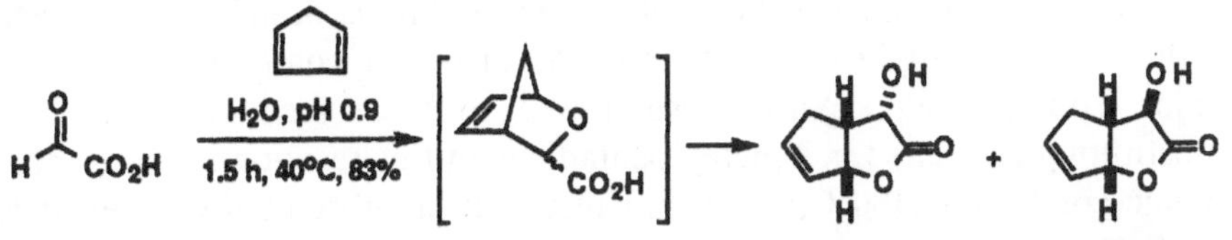

Product formation was postulated to arise from the spontaneous rearrangement of the initially formed hetero Diels–Alder adducts **127**. The similar reaction when performed with an alkyl glyoxylate in the absence of water (neat) or in organic solvent (toluene) inevitably leads to cyclopentadiene dimerization instead of the expected cycloaddition.

The reaction of cyclohexadiene with aqueous glyoxylic acid at pH 1 is complete after 2 days at 90°C, providing α–hydroxy-γ-lactones **131** and **132** (60/40 ratio) in 85% yield [41]:

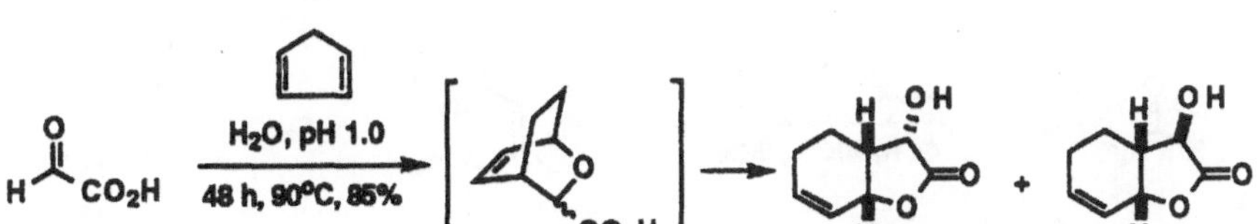

The reaction of butyl glyoxylate with cyclohexadiene (neat) requires 21 h at 120°C to produce the corresponding cycloadducts in 57% yield (endo/exo 9/1) [43]. The free acids (**130**), isolated by hydrolysis of the Diels–Alder adducts obtained by using an alkyl glyoxylate dienophile under non-aqueous conditions, were shown to rearrange spontaneously at room temperature in water to the corresponding α-hydroxy-γ-lactones **131** and **132**. In the absence of water, this rearrangement requires high temperatures (150°C) to proceed [44].

The hetero Diels–Alder reaction also proceeds with acyclic dienes [45]. For example, reaction of 2-methylpentadiene with aqueous glyoxylic acid furnishes, after 1.5 h at 100°C, a near-quantitative yield of dihydropyrans 133 and 134 in a 64/36 ratio:

133; R = H
135; R = Me

134; R = H
136; R = Me

Ratio   64 : 36

The product ratio reflects thermodynamic conditions, as evidenced by each of the purified methyl esters (135 and 136), providing compound 133 as the major isomer when separately submitted to these reaction conditions. Substituting isoprene for 2-methylpentadiene in this reaction requires more drastic conditions (100°C, 18 h) and results in a substantially lower yield (28%) of the corresponding dihydropyrans.

Lubineau et al. [45] have also demonstrated Lewis-acid catalysis in the above equation by the use of water-tolerant lanthanide triflates as reported by Kobayashi [9, 10]. Adducts 133 and 134 are obtained quantitatively after 12 h at 60°C in the presence of 0.1 equiv of $Yb(OTf)_3$ or $Nd(OTf)_3$, whereas without catalyst under the same conditions a 55% yield is obtained.

Other oxo-heterodienophiles, such as pyruvaldehyde and glyoxal, both conveniently available as aqueous solutions, can be used to form substituted dihydropyrans [45]. Reaction of aqueous pyruvaldehyde with 2-methylpentadiene gives, in 96% yield, cycloadducts 137 and 138 in a 47/53 ratio after 48 h at 100°C :

138   137

Ratio   53 : 47

Reaction of aqueous glyoxal with 2-methyl pentadiene after 60 h at 100°C gives, in 36% yield, the cycloadduct derivatives 143 and 144 in a 1/1 ratio after reduction ($NaBH_4$) and acylation (AC_2O, pyr.) of the initially formed dihydropyran hetero Diels–Alder products, 139 and 140.

The oxo Diels–Alder reaction is also possible with ketones (i.e. pyruvic acid) [45]. Aqueous pyruvic acid and 2-methylpentadiene at 100°C for 6 h react to form cycloadducts 145 and 146 in 74% yield in a 1/2 ratio.

Interestingly, the same reaction run in the absence of water (neat or in

R = CHO   **139**   **140**

R = CH₂OH   **141**   **142**

R = CH₂OAc   **143**   **144**

Ratio   1 : 1

145   **146**

Ratio   1 : 1

toluene) gives **145** and **146** but with a reverse isomeric selectivity (2/1). The stereoselectivity difference was shown to be a consequence of thermodynamic control in water versus kinetic control in toluene. It is noteworthy that the oxo hetero Diels–Alder examples alluded to above all proceed in good to excellent yields despite the very low solubility of the diene components and extensive hydration of the dienophilic carbonyl moiety.

2.5.2 Synthetic exploitation of the aqueous oxo Diels–Alder reaction

The development of efficient and practical aqueous oxo Diels–Alder methodology allowing for convenient access to substituted dihydropyrans and α–hydroxy-γ-lactones set the stage for further synthetic exploitation. The following is a brief account of synthetic applications of the aqueous oxo Diels–Alder reaction.

The total synthesis of sesbanimide A (**149a**) and B (**149b**) reported by Grieco *et al.* [42] commenced with bicyclic lactone **128** obtained through the reaction of cyclopentadiene and aqueous glyoxylic acid in 40% yield (see above). Elaboration of **128** to unsaturated δ-lactone **147** defined the relative stereochemistry at C(7), C(8) and C(9) in the natural product, and set the stage for a novel lithium perchlorate–ether promoted conjugate addition of 1-methoxy-1-(*t*-butyldimethylsiloxy)ethene to **147**, thus providing lactone **148** in 65% yield (Scheme 2.20). Further manipulation of **148** completed the total synthesis of sesbanimide A (**149a**) and B (**149b**).

Roberts *et al.* [46] have enzymatically resolved bicyclic lactone **128** using *Pseudomonas fluorescens* lipase and vinyl acetate. The enantio-pure ester, (–)-**151** (>95% *ee*), obtained by this process is converted in three steps into diol **153** via triol **152** in 50% overall yield without purification of the intermediates

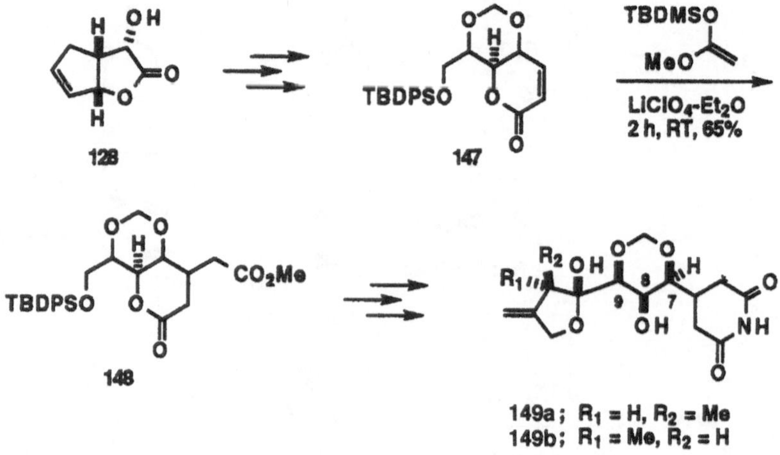

Scheme 2.20

(Scheme 2.21). Ready access to diol **153** provided the starting point for the formal synthesis of the anti-HIV agent carbovir (−)-**154** [47].

Roberts *et al.* [48] have used the other diol enantiopode **155**, derived from (+)-**150**, to construct the hydroxylactone moiety (**156**) of mevinic acids such as compactin and mevinolin:

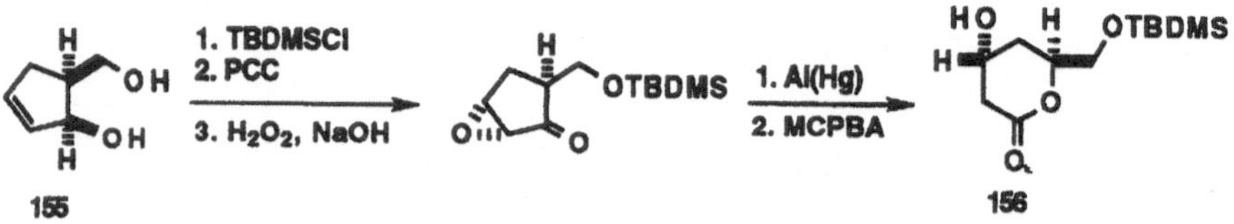

Racemic diol **157**, derived from **128**, is also used in the synthesis of the antiviral carbocyclic nucleoside, 9-(c-5-hydroxy-c-4-hydroxymethylcyclo-pent-2-en-r-1-yl)-9H-adenine (Epinor-BCA) [49]. Diol **157** is elaborated to

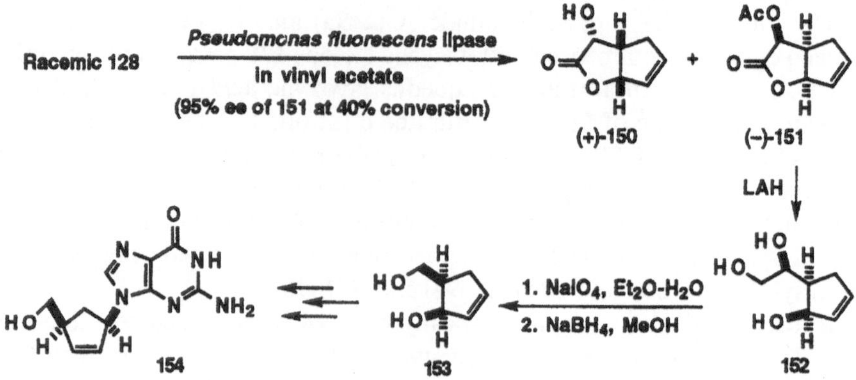

Scheme 2.21

the protected cyclopentene derivative, **158**, which provided the carbocyclic backbone of BCA (**159**):

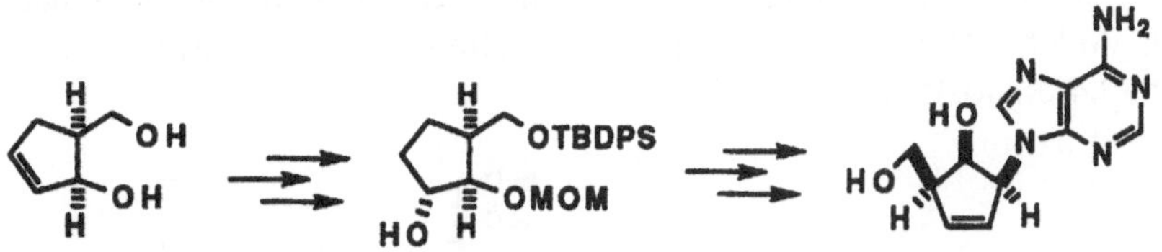

Racemic bicyclic lactone **128** has also been efficiently resolved through treatment with (*S*)-*O*-acetyllactyl chloride in pyridine followed by separation of the resultant diastereomers, **160** and **161**, by silica gel chromatography [50]. Ester hydrolysis provides enantiopure **150** and **162** (Scheme 2.22).

Lubineau *et al.* [51] exploit the aqueous oxo cycloaddition in the synthesis of the monosaccharide 3-deoxy-D-manno-2-octulosonic acid (KDO) **168** and related derivatives. The strategy involves reacting the sodium salt of glyoxylic acid with the water-soluble diene, **163**, obtained from D-glyceraldehyde. Exposure of the aqueous reaction mixture at pH 6 to 100°C for 18 h gives, after esterification, a 54% yield of an inseparable mixture of four cycloadducts, **164**, **165**, **166** and **167**, in a 42/32/19/7 ratio (Scheme 2.23). The major isomers, **164** and **165**, are described to arise from endo-*si* and exo-*si* transition states; the minor isomers, **166** and **167**, result from endo-*re* and exo-*re* transition states, respectively. The reaction performed in the absence of water (neat), using the benzyl-protected diene of **163**, requires a reaction temperature 20°C higher, resulting in comparable yield and stereoisomeric product ratios. However, the non-aqueous conditions also require a tedious preparation of the alkyl glycolate dienophile. The major isomer, **164**,

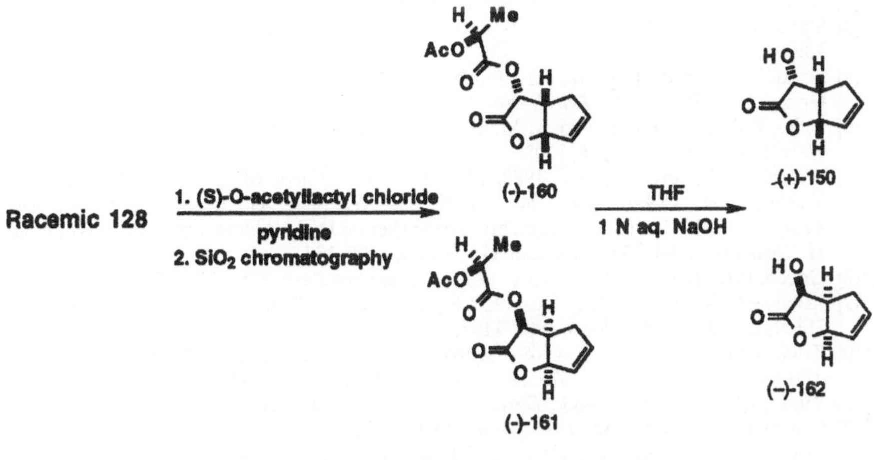

Scheme 2.22

contains the eight-carbon backbone required for further elaboration to KDO (**168**) and related saccharide analogs. The analogous seven-carbon backbone of the heptulosonic saccharide class is also accessible via aqueous hetero Diels–Alder methodology through the reaction of sodium glyoxylate with penta-2,4-dienol [52].

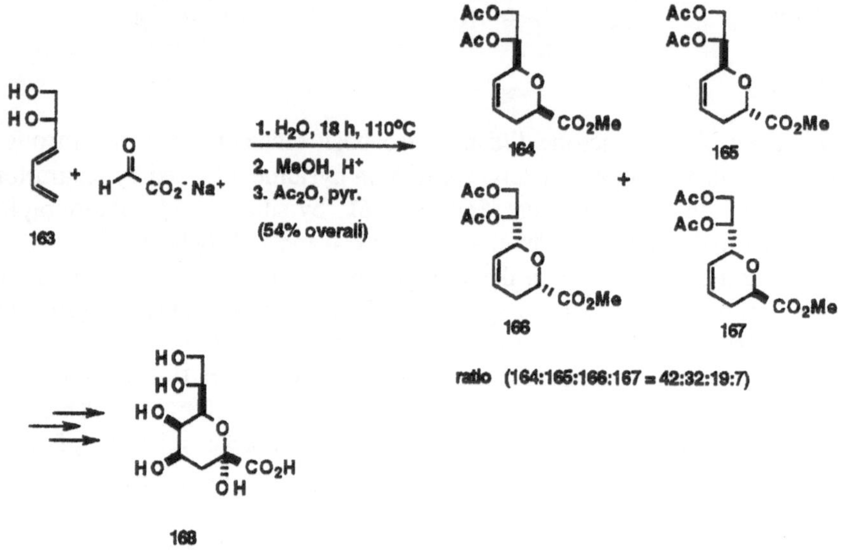

Scheme 2.23

References

[1] Larsen, S.D.; Grieco, P. A., *J. Am. Chem. Soc.*, 1985, **107**, 1768. For review of Diels–Alder reactions in water see: Li, C.H., *Chem. Rev.*, 1993, **93**, 2023; Lubineau, A.; Augé, J.; Queneau, Y., *Synthesis*, 1994, **9**, 741.

[2] Hanley, J.A.; Forsyth, D.A., *J. Labelled Compd. Radiopharm.*, 1990, **28**, 307.

[3] Staninets, V.I.; Mironova, D.F.; Iksanova, S.V.; Sinitsa, A.D., *Ukr. Khim. Zh.*, 1990, **56**, 1321.

[4] Skvarchenko, V.R.; Lapteva, V.L.; Gorbunova, M.A., *Zh. Org. Khim.*, 1990, **26**, 2588.

[5] Grieco, P.A.; Bahsas, A., *Tetrahedron Lett.*, 1988, **29**, 5855.

[6] Villar, P.E.; Boelsterli, J.; Cid, M.M.; France, J.; Fuchs, B.; Walkinshaw, M.; Weber, P.-H., *Helv. Chim. Acta.*, 1993, **76**, 1203.

[7] Waldman, H., *Angew. Chem.*, 1988, **100**, 307; *Angew. Chem. Int. Ed. Engl.*, 1988, **27**, 274; Waldman, H., *Liebigs Ann. Chem.*, 1989, 231; Waldman, H.; Braun, M, *Liebigs Ann. Chem.*, 1989, 1045. For reviews on asymmetric hetero Diels–Alder reactions see: Waldman, H., *Synthesis*, 1994, 535; Waldman, H., *Synlett*, 1995, 133.

[8] Grieco, P.A.; Larsen, S.D.; Fobare, W.F., *Tetrahedron Lett.*, 1986, **27**, 1975.

[9] Kobayashi, S.; Ishitani, H.; Nagayama, S., *Chem. Lett.*, 1995, 423; Kobayashi, S.; Ishitani, H.; Nagayama, S., *Synthesis*, 1995, 1195.

[10] Kobayashi, S., *Synlett*, 1994, 689; Kobayashi, S.; Hachiya, I.; Araki, M.; Ishitani, H., *Tetrahedron Lett.*, 1993, **34**, 3755; Kobayashi, S.; Hachiya, I., *J. Org. Chem.*, 1994, **59**, 3590.

[11] Yu, L.; Chen, D.; Wang, P.G., *Tetrahedron Lett.*, 1996, **37**, 2169.

[12] Grieco, P.A.; Larsen, S.D., *Org. Synth.*, 1990, **68**, 206.

[13] Roberts, S.M.; Smith, C.; Thomas, R.J., *J. Chem. Soc., Perkins Trans. 1*, 1990, 1493; Maurya, R.; Pittol, C.A.; Pryce, R.J.; Roberts, S.M.; Thomas, R.J.; Williams, J.O., *J.*

Chem. Soc., Perkins Trans. 1, 1992, 1617.

[14] Chao, S.; Kunng, F.A.; Gu, J.M.; Mariano, P.S., *J. Org. Chem*, 1984, **49**, 2708; Baxter, E.W.; Labaree, D.; Chao, S.; Mariano, P.S., *J. Org. Chem*, 1989, **54**, 2893.

[15] Fray, A.H.; Augeri, D.J.; Kleinman, E.F., *J. Org. Chem*, 1988, **53**, 896.

[16] Carroll, F.I.; Abraham, P.; Chemburkar, S.; He, X.; Mascarella, S.W.; Kwon, Y.W.; Triggle, D.J., *J. Med. Chem.*, 1992, **35**, 2184.

[17] Ramberg, P., Master's Thesis, Indiana University, 1987.

[18] Kunng, F.A.; Gu, J.M.; Chao, S.; Chen, Y.; Mariano, P.S., *J. Org. Chem*, 1983, **48**, 4262; Chao, S.; Kunng, F.A.; Gu, J.M.; Mariano, P.S., *J. Org. Chem*, 1984, **49**, 2708.

[19] Grieco, P.A.; Larsen, S.D., *J. Org. Chem.*, 1986, **51**, 3553.

[20] Grieco, P.A.; Parker, D.T., *J. Org. Chem.*, 1988, **53**, 3658.

[21] Weinreb, S.M., *Acc. Chem. Res.*, 1985, **18**, 16; Bremmer, M.L.; Khatri, N.A.; Weinreb, S.M., *J. Org. Chem.*, 1983, **48**, 3661.

[22] Grieco, P.A.; Parker, D.T., *J. Org. Chem.*, 1988, **53**, 3325.

[23] Grieco, P.A.; Larsen, S.D., Unpublished results, Indiana University.

[24] Chiu, C.K.-F., *Synth. Commun.*, 1996, **26**, 577.

[25] Pombo-Villar, E.; Supavilai, P.; Weber, H.P.; Boddeke, H.W.G.M., *Bioorg. Med. Chem. Lett.*, 1992, **2**, 501.

[26] Cid, M.M.; Eggnauer, U.; Weber, H.P.; Pombo-Villar, E., *Tetrahedron Lett.*, 1991, **32**, 7233.

[27] Roberts, S.M.; Smith, C.; Thomas, R.J., *J. Chem. Soc., Perkins Trans. 1*, 1990, 1493.

[28] Cid, M.M.; Pombo-Villar, E., *Helv. Chim. Acta*, 1993, **76**, 1591.

[29] Shealy, Y.F.; Clayton, J.D., *J. Am. Chem. Soc.*, 1969, **91**, 3075; Just, G.; Reader, G., *Tetrahedron Lett.*, 1973, 1521.

[30] Kobayashi, T.; Ono, K.; Kato, H., *Bull. Chem. Soc. Jpn*, 1992, **65**, 61.

[31] Bourgeois-Cury, A.; Doan, D.; Gore, J., *Tetrahedron Lett.*, 1992, **33**, 1277.

[32] Kobayashi, T.; Fujieda, H.; Murakami, Y.; Nakamura, T.; Ono, K.; Yamamoto, S.; Kato, H., *Bull. Chem. Soc. Jpn*, 1994, **67**, 3082.

[33] Braillon, B.; Lasne, M.-C.; Ripoll, J.L., *Nouv. J. Chim.*, 1982, **6**, 121.

[34] Grieco, P.A.; Parker, D.T.; Fobare, W.F.; Ruckle, R., *J. Am. Chem. Soc.*, 1987, **109**, 5859.

[35] Grieco, P.A.; Clark, J.D., *J. Org. Chem.*, 1990, **55**, 2271.

[36] Grieco, P.A.; Bahsas, A., *J. Org. Chem.*, 1987, **52**, 5746.

[37] Grieco, P.A.; Carroll, W.A., *Tetrahedron Lett.*, 1992, **33**, 4401.

[38] Parker, D.T., Unpublished results and Doctoral Thesis, Indiana University, 1989.

[39] Carroll, W.A.; Grieco, P.A., *J. Am. Chem. Soc.*, 1993, **115**, 1164.

[40] Wenkert, E., *J. Am. Chem. Soc.*, 1962, **84**, 98; Scott, A.I., *Bioorg. Chem.*, 1974, **3**, 398; Kutney, J.P., *Heterocycles*, 1977, **7**, 593.

[41] Lubineau, A.; Augé, J.; Lubin, N., *Tetrahedron Lett.*, 1991, **32**, 7529.

[42] Grieco, P.A.; Henry, K.J.; Nunes, J.J.; Matt, J.E., Jr., *J. Chem. Soc., Chem. Commun.*, 1992, 368.

[43] Achmatowicz, O., Jr.; Jurczak, J.; Pyrek, J.S., *Rocz. Chem.*, 1975, **49**, 1831.

[44] Achmatowicz, O., Jr.; Jurczak, J.; Pyrek, J.S., *Tetrahedron*, 1976, **32**, 2113.

[45] Lubineau, A.; Augé, J.; Grand, E.; Lubin, N., *Tetrahedron*, 1994, **50**, 10265.

[46] Hibbs, D.E.; Hursthouse, M.B.; Knutsen, L.J.S.; Abdul Malik, K.M.; Olivo, H.F.; Roberts, S.M.; Varley, D.R.; Xiong, H., *Acta Chem. Scand.*, 1995, **49**, 122.

[47] MacKeith, R.A.; McCague, R.; Olivo, H.F.; Palmer, C.F.; Roberts, S.M., *J. Chem. Soc., Perkins Trans. 1*, 1993, **14**, 313.

[48] McCague, R.; Olivo, H.F.; Roberts, S.M., *Tetrahedron Lett.*, 1993, **34**, 3785.

[49] Toyota, A.; Katagiri, N.; Kaneko, C., *Heterocycles*, 1994, **38**, 27.

[50] Burlina, F.; Clivio, P.; Fourrey, J.-L.; Riche, C.; Thomas, M., *Tetrahedron Lett.*, 1994, **35**, 8151.

[51] Lubineau, A.; Augé, J.; Lubin, N., *Tetrahedron Lett.*, 1993, **49**, 4639.

[52] Lubineau, A.; Queneau, Y., *J. Carbohydr. Chem.*, 1995, **14**, 1295.

3 Claisen rearrangements in aqueous solution

J.J. GAJEWSKI

3.1 Introduction

The thermal rearrangement of allyl vinyl ethers to γ,δ-unsaturated carbonyl compounds was first described by Claisen in 1912 [1]. The rearrangement of allyl phenyl ethers to o-allyl phenols was described soon thereafter [2]. Claisen later proposed that a cyclic mechanism was involved in both reactions [3]. The aromatic Claisen rearrangement was subsequently demonstrated to be intramolecular and to proceed with inversion of the allyl group; that is, the α and γ carbons of the allyl group interchange [4]. The rearrangements represent an important tool in the arsenal of synthetic organic chemistry, and extensive reviews provide a wealth of data on the reactions [5]. Both the aliphatic and aromatic Claisen rearrangements involve a 3,3-sigmatropic shift [6]. That is, a bond is broken and a new bond is formed to an atom that is three atoms away along a chain from one of the atoms in the initial bond, and the new bond is formed three atoms away along the chain from the other atom that was part of the original bond. In the case of the aromatic Claisen rearrangement, a subsequent acid- or base-catalyzed tautomerization of the dienone intermediate must occur (Scheme 3.1).

In the case of the aliphatic variant, asymmetry at the terminal methylene carbons is transformed into asymmetry at the two new saturated centers in a sense that suggests a chair-like transition state (Scheme 3.2) [7a,b]. In the case of the aromatic Claisen rearrangement, the initially formed dienone undergoes tautomerization to the phenol, so the only stereochemistry observable is the transfer of chirality to a product from an optically active starting material with a specific double-bond geometry. This experiment reveals that the aromatic Claisen rearrangement also proceeds with stereochemistry appropriate for a chair-like transition state (Scheme 3.2) [7c–e].

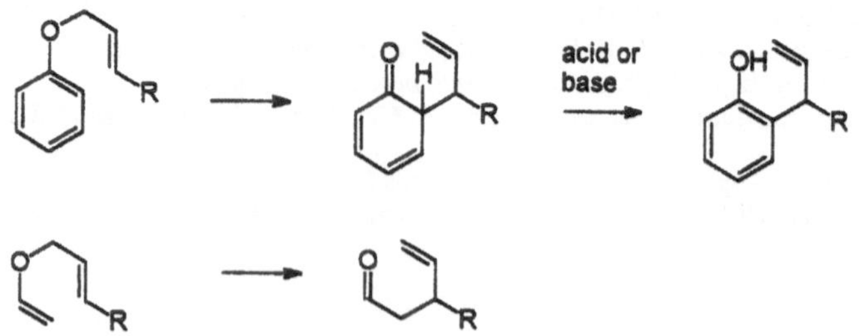

Scheme 3.1

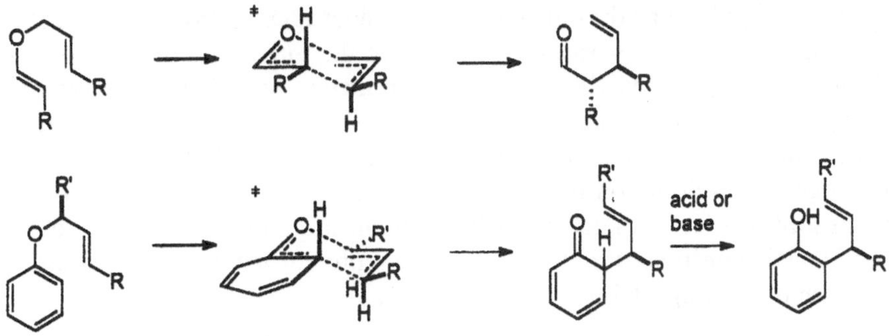

Scheme 3.2

3.2 Aromatic Claisen rearrangement

3.2.1 Solvent effects

Allyl phenyl ethers undergo the rearrangement at temperatures around 200°C in a first-order process in diphenyl ether solvent [8]. However, the rate constants drift upward by 10% at 80% reaction, possibly due to formation of the phenol product, which might be a catalyst for the reaction. Further, in the absence of diphenyl ether, the rate constants increased by a factor of 2 at roughly 80% reaction. Perhaps confirming the speculation that phenol is a catalyst, the rate constants obtained upon addition of dimethylamine did not change from the initial rate constants in diphenyl ether solvent alone. However, small amounts of acetic acid had no effect on the reaction. Oxygen also had no effect on the reaction.

Both White [9] and Goering [10] examined the effect of substituents on the rate of the aromatic Claisen rearrangement and found small negative ρ^+ values (–0.6). This is not consistent with formation of a phenoxide–allyl cation pair despite the charge-affinity relationships between the atoms of the starting material that would suggest this. Both White and Goering also discovered a slight (factors of 1.4×) rate acceleration in diethylene glycol monoethyl ether (DGME) solvent relative to nonhydroxylic polar solvents like sulfolane and benzonitrile, and the rate in these polar solvents was roughly twice as fast as in hydrocarbon solvents. Goering found, however, that the rates in ethylene glycol were roughly five times faster than in DGME, so hydrogen-bond donation to the solute was identified as a contributor to the reaction mechanism. Phenol was also used as a solvent, and it was found to be effective in promoting the reaction relative to DGME by a factor of 12.

3.2.2 Effect of an aqueous environment

In 1970 White provided rate data on the aromatic Claisen rearrangement in a still wider range of solvents including 28.5% ethanol–water where the rate in water was similar to that in phenol [11]. He suggested that hydrogen-bond

donation might not be the only factor to consider since hydrogen-bond donation might have been expected to track with the pK_a values of the solvents.
White also determined the activation parameters for all the solvents and
found a good correlation between the enthalpies and entropies of activation,
which suggests that the mechanism is the same in all the solvents, so
hydrogen-bond donation might not be a factor. The conclusion was that a
polar solvent effect with no change in mechanism was involved. Regardless of
mechanistic interpretation, this appears to represent the first reported use of
water in promoting Claisen rearrangements.

3.3   Aliphatic Claisen rearrangement

3.3.1   *Mechanism*

The first kinetic measurements of the aliphatic Claisen rearrangement are due
to Schuler and Murphy, who found that

$$\log k \, (\mathrm{s}^{-1}) = 11.70 - 30\,600/2.3RT$$

in the gas-phase pyrolysis of allyl vinyl ether [12]. The relatively low pre-
exponential term and the stereochemistry [6] of the reaction suggest that a
chair-like, concerted transition state is involved. However, the questions of
the extent of bonding between the two allyl-like fragments and the extent of
charge transfer have occupied experimentalists and theorists for some time.
The magnitude of the changes in bonding was addressed by a determination
of secondary-deuterium kinetic isotope effects at C(4) and C(6) of allyl vinyl
ether in its pyrolysis at 160°C, and these were compared with values expected
for either complete C–O bond cleavage or complete terminal C–C bond
making [13]. This gave an estimate of 33% bond cleavage and 15% bond
making. A subsequent determination of heavy-atom kinetic isotope effects
and comparison to force-field calculations of the isotope effects based on
bond orders led to roughly the same values. Thus the transition state more
resembles reactant than product, and it more resembles a two allyl-like
radical pair than an oxacyclohexane 1,4-diyl (Scheme 3.3) (see later).

3.3.2   *Effect of an aqueous environment*

While numerous examples of the Claisen rearrangement occur in the literature, the reported conversion of chorismate to prephenate appears to be the
first example of the use of pure water as a medium for the rearrangement [14]:

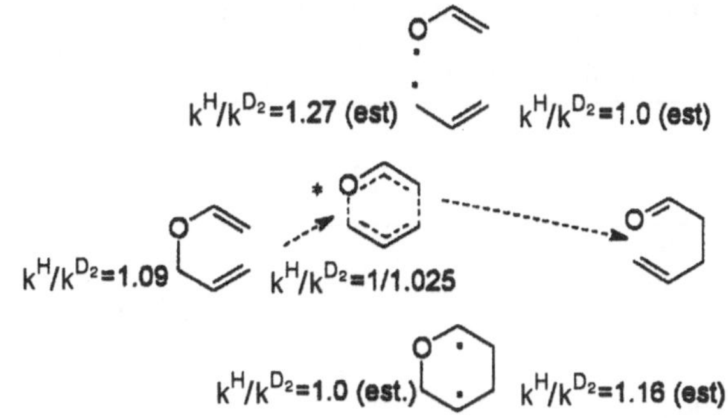

Scheme 3.3

This conversion, which is also catalyzed by the enzyme chorismate mutase, occurs in the biosynthetic pathway from shikimate to phenylalanine. Unfortunately, no comparison of the rate of the rearrangement in different solvents was made to provide insight into the effect of water on the reaction, but later Knowles examined the rate of the rearrangement in water and in methanol and found that the reaction was 100 times faster in water [15]. This would appear to be an extraordinary rate difference for a Claisen rearrangement, but subsequent experiments have revealed that it is not unusual. The origin of the rate effect, however, is still a matter of controversy, but it would appear that the combination of increased hydrogen bonding to the transition state and destabilization of the ground state by hydrophobic effects are responsible (*vide infra*).

An early report by Ponaras lists the use of water as a cosolvent for the Claisen rearrangement of various 2-aminoallyl vinyl ether derivatives [16]. Here, however, no specific rate effect was noted, nor in the table of data provided is there a comparison of rates in different solvents at the same temperature to assess the effect of water.

Another report of a significant effect of water on the rate of a simple aliphatic Claisen rearrangement came from the Carpenter–Ganem collaboration at Cornell [17]. Here both allyl vinyl ether itself and 2-hepta-3,5-dienyl vinyl ether were found to undergo the 3,3-shift with rate constants that increased in solvents with increasing values of the solvent E_T parameter, a solvatochromically derived measure of solvent polarity. The rates in 2/1 methanol–water are roughly 40 times those in acetone solvent, and in acetone the reactions were only 1.4–3.3 times faster than in cyclohexane solvent (Table 3.1). The Cornell group suggested that the rearrangements had increased dipolar character in the transition state, which would rationalize this solvent effect, but, as will be pursued later, the reactions are faster in aqueous methanol than might be expected based on the correlation with E_T values.

Simultaneous with publication of the efforts of the Cornell group was a

Table 3.1 Relative rate constants k_{rel} for the Claisen rearrangement of allyl vinyl ether (AVE) and 2-hepta-3,5-dienyl vinyl ether (HDVE) in various solvents [17]

	k_{rel}	
Solvent	AVE	HDVE
cyclohexane		1.0
di-n-butyl ether	1.0	
benzene		2.6
diethyl ether		2.6
acetone	1.4	3.3
dimethylsulfoxide		5.9
2-propanol		16
ethanol	4.0	20
methanol		28
methanol–water (2/1)	58	117

combined publication from the Curran and Coates groups [18]. This work described rate effects due to alkoxy substitution and solvent changes from benzene to methanol to 80% aqueous ethanol (80/20 ethanol–water) that were interpreted in terms of increased hydrogen bonding to the transition state. Here up to 68-fold rate accelerations were observed with the 4- and 6-alkoxy-substituted allyl vinyl ethers shown below:

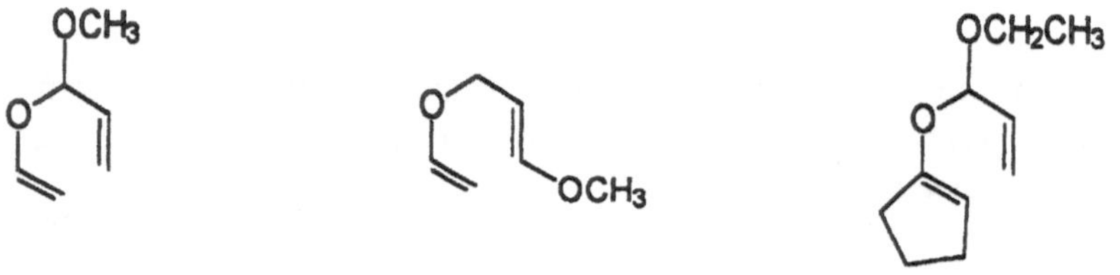

3.4 Synthetic exploitation of aqueous Claisen rearrangements

3.4.1 Synthetically important reactions

As a result of the report from the Cornell group and of previous experience with rate accelerations of Diels–Alder reactions in water solvent [19], the Grieco group examined some Claisen rearrangements in aqueous solvent systems [20]. These studies provided dramatic results on particularly 'reluctant' Claisen rearrangement systems. Thus the following acid could be induced to undergo the Claisen rearrangement at reflux in aqueous base for 5 h to provide 82% yield of the aldehyde:

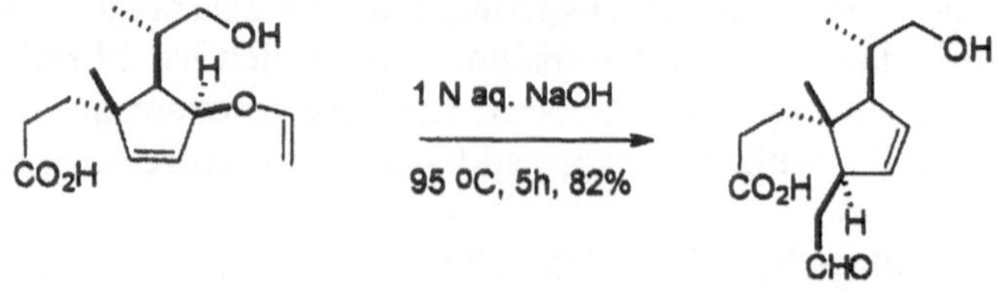

This aldehyde is a key intermediate in the synthesis of the Inhoffen–Lythgoe diol, which is one possible precursor to vitamin D and its derivatives [21]. The aldehyde could not be formed by pyrolysis of the corresponding ester in decalin at 95°C. Other procedures that accomplished the rearrangement required heating in excess of 220°C [22].

Another important example of the use of an aqueous medium to effect the Claisen rearrangement is the formation of the aldehyde below from the vinyl ether precursor in the synthesis of aphidicolin. This reaction was accomplished upon heating in 2.5:1 water–methanol solutions at 80°C for 24 h [20]:

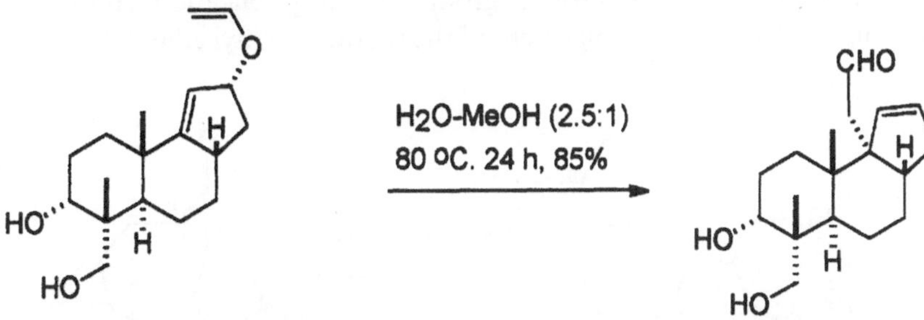

Previous efforts by McMurry accomplished the same conversion by heating the vinyl ether, with the diol protected as the acetonide, at 220°C in a base-washed silylated sealed glass tube containing toluene and sublimed sodium *t*-pentoxide, but the yield was lower, 60% [23].

Perhaps the most spectacular accomplishment of the Grieco group in the utilization of the aqueous solution methodology is the rearrangement of the allyl vinyl ether precursors to the *ctct*- and *ctcc*-[4.5.5.5]fenestrenes (Scheme 3.4) [20]. It is the latter reaction, which generates a *trans* ring fusion between

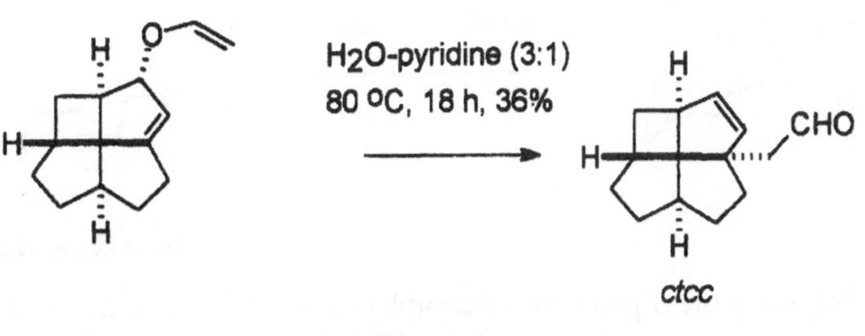

Scheme 3.4

adjacent five-membered rings, that is remarkable. Unlike the Claisen rearrangement to give the *ctct* structure, which is less strained by roughly 7 kcal mol^{-1}, the reaction to give the *ctcc* product provides no strain relief. Yet the latter rearrangement proceeds in 36% yield upon heating at 80°C for 18 h in a 3/1 water–pyridine solution. It is the exothermicity of the formation of a C=O double bond that drives the rearrangement, but it is the water that provides the critical rate acceleration to promote a rapid rearrangement at temperatures that do not destroy the precursor or product [24].

Nearly simultaneous with reports of the aqueous-solution Claisen rearrangement from the Grieco group was a publication from Broka revealing the Claisen rearrangement of the following vinyl ether [25]:

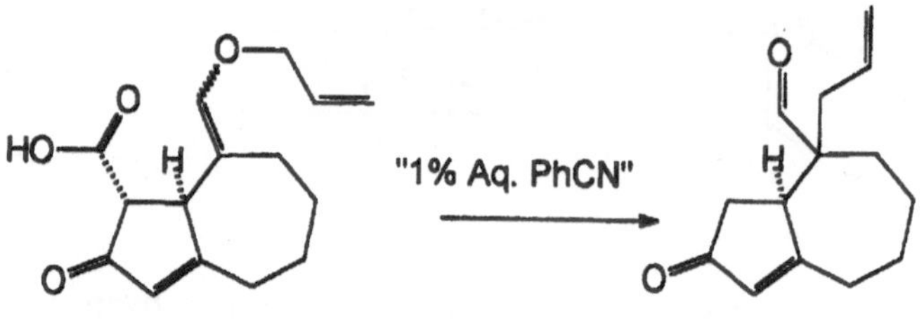

The paper describes the rearrangement as occurring in a 1% aqueous benzonitrile solution. While this might indicate that only 1% of water was present, past usage of this terminology could also suggest that there was present only 1% benzonitrile with the rest being water. Indeed, the *Beilstein* Database records that this reaction was performed in water solvent. However, a careful reading of the 'Discussion' section of the paper indicates that the major component of the solvent system is benzonitrile. The purpose of the water is unclear, although accompanying the rearrangement is a decarboxylation of a carboxylic acid.

Another example of a striking rate acceleration in a Claisen rearrangement conducted in water is that of 6-β-glycosylallyl vinyl ether:

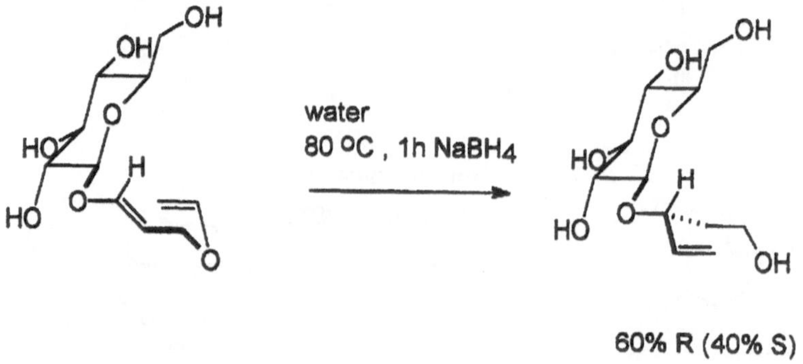

60% R (40% S)

The rearrangement proceeds to completion in 1 h at 80°C in water solvent but requires 13 days at 80°C in toluene [26]. Sodium borohydride was added to the reaction mixture to reduce the aldehyde as it was formed. Interestingly, when the D sugar was used, a 60/40 ratio of (R) to (S) isomers was produced in both water and toluene.

A later report by the same authors described the Claisen rearrangement of the α-anomer in water at 60°C for 2.5 h, but no reaction was observed in toluene even upon heating to higher temperatures, which resulted in destruction of the material [27]:

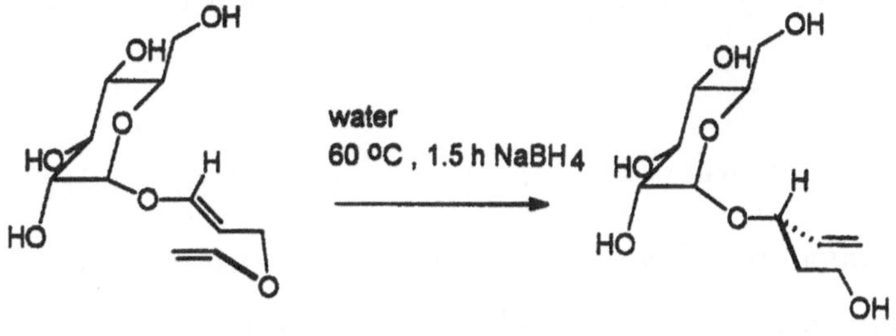

60% S (40% R)

This rearrangement also proceeded with asymmetric induction to provide a 60/40 mixture of (S) and (R) diastereomers. The transition-state-like geometries drawn for the previous two reactions depict the stereochemistry, but it must be recognized that a 60/40 mixture represents a free-energy difference of only 0.26 kcal mol^{-1} at 60°C, so any rationalization is tenuous at best.

3.4.2   Failures of the aqueous solvent methodology

A major difficulty with the use of water as a solvent for the Claisen rearrangement is the solubility of the substrate. Hydrocarbon-like materials are not easily water-soluble even with addition of cosolvent. Two examples by the Fukumoto group serve to illustrate the point:

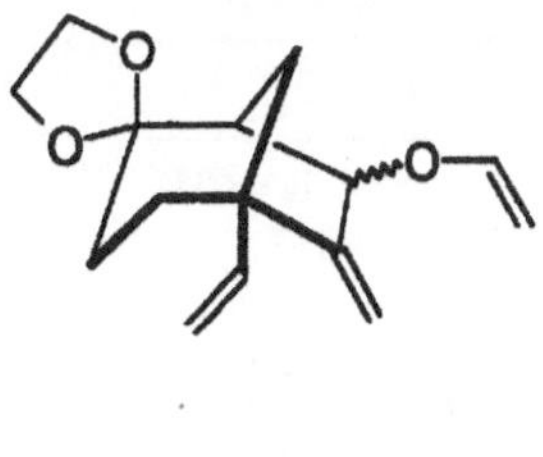

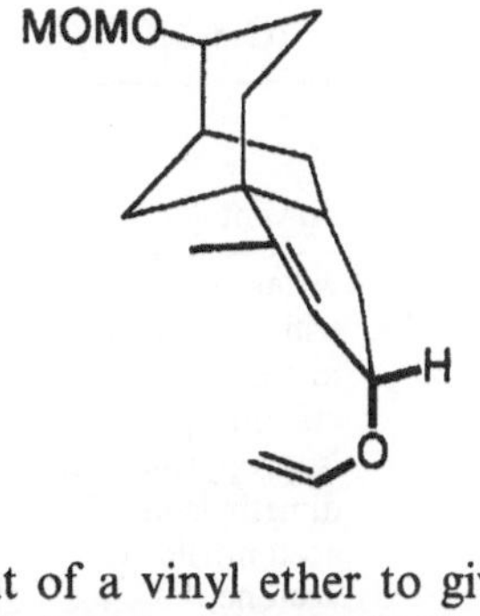

The first is an attempted Claisen rearrangement of a vinyl ether to give a precursor to aphidicolin [28], and the second is an attempted Claisen rearrangement of a vinyl ether to give an intermediate in the synthesis of stemodin [29].

Another notable failure is the attempt by Keese to effect the Claisen rearrangement to give a *trans*-fused bicyclo[3.3.0]octenyl system [30]:

In this case hydrolysis of the vinyl ether occurred before the 3,3-sigmatropic shift.

3.5   Water as a solvent

3.5.1   Quantification of the effect of water on Claisen rearrangements

In an effort to assess the effect of pure water on the rate of the Claisen rearrangement, Brandes prepared an allyl vinyl ether with a pentamethylene chain at C(4) terminating in a carboxyl group (Table 3.2) [31]. This material was soluble as its sodium salt in water and as its methyl ester in cyclohexane. Brandes then determined the rate of rearrangement of both of these materials in a large number of pure solvents and in an even larger number of mixtures of solvents. In those solvent systems where both the carboxylate and the ester were soluble rate constants that were within a factor of 2 of one another were determined for the two compounds, indicating that the appended group had little effect on the rate. Thus the Claisen rearrangement could be examined

Table 3.2 Relative rate constants k_{rel} for the Claisen rearrangement of a soluble allyl vinyl ether derivative in various solvents[a]

Solvent	k_{rel}	
	R = Na	R = CH$_3$
water	214	
trifluoroethanol	31	56
methanol	9.4	8.6
ethanol		6.1
isopropyl alcohol		5.0
dimethylsulfoxide		3.2
acetonitrile		3.1
acetone		2.1
benzene		2.0
cyclohexane	58	1.0

[a]Reference [31].

over a wide range of solvents: nonpolar, highly protic like trifluoroethanol (TFE), dipolar aprotic like dimethylsulfoxide (DMSO), and highly self-associative like water.

The results of Table 3.2 are not unlike those reported previously by the Cornell group, but they contain the additional solvents, DMSO, ethanol, methanol and, importantly, TFE and nearly pure water. The paper describes a correlation of the logarithm of the rate constant with Grunwald–Winstein Y value, that is, the logarithm of the relative rate of solvolysis of t-butyl chloride in a solvent relative to that in 80% aqueous ethanol (v/v), an 80/20 mixture of ethanol and water by volume [32]. However, the slope of the correlation (m = 0.31) is much less than in any solvolysis reaction. Thus, it was concluded that there was little increase in polar character in the Claisen rearrangement transition state. However, this correlation established a link between the Claisen rearrangement and the solvolysis reaction despite the fact that there was a large difference in the charge buildup in the transition state in the two reactions.

3.5.2 Origin of the solvent effect on solvolysis (S_N1) reactions

Concern for the origin of the similarities in the Claisen rearrangement and the solvolysis reaction led to the recognition that the effect of polar solvents on the S_N1 reaction is mostly on the ground state! This is most dramatically illustrated in work by Abraham, who determined the free energy of transfer of t-butyl chloride between a wide range of solvents all relative to dimethylformamide (DMF) [33]. This work found a free-energy difference of $5.5 \, \text{kcal mol}^{-1}$ between cyclohexane and water, but the difference between methanol and water was almost as large, $4.6 \, \text{kcal mol}^{-1}$. The solvolysis rate difference in these two solvents corresponds to an energy difference of $6.2 \, \text{kcal mol}^{-1}$ so that ground-state destabilization is the major factor. This relatively recent quantification of solvent effects on the solvolysis reaction followed earlier work by Winstein, who in 1957 came to the conclusion that ground-state effects dominated the rate difference between methanol and water [34]. This was also based on a determination of the free energy of transfer of the reactant between the two solvents, and the quantification of the effects on the ground state and transition state was nearly the same. This important effect on the classic S_N1 reaction seems to have been lost to the authors of organic chemistry textbooks [35].

Abraham also found that the free energies of transfer of t-butyl chloride between solvents correlated with the cohesive energy density of the solvent. This attribute of the solvent was introduced by Hildebrand and is defined as the heat of vaporization of the solvent divided by its molar volume, all minus an RT term [36]. It is reasonable that this correlation represents causality since the interaction of any solute with a solvent requires disruption of solvent–solvent interaction. Subsequent work in our laboratory suggests that similar effects make an important contribution to the accelerated rates of the Claisen rearrangement in aqueous media.

3.6   Multiparameter correlation-factor analyses

3.6.1   General multiparameter correlation analysis of solvent effects

Multiparameter correlation analyses of the effect of solvent on rates and equilibria were introduced by Koppel and Palm [37], but these have been utilized extensively by Taft and his students who recognized the contribution of five factors [38]. These are: (i) the ability of the solvent to stabilize an ion or dipole by dielectric effects (π^*); (ii) the polarizability of the solvent; (iii) the hydrogen-bond donation (α); (iv) the acceptor ability (β); and (v) the cohesive energy density of the solvent (δ^2). In the case of the solvolysis reaction of t-butyl chloride, Taft's group found that the logarithm of the rate constant in 21 solvents from water to diethyl ether could be correlated with four factors [39]:

$$\log k = -14.60 + 5.10\pi^* + 4.17\alpha + 0.73\beta + 0.048\delta^2/100 \qquad r = 0.9973$$

However, when the free energies of transfer of t-butyl chloride in these same solvents were determined, the data provided a good correlation with the same four factors but only the cohesive energy density term was significant [33]:

$$\Delta G_t = -0.856 - 0.542\pi^* - 0.032\alpha - 0.308\beta + 1.130\delta^2/100 \qquad r = 0.9857$$

When the free energies of transfer of the ground state were subtracted from the free energies of activation, the free energies of transfer of the transition state were obtained, and these correlated well with the same four factors:

$$\delta\Delta G_t^{\ddagger} = 8.25 - 6.96\pi^* - 5.70\alpha - 0.99\beta - 0.65\delta^2/100 \qquad r = 0.9973$$

However, the dominant factors affecting the transition state are the ability of the solvent to stabilize dipoles and the ability of the solvent to donate a hydrogen bond.

There has been some concern that the Taft π^* parameter, which is derived from the solvent effect on the wavelength of ultraviolet absorptions modified by experience, is too encompassing a parameter and includes hydrogen-bonding and cohesive-energy-density contributions. So Beak [40] utilized the Kirkwood–Onsager [41] function, $(\varepsilon-1)/(2\varepsilon+1)$, as a parameter instead of π^* to characterize various tautomeric equilibria. As a result not only of concern about π^* but also of concern about the origin of the α and β parameters, Gajewski attempted to use the Kirkwood–Onsager function and the cohesive energy density as parameters along with hydrogen-bond donor and acceptor parameters derived from the transfer of chloride ion and potassium ion, respectively, in bulk solvents [42]. This equation was called the KOMPH equation, after Kirkwood, Onsager, Marcus, Parker and Hildebrand. Marcus and Parker have pioneered the measurement and analysis of single-ion-transfer free energies between various solvents and water [43]. The Taft hydrogen-bond parameters are derived from spectroscopic wavelength shifts of a dilute solution of the solvent in a nonpolar solvent with added base or acid so that they measure single- or perhaps two-molecule interactions as opposed to solute interaction with bulk solvent. However, after Gajewski updated the hydrogen-bond parameters to exclude the contribution of dielectric and cohesive-energy-density effects (the KOMPH2 model), the hydrogen-bond donor parameters were found to be similar from both sources (bulk and single molecule), that is they correlate with one another with $r = 0.982$ [44]. Unfortunately, the hydrogen-bond acceptor parameters correlate less well: $r = 0.902$.

When the KOMPH2 model is used to correlate the solvent-induced rate effects in the solvolysis of t-butyl chloride in solvents ranging from acetone to water, only the hydrogen-bond donor parameter and the cohesive-energy-density term were significant [44]:

$$\ln k = 46.1\alpha' + 11.4\delta^2 - 6.23 \qquad (r = 0.979, \text{SD} = 1.22, \text{range} = 19.0)$$

Analysis of the relative contributions of the two terms using the transfer data provided by Abraham reveals that the hydrogen-bonding term is a transition-state effect and the cohesive-energy-density term is a ground-state effect. Further, the coefficient of the α' parameter suggests that hydrogen bonding to the developing chloride ion is roughly half that to the chloride ion itself, and the coefficient of the cohesive-energy-density term indicates a volume difference between ground and transition states of roughly $11 \, \text{cm}^3 \, \text{mol}^{-1}$, which is roughly half of the volume of activation measured for solvolysis of t-butyl chloride in a variety of hydroxylic solvents [45]. It should be further recognized that these correlations provide a measure of the relative contribution of each factor to the overall effect produced by any of the solvents on the reaction. Thus in trifluoroethanol (TFE) solvent, a solvent with low self-association but strong hydrogen-bond donor abilities, the major contribution to the high solvolysis rates observed in this solvent is the hydrogen-bond donor ability, which is a transition-state effect providing stabilization of the chloride ion. However, in water solvent, the hydrogen-bond donor effect is only slightly more than the effect of the cohesive energy density of the water, which destabilizes the ground state. Relative to methanol solvent, the transition-state stabilization by TFE ($-3.2 \, \text{kcal} \, \text{mol}^{-1}$) is substantially more than the transition-state stabilization by water ($-1.65 \, \text{kcal} \, \text{mol}^{-1}$), but this is substantially less than the ground-state destabilization by water ($+4.6 \, \text{kcal} \, \text{mol}^{-1}$), again, relative to methanol (see Figure 3.1).

It may be of some concern that the solvent dielectric properties are nowhere in evidence in this correlation, but that is the result of the solvents chosen. All have a value of the Kirkwood–Onsager (KO) parameter between 0.44 and 0.48. Only if data in solvents of very low dielectric constant are included (e.g. benzene, ether, THF, ethyl acetate) is there a contribution from the KO function [44]:

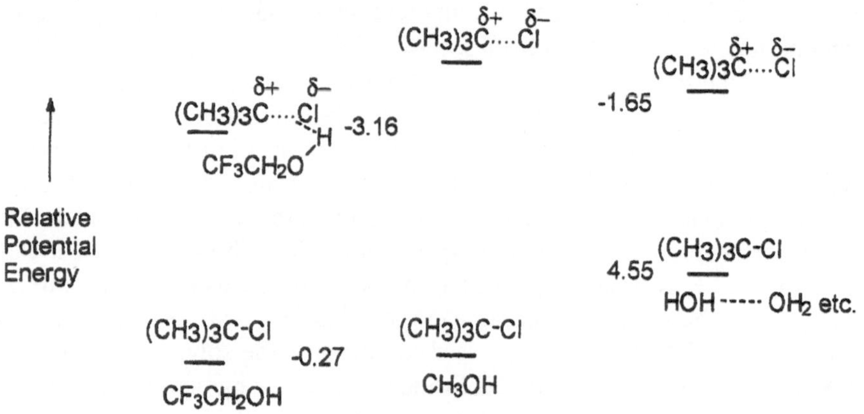

$$\overset{\delta+}{(CH_3)_3C}\cdots\overset{\delta-}{Cl}$$

$$\overset{\delta+}{(CH_3)_3C}\cdots\overset{\delta-}{Cl} \quad -1.65$$

$$\overset{\delta+}{(CH_3)_3C}\cdots\overset{\delta-}{Cl}_{H} \quad -3.16$$
$$CF_3CH_2O$$

Relative
Potential
Energy

$$(CH_3)_3C\text{-}Cl \quad 4.55$$
$$HOH\text{-----}OH_2 \text{ etc.}$$

$$(CH_3)_3C\text{-}Cl \quad -0.27$$
$$CF_3CH_2OH$$

$$(CH_3)_3C\text{-}Cl$$
$$CH_3OH$$

Figure 3.1 Solvent effects on the ground state and transition state in the S_N1 reaction of t-butyl chloride (in kcal mol^{-1} relative to methanol).

$$\ln k = 36.0(\varepsilon-1)/(2\varepsilon+1) + 39.9\alpha' + 11.7\delta^2 - 22.8$$

$$(r = 0.985, SD = 1.12, range = 25.4)$$

The dielectric term corresponds to a dipole moment change of roughly 10 D if a cavity radius of 4 Å is assumed. However, there might be some concern about the nature of the solvolysis reaction in these very nonpolar solvents. Indeed, even in acetone, the reaction is probably dominated by rate determining ion-pair formation.

3.6.2 Multiparameter correlations of solvent effects on Claisen rearrangements

In order to put into perspective the effects of solvent on the aliphatic Claisen rearrangement, it should be noted that the large change in dielectric constant and basicity between cyclohexane and DMSO provides but a small rate difference (a factor of 4), which indicates little change in dipolar character or Lewis acidity from ground state to transition state. Hydroxylic solvents have a modest rate effect, but even in bulk TFE, a solvent in which the chloride ion is more stable by 2.5 kcal mol^{-1} relative to water solvent [43] because of better hydrogen-bond donation, there is a rate acceleration that is less than that found in pure water. This suggests that ground-state destabilization is more important than hydrogen-bond donation to the transition state, and application of the multiparameter factor analysis KOMPH2 equation provides a good correlation that supports this view [44]:

$$\ln k = 9.43\alpha' + 3.27\delta^2 - 0.29 \qquad (r = 0.969, SD = 0.40, range = 5.37)$$

In TFE solvent, hydrogen-bond donation is the major factor, but in water the hydrogen-bond donation and the cohesive energy density are nearly equal contributing factors. Again it is useful to compare the effects in these two solvents to that in methanol, where the hydrogen-bonding effect is assumed to be a transition-state effect, but the cohesive-energy-density effect is assumed to be a ground-state effect (see Figure 3.2).

Just why there should be any cohesive-energy-density effect in the Claisen rearrangement can be attributed to the volume difference between the reactant and the transition state. Brower found a negative activation volume of $-10 \, cm^3 \, mol^{-1}$ for the rearrangement as the result of examining the rate of the rearrangement as a function of external pressure [46]. The cyclic nature of the transition state relative to the reactant state can easily account for the volume difference. However, in aqueous solution, the favored conformations of the reactant probably minimize the potential energy in this solvent, which might be expected to result in a structure more resembling the transition state. The combination of hydrophobic interactions forcing more transition-state-like structure on the ground state is the likely source of the cohesive-energy-density effect.

3.7 Other developments in Claisen rearrangements in aqueous solution

3.7.1 Experimental examination of transition-state structure

Secondary-deuterium kinetic isotope effects (KIE) have been used to probe the extent of bonding changes in the Claisen rearrangement in the gas phase [13]. The question of how the transition-state structure might change in aqueous solution was addressed by Brichford, who determined the C(4) and C(6) KIE in *m*-xylene solution and in aqueous methanol solutions at 100°C [47]. These KIEs are given in Figure 3.3 along with the values for the gas-

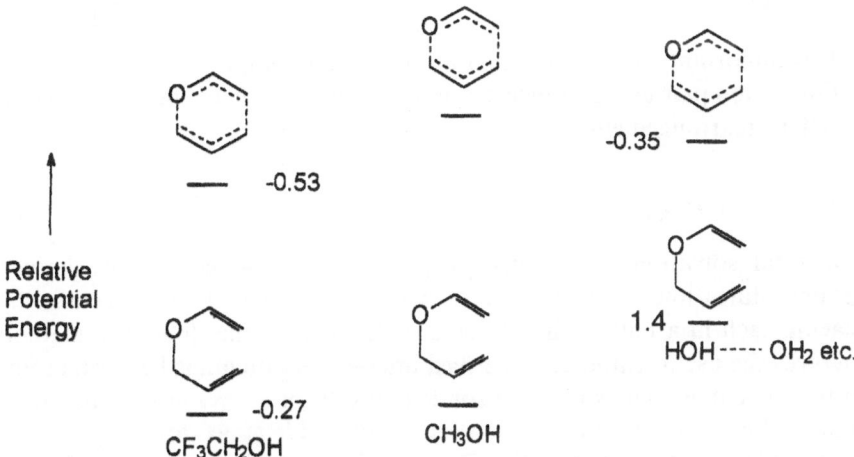

Figure 3.2 Solvent effects on the ground state and transition state in the Claisen rearrangement reaction (in $kcal \, mol^{-1}$ relative to methanol).

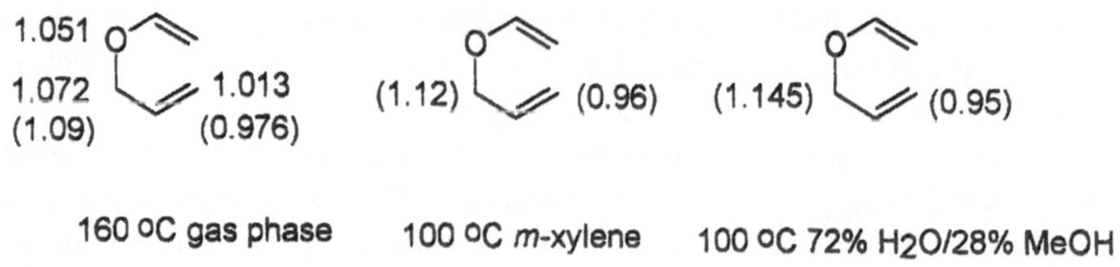

Figure 3.3 Experimental heavy-atom, $^{16}O/^{18}O$, $^{12}C/^{14}C$ and H/D$_2$ kinetic isotope effects in the Claisen rearrangement.

phase rearrangement at 160°C and the heavy-atom isotope effects determined for the gas-phase reaction by a Shine–Saunders collaboration [48]. An analysis of the heavy-atom isotope effects reinforces the original interpretation of the bonding changes occurring in the rearrangement, and the deuterium KIE values at 100°C are reasonable extrapolations of the higher-temperature values. It should also be noted that the differences in the deuterium KIE values between the hydrocarbon solvent and the highly aqueous medium are not very different; however, the experimental error is roughly 0.04 at C(4) and 0.02 at C(6), so it might be argued that the transition states could be different.

The concern for the possibility of generation of an ion pair that might be induced by the aqueous solvent system led to an examination of the solvolysis of 1,1-dideuterioallyl mesylate in an aqueous methanol solvent system. Here, however, no allylic rearrangement was observed in the formation of the substitution products; thus, there was no rearrangement in the starting mesylate as well [47]:

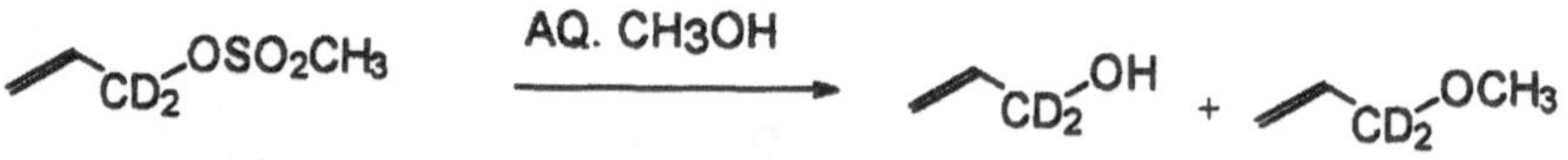

It is important to note here that a deliberate attempt to generate an allyl cation–anion pair using a much more stable anion than an enolate ion led to no allylic rearrangement!

3.7.2   Theoretical approaches

The total solvation of various gas-phase quantum-mechanically-derived ground states and transition states has been evaluated by Jorgensen by placing each in a bath of hundreds of solvent molecules [49]. The method involves the use of empirically derived interaction potentials between atoms of the solvent as well as between solute and solvent atoms without polarization of the solute. This allows evaluation of stable states of the system by Monte Carlo methods including Metropolis sampling procedures. In this way a 3.85 kcal mol^{-1} energy benefit for doing the Claisen rearrangement in water versus the gas phase was calculated. The quantum-mechanical calculation was at the RHF/6-31G* level, and this revealed no dipole change from ground to transition state, although there could be changes in higher moments [50]. Thus the energy change could not come from a generalized Kirkwood–Onsager dielectric effect. The solvent-dynamics calculations did reveal an increase in hydrogen bonding from reactant to transition state in the aqueous environment, and apparently the entire rate effect in water was attributed to this factor. No mention was made of solvent self-association. However, it is important to recognize that it is difficult to separate hydrogen-bonding and hydrophobicity contributions in any of these calculations without reference to a calculation on a nonpolar model, e.g. the Cope rearrangement.

In a related effort, Gao [51] started with the gas-phase reaction coordinate and performed quantum-mechanical/molecular-mechanical calculations. In these calculations, the solute was modeled using the AM1 semiempirical Hamiltonian and Jorgensen's van der Waals parameters, with the quantum-mechanical Hamiltonian being perturbed by surrounding water molecules, which were treated classically by a Jorgensen water–water potential (TIP3P). Thus changes in transition-state structure resulting from immersion in water could be calculated. The rate factor calculated for the reaction in water relative to the gas phase was 368 at 25°C. This corresponds to $3.5\,kcal\,mol^{-1}$ and negative charge buildup on the oxygen was identified as the major contributor.

In a different approach, Gao started with the RHF/6-31G* transition state and systematically varied the average distance between the two three-atom fragments using an AM1 calculation coupled with a Jorgensen-like Monte Carlo simulation. He found a monotonic increase in dipole moment in both the gas phase and in water and an increase in free energy of aquation with increasing distance [52]. The correlation of energy and dipole moment with distance was then applied to the geometries resulting from RHF/6-31G* calculations. A 1.46 D change in dipole moment from reactant to transition state in an aqueous environment was estimated. If this dipole moment change were associated with charge changes at an average distance of 2 Å, the increase in charge at each end of the dipole is roughly 0.15 of an electron. Thus, any depiction of the transition state with $\delta+$ and $\delta-$ might be better represented as $\delta\delta\delta+$ and $\delta\delta\delta-$, where each δ represents division of the charge by a factor of 2. Just what this charge shift leading to a dipole moment change of 1.46 D means in terms of energy can be addressed by a classical continuum dielectric model using the Kirkwood–Onsager approach, which gives $1\,kcal\,mol^{-1}$ stabilization for a cavity radius of 2.5 Å. This is the effect due simply to solvent polarity and corresponds to a rate factor of 5 at 60°C. The effect on hydrogen bonding is more difficult to assess; however, some consideration should be given to the very weakly basic character of aldehydic oxygen, which suggests that, even if the carbonyl group were fully formed in the transition state, there would be little hydrogen bonding to it.

In still another approach, Cramer and Truhlar have pursued a combined

AM1–Born solvation continuum approach to the question of water acceleration of the Claisen rearrangement. The most recent version [53], like its
predecessors, also uses an exposed surface calculation as a way of modeling
the hydrophobic effect. The results suggest that hydrophobic packing is relatively unimportant in water and that electric polarization and first-hydration-
sphere hydrophilic effects dominate the calculated acceleration, which when
using the Houk CASSCF/6-31G* geometry provided a rate factor of 1400. It
is remarkable that the factors of solvent self-association and hydrogen
bonding appear to be absent from this model, yet it seems to approximate the
experimental observations.

Continuum solvent models, i.e. solvent continuum reaction field (SCRF)
equations derived from Kirkwood–Onsager approaches to solvation [41], are
included in the Gaussian programs, and Hillier has used these to assess solvation in the Claisen rearrangement. Using the RHF/6-31G* transition state he
found little (0.8 kcal mol^{-1}) stabilization of the transition state relative to the
reactant state [54]. However, when a polarizable continuum model (PCM)
[55] was used, a 2.8 kcal mol^{-1} net stabilization was calculated, and additional
calculations of dispersion and cavity effects provided another 0.6 kcal mol^{-1}
of solvation. Here, again, a model that minimizes hydrophobic effects
appears to do well in providing an energy difference in aqueous solution
which is similar in magnitude to that determined experimentally.

In a more recent study, Hillier placed two water molecules around the
oxygen of allyl vinyl ether and its transition state for formation of 4-pentenal
in an MP2/RHF/6-31G* calculation [56]. When the SCRF model was used,
no net decrease in activation free energy was obtained at the $l = 1$ level (atomic
monopoles) and lack of convergence accompanied attempts to use higher
terms in the multipole expansion. However, the PCM model provided a net
energy decrease of 4.3 kcal mol^{-1}, which corresponds favorably to experiment. Somewhat disconcerting, however, were the calculated kinetic isotope
effects ($k^H/k^{C(4)D_2} = 1.149$ and $k^H/k^{C(6)D_2} = 0.919$), which differed from the experimental values, but not by as much as the authors suggested, since they
referred to the less aqueous result cited in reference [47] (see Figure 3.3).
Nonetheless, the small contribution of solvent-cavity or cohesive energy-
density effects in these calculations is of some concern.

3.8 Summary

The Claisen rearrangement of allyl vinyl ethers is an effective synthetic tool
for introducing a γ, δ-unsaturated carbonyl system with transfer of asymmetry from the double bonds to the newly formed α, β carbon–carbon bond.
The reaction is intramolecular and highly exothermic so that relatively
unstable unsaturated systems can be prepared. The only drawback is the high
temperature necessary to effect the reaction, but its rate can be enhanced by

factors of up to 200 by the use of water as a solvent instead of a hydrocarbon solvent. A rate factor of 200 corresponds to a free energy of $4.6\,\text{kcal}\,\text{mol}^{-1}$ at 160°C, and for a reaction with activation parameters like that of the aliphatic Claisen rearrangement this could lower the temperature necessary to effect the reaction by roughly 70°C!

The origin of the rate acceleration in aqueous solution is controversial. Solvent-effect calculations using various approaches including discrete and continuum solvent models [50–53] suggest either increased hydrogen-bond donation to the transition state or increased polarity of the transition state, which responds to the dielectric constant of the medium. However, the experimental evidence is consistent with destabilization of the ground state relative to the transition state owing to solvent self-association and hydrogen-bond donation to stabilize the transition state, with both contributing nearly equally in water solvent [42, 44]. The former factor results from the negative activation volume for the Claisen rearrangement.

Acknowledgements

I thank the National Science Foundation and the Department of Energy for support of different aspects of this research. I also thank Professor Paul A. Grieco and Dr Ellen B. Brandes for their efforts in this area.

References

[1] Claisen, L., *Berichte*, 1912, **45**, 3157.
[2] Claisen, L., Eisleb, O., *Annalen*, 1913, **401**, 21.
[3] Claisen, L., Tietze, E., *Berichte*, 1925, **58**, 275.
[4] Hurd, C.D.; Schmerling, L., *J. Am. Chem. Soc.*,1937, **59**, 107.
[5] Wipf, P., *Comprehensive Organic Synthesis*, Trost, B.M.; Fleming, I.; Paquette, L.A., Eds.; Pergamon Press, New York, 1991, vol. 5. p. 827; Ziegler, F.E., *Chem. Rev.*, 1988, **88**, 1423; Rhoads, S.J.; Raulins, N.R. *Org. React.*, 1975, **22**, 1.
[6] Woodward, R.B.; Hoffmann, R.J., *Angew. Chem. Int. Ed. Engl.*, 1969, **8**, 781.
[7] (a) Vitorelli, P.; Winkler, T.; Hansen, H.-J.; Schmidt, H., *Helv. Chim. Acta*, 1968, **51**, 1457; (b) Hill, R.K.; Edwards, A.G., *Tetrahedron Lett.*, 1964, 3239; (c) Marvel, E.N.; Stephenson, J.L., *J. Org. Chem.*, 1960, **25**, 676; (d) Hart, H., *J. Am. Chem. Soc.*, 1954, **76**, 4033; (e) Alexander, E.R.; Kluiber, R.W., *J. Am. Chem. Soc.*, 1951, **73**, 4304.
[8] Kincaid, J.F.; Tarbell, D.S, *J. Am. Chem. Soc.*, 1939, **61**, 3085.
[9] White, W.N.; Gwynn, D.; Schlitt, R.; Girard, C; Fife, W., *J. Am. Chem. Soc*, 1958, **80**, 3271.
[10] Goering, H.L.; Jacobson, R.R., *J. Am. Chem. Soc.*, 1958, **80**, 3277.
[11] White, W.N.; Wolfarth, E.F., *J. Org. Chem.*, 1970, **35**, 2196.
[12] Schuler, F W.; Murphy, G.W., *J. Am. Chem. Soc.*, 1950, **72**, 3155.
[13] Gajewski, J.J.; Conrad, N.D. *J. Am. Chem. Soc.*, 1979, **101**, 2747.
[14] Andrews, P.R.; Smith, G.D.; Young, *I.G., Biochemistry*, 1973, **12**, 3492.
[15] Copley, S.D.; Knowles, J.R., *J. Am. Chem. Soc.*, 1987, **109**, 5008.
[16] Ponaras, A.A., *J. Org. Chem.*, 1983, **48**, 3866.
[17] Gajewski, J.J.; Jurayj, J.; Kimbrough, D R.; Gande, M.E.; Ganem, B.; Carpenter, B.K., *J. Am. Chem. Soc.*, 1987, **109**, 1170. (This was a paper resulting from two separate efforts at Cornell and Indiana, which were forced into combination by the editors of the journal.)

[18] Coates, R.M.; Rogers, B.D.; Hobbs, S.J.; Peck, D.R.; Curran, D.P., *J. Am. Chem. Soc.,* 1987, **109**, 1160. (This paper, like reference [17], was a combination of work in two separate laboratories.)

[19] Rideout, D.; Breslow, R., *J. Am. Chem. Soc.,* 1980, **102**, 5613; Grieco, P.A.; Garner, P.; He, Z., *Tetrahedron Lett.,* 1983, **24**, 1897.

[20] Grieco, P.A.; Brandes, E.B.; McCann, S.; Clark, J.D., *J. Org. Chem.,* 1989, **54**, 5849.

[21] Zhu, G.D.; Okamura, W.H., *Chem. Rev.,* 1995, **95**, 1877–1952.

[22] Trost, B.M.; Bernstein, P.R.; Funschulling, P.C., *J. Am. Chem. Soc.,* 1979, **101**, 4378; Grieco, P.A.; Takigawa, T.; Moore, D.R., *J. Am. Chem. Soc.,* 1979, **101**, 4380; Brandes, E.B.; Grieco, P.A.; Garner, P., *J. Chem. Soc., Chem. Commun.,* 1988, 500.

[23] McMurry, J.E.; Andrus,A.; Ksander, G.M.; Musser, J.J.; Johnson, M.A., *Tetrahedron,* 1981, **37** (Suppl. No. 1), 319.

[24] Wang, J; Thommen, M.; Keese, R., *Acta Crystallog. C,* 1996, **52**, 2311–13. (For a crystal structure of a derivative of this material to compare with that of the *ctct* material of reference [21]).

[25] Broka, C.A., *J. Org Chem.,* 1988, **53**, 575–83.

[26] Lubineau, A.; Augé J.; Bellanger, N; Caillebourdin, S., *Tetrahedron Lett.,* 1990, **31**, 4147.

[27] Lubineau, A.; Augé, J.; Bellanger, N; Caillebourdin, S., *J. Chem. Soc, Perkin Trans. 1,* 1992, **13**, 1631–6.

[28] Toyota, M.; Nishikawa, Y.; Fukumoto, K., *J. Tetrahedron,* 1994, **50**, 11153–66.

[29] Toyota, M.; Seishi, T.; Fukumoto, K. *Tetrahedron,* 1994, **50**, 3673–86.

[30] Bourgin, D.; Buchel, R.; Gerber, P; Keese, R., *Tetrahedron Lett.,* 1994, **20**, 3267–8.

[31] Brandes, E B.; Grieco, P.A.; Gajewski, J.J., *J. Org. Chem.,* 1989, **54**, 515.

[32] Grunwald, E.; Winstein, S. J., *J. Am. Chem. Soc.,* 1948, **70**, 846.

[33] Abraham, M.H.; Grellier, P.L.; Nasehzadeh, A.; Walker, R.A.C., *J. Chem. Soc., Perkin Trans. 2,* 1988, 1717.

[34] Winstein, S.; Fainberg, A.H., *J. Am. Chem. Soc.,* 1957, **79**, 5937; Smith, S.G.; Fainberg, A.H.; Winstein, S., *J. Am. Chem. Soc.,* 1961, **83**, 618.

[35] See, however, Harris, J.M.; Wamser, C.C., *Fundamentals of Organic Reaction Mechanisms;* Wiley, New York, 1976, p. 143.

[36] Hildebrand, J.H.; Prausnitz, J.M; Scott, R.L., *Regular and Related Solutions;* Van Nostrand-Reinhold, Princeton, NJ, 1970.

[37] Koppel, I.A.; Palm, V.A. in *Advances in Linear Free Energy Relationships,* Chapman, N.B.; Shorter, J., Eds.; Plenum, London, 1972, ch. 5.

[38] Araham, M.H.; Grellier, P.L.; Abboud, J.-L.M.; Doherty, R.M.; Taft, R.W., *Can. J. Chem.,* 1988, **66**, 2673 (Review).

[39] Abraham, M.H.; Doherty, R.M.; Kamlet, M.J.; Harris, J.M.; Taft, R.W., *J. Chem. Soc., Perkin Trans. 2,* 1987, 913.

[40] Mills, S.G.; Beak, P., *J. Org. Chem.,* 1985, **50**, 1216 and references therein.

[41] Kirkwood, J.G., *J. Chem. Phys.,* 1934, **2**, 351; Onsager, L., *J. Am. Chem. Soc.,* 1936, **58**, 1486.

[42] Gajewski, J.J., *J. Org. Chem.,* 1992, **57**, 5500.

[43] Cox, B.G.; Hedwig, G.R.; Parker, A.J.; Watts, D.W., *Aust. J. Chem.,* 1974, **27**, 477; Marcus, Y., *Pure Appl. Chem.,* 1983, **55**, 977.

[44] Gajewski, J.J.; Brichford, N.L., in *Structure and Reactivity in Aqueous Solution,* ACS Symp. Ser. 568, Cramer, C J., Truhlar, D.G., Eds.; American Chemical Society, Washington DC, 1994, pp. 229–42 .

[45] LeNoble, W.J., *Chem. Rev.,* 1989, **89**, 549.

[46] Brower, K.R., *J. Am. Chem. Soc.,* 1961, **83**, 4370. For similar results with Cope and aromatic Claisen rearrangement, see Walling, C.; Naiman, M., *J. Am. Chem. Soc.,* 1962, **84**, 2628.

[47] Gajewski, J.J.; Brichford, N.L., *J. Am. Chem. Soc.,* 1994, **116**, 3165.

[48] Kupczyk-Subotkowska, L.; Saunders, W.H., Jr.; Shine, H.J.; Subotkowski, W.J., *J. Am. Chem. Soc.,* 1993, **115**, 5957.

[49] Severance, D.L.; Jorgensen, W.L., *J. Am. Chem. Soc.,* 1992, **114**, 10966; Severance, D.L.; Jorgensen, W.L., in *Structure and Reactivity in Aqueous Solution,* ACS Symp. Ser. 568, Cramer, C.J., Truhlar, D.G., Eds.; American Chemical Society, Washington, DC, 1994, pp 243–59.

[50] Yoo, H.Y.; Houk, K.N., *J. Am. Chem. Soc.*, 1994, **116**, 12047.

[51] Gao, J., *J. Am. Chem. Soc.*, 1994, **116**, 1563.

[52] Sehgal, A.; Shao, L.; Gao, J., *J. Am. Chem. Soc.*, 1995, **117**, 11337.

[53] Storer, J.W.; Giesen, D.J.; Hawkins, G.D.; Lynch, G.C.; Cramer, C.J.; Truhlar, D.G.; Liotard, D.A., in *Structure and Reactivity in Aqueous Solution*, ACS Symp. Ser. 568, Cramer, C.J., Truhlar, D.G., Eds.; American Chemical Society, Washington, DC, 1994, pp 24–49.

[54] Davidson, M M.; Hillier, I.H.; Hall, R.J.; Burton, N.A., *J. Am. Chem. Soc.*, 1994, **116**, 9294.

[55] Orozco, M.; Jorgensen, W.L.; Luque, F.J., *J. Comput. Chem.*, 1993, **14**, 1498.

[56] Davidson, M M.; Hillier, I.H., *J. Phys. Chem.*, 1995, **99**, 6748.

4 Carbonyl additions and organometallic chemistry in water

A. LUBINEAU, J. AUGÉ and Y. QUENEAU

4.1 Introduction

This chapter is devoted to aqueous carbonyl additions, with the focus on organometallic additions. Thus Barbier-type reactions in water, the success of which is increasing, are described at length in section 4.2. The subsequent section deals with conjugate additions, and section 4.4 deals with cross-aldol reactions, with special attention being paid to organometallic additions. Section 4.5 is devoted to organometallic pinacol couplings in water. Some miscellaneous reactions in which the carbonyl group is the reactive site are reviewed briefly at the end of the chapter.

4.2 Barbier-type alkylation reactions

The Barbier reaction is that between an alkyl halide and a carbonyl compound in the presence of magnesium (Barbier, 1898). In Barbier-type reactions, an organometallic intermediate species is supposed to be produced in the presence of the substrate. As this organometallic intermediate is usually highly reactive with water, Barbier-type reactions were previously conducted after careful exclusion of water.

As long ago as 1977, a Barbier allylation mediated by zinc was accomplished in refluxing ethanol containing 5% of water, unfortunately giving poor yields (Killinger *et al.* 1977). Luche and Damiano (1980) observed that ultrasonic waves were able to promote the formation of lithium or magnesium organometallic reagents even in the presence of aqueous solvents. These salient features have paved the way to a new field of investigation, which turned out to be extremely fruitful in the case of the allylation of carbonyl compounds mediated by tin, zinc and indium in aqueous media, as outlined in a recent review (Li, 1996).

4.2.1 Allylation of carbonyl compounds mediated by tin

The aqueous allylation reaction mediated by tin was inspired by the observation that allylation of benzaldehyde with diallyltin dibromide seemed to be accelerated by the addition of water (Nokami *et al.*, 1983). Acidic conditions (HBr, AcOH) were needed to perform the heterogeneous reaction using metallic tin, allylic bromides and carbonyl compounds in a mixture of ether and water. Metallic aluminum as an additive was recommended for the allylation of ketones and the less reactive aldehydes (Nokami *et al.*, 1983). The procedure was then applied to intramolecular reactions to synthesize five- or six-membered rings (Nokami *et al.*, 1984):

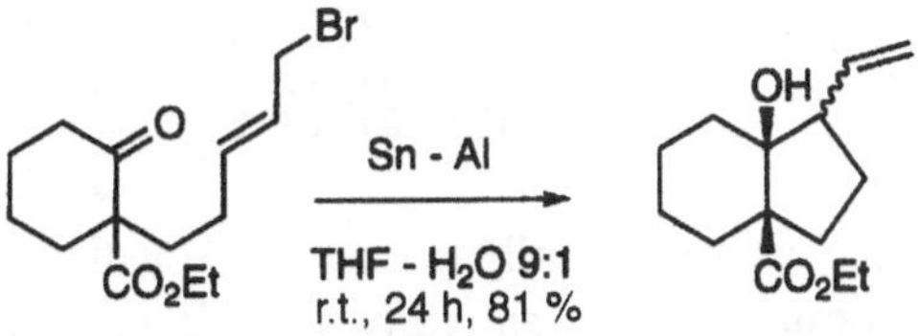

With functionalized allylic bromides, the same procedure allowed preparation of bromo-homoallylic alcohols and 1,3-ketoacetates (Mandai *et al.*, 1984) or α-methylene-γ-butyrolactones (Nokami *et al.*, 1986).

An improvement of the allylation reaction was achieved with the use of ultrasonic waves in a water–tetrahydrofuran (5/1) mixture without the help of acidic co-reagents (Pétrier *et al.*, 1985). Noteworthy are the sonicated reactions involving commercially available aqueous solutions of aldehydes, such as formaldehyde, and those involving aldehydes containing free hydroxyl groups (Einhorn and Luche, 1987):

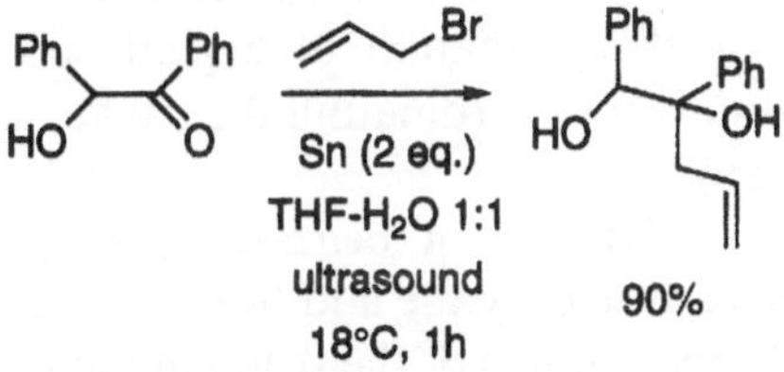

When applied to carbohydrates, the sono-allylation proceeded with useful diastereoselectivity (*threo*-selectivity) and allowed preparation of higher-carbon sugars from water-soluble substrates directly in aqueous ethanol without protection (Schmid and Whitesides, 1991). Later, the authors showed that the allylation of aldoses could be advantageously carried out by heating the reaction mixture; for example, the allylation of D-arabinose required 16–20 h under sonication, but only 2 h under reflux (Kim *et al.*, 1993):

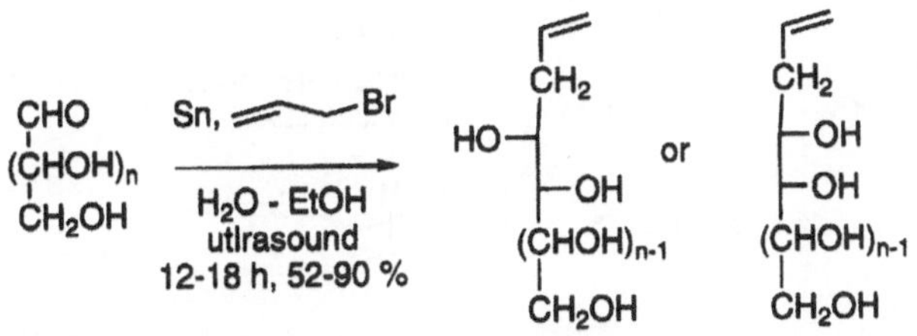

Of interest was the use of a suspension of tin powder in a saturated aqueous NH_4Cl/THF solution at 60°C (Zhou *et al.*, 1992).

Active zero-valent tin could also be generated by tin(II) chloride–aluminum in aqueous THF (Uneyama *et al.*, 1985), providing excellent *anti*-selectivity in the reaction of cinnamyl chlorides with aldehydes (Uneyama *et al.*, 1986a):

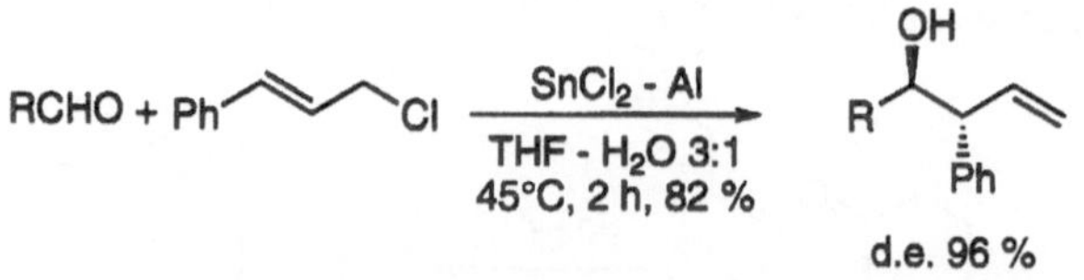

$$RCHO + Ph \diagup\diagdown Cl \xrightarrow[\substack{THF - H_2O\ 3:1 \\ 45°C,\ 2\ h,\ 82\ \%}]{SnCl_2 - Al} \begin{array}{c} OH \\ R \diagup\diagdown\diagup \\ \vdots \\ Ph \end{array}$$

d.e. 96 %

In the carbonyl allylation with allylic bromides mediated by tin(II) halide, organotin(IV) compounds were generally postulated (Mukaiyama *et al.*, 1980). Tin(II) chloride in the acidic aqueous medium $MeOCH_2CH_2OH/H_2O/AcOH/HCl$ (Uneyama *et al.*, 1986b) or tin(II) chloride with Amberlyst 15 in THF/H_2O (Talaga *et al.*, 1990) served as the promotor in allylation reactions yielding α-methylene-γ-lactones.

When carried out with (*E*)-rich 1-bromo-2-butene in THF/H_2O solution, the $SnCl_2$ Barbier-type allylation of aldehydes provided homoallylic alcohols, mainly with the *anti* configuration. By contrast, *syn* allylation occurred with tin(II) iodide (Scheme 4.1). The *syn*-selectivity, suggesting an acyclic antiperiplanar transition state, was even improved by the addition of tetrabutyl-ammonium bromide in the aqueous medium. This enhancement of selectivity could be explained by the formation of a pentacoordinated tin species, preventing the tin atom from coordinating the oxygen atom of the aldehyde (Masuyama *et al.*, 1996).

Palladium-catalyzed allylation of benzaldehyde by (*E*)-but-2-enol with tin(II) chloride was accelerated by the addition of water in any solvent used. Furthermore, the selectivity (*anti*) of the reaction increased with the amount of water, which was interpreted by the rupture of the coordination of the cosolvent to the tin(IV) species (Masuyama *et al.*, 1989). Functionalized

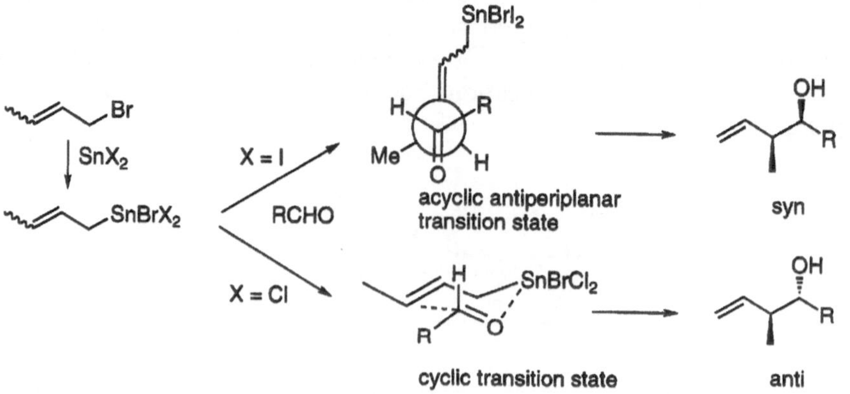

Scheme 4.1

allylic alcohols, or even more easily their methylcarbonate derivatives, upon exposure to aldehydes in a mixture of 1,3-dimethylimidazolidinone and water, underwent the same type of reaction leading to α-methylene-γ-lactones in moderate yields (Masuyama *et al.*, 1991):

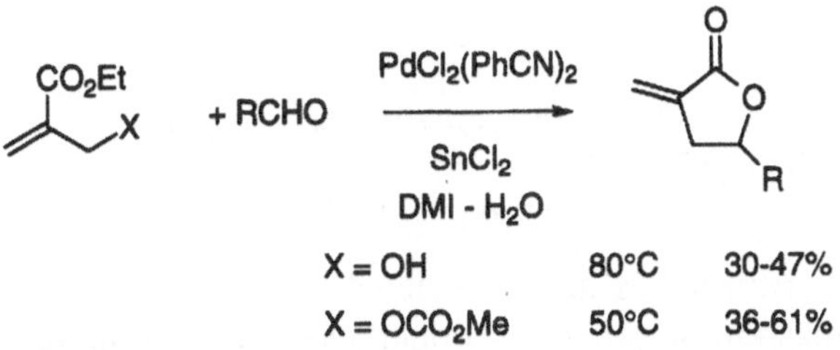

| X = OH | 80°C | 30-47% |
| X = OCO$_2$Me | 50°C | 36-61% |

As organotin(IV) intermediates are presumed to be involved in the reaction, it is indeed of great interest to examine the use of preformed organotin(IV) compounds as synthons in aqueous media. Thus the addition of allyldibutyltin chloride or allylbutyltin dichloride to aldehydes or ketones in water at 25°C led to the corresponding homoallylic alcohols in excellent yields (Boaretto *et al.*, 1985a). The reaction was then extended to other organotin monochlorides having a 2,3-unsaturated organic group bound to the tin atom. Crotyl-, 1-methylallyl- and cyclohex-2-enyl-stannation could occur in high yields (80–100%), but the *syn/anti*-selectivity, though slightly enhanced compared to reactions in a solvent-free mixture, was usually low (Furlani *et al.*, 1988). The regio- and stereochemical course of the allylation of aldehydes with crotyltin mono-, di- and tri-chlorides was studied in aqueous perchloric acid; with increasing acid concentration, the γ-selectivity was gradually decreased from the 100% value observed in water (Marton *et al.*, 1989). With tetraallyltin in acidic aqueous media, the allylation displayed exclusive chemoselectivity toward aldehydes (Yanagisawa *et al.*, 1993).

A related reaction, i.e. the condensation of allyltributylstannane with formaldehyde-generated immonium salts in aqueous media, gave, in excellent yields, bis(homoallylamines) with primary amines, and tertiary homoallylamines with secondary amines (Grieco and Bahsas, 1987):

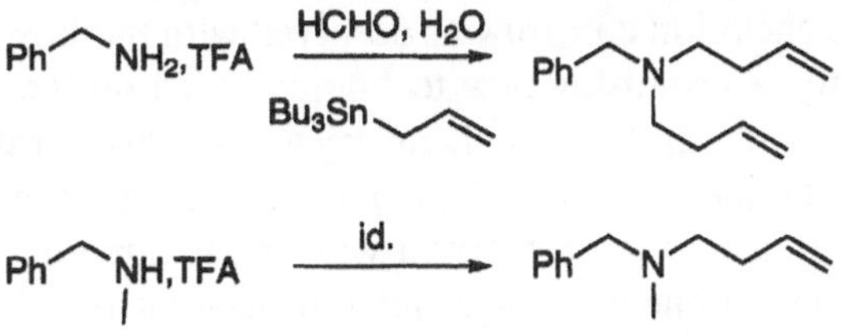

4.2.2   Allylation of carbonyl compounds mediated by zinc

Unlike the metallic tin-mediated allylation of carbonyl compounds, which requires either acidic medium, aluminum additive, ultrasonic radiation or heating, the metallic zinc-mediated allylation proceeds at room temperature, without metallic additive or sonication, in a mixture of tetrahydrofuran and saturated ammonium chloride solution (Pétrier and Luche, 1985). The chemoselectivity toward aldehydes relative to ketones turned out to be under kinetic control (Pétrier et al., 1985). Allyl bromide or chloride could be used in the reaction:

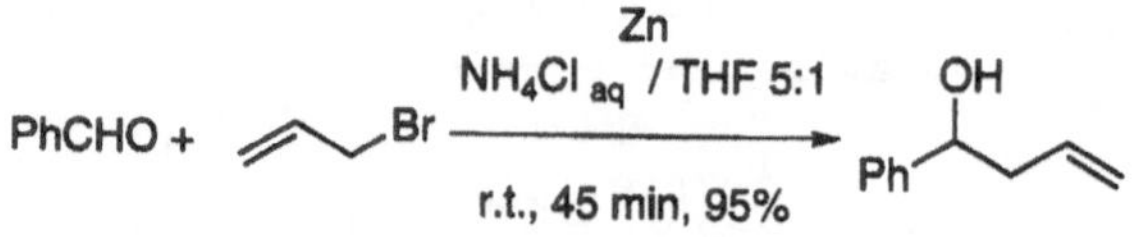

Substituted allyl bromides always reacted at the more-substituted carbon atom. Poor diastereoselectivities, nonetheless, were obtained with crotyl bromide or 3-chloro-1-butene. The scope and limitations of the reaction were carefully investigated. Cinnamyl bromide yielded phenylpropene and the aldehydic co-reactants remained unchanged; in the absence of the halide, some ketones dimerized to the pinacol in a slow, low-yield reaction (Einhorn and Luche, 1987).

With functionalized allyl bromides, such as ethyl (2-bromomethyl)acrylate, the zinc-mediated allylation in a saturated aqueous NH_4Cl/THF mixture under reflux provided α-methylene-γ-butyrolactones in good yield (Mattes and Benezra, 1985):

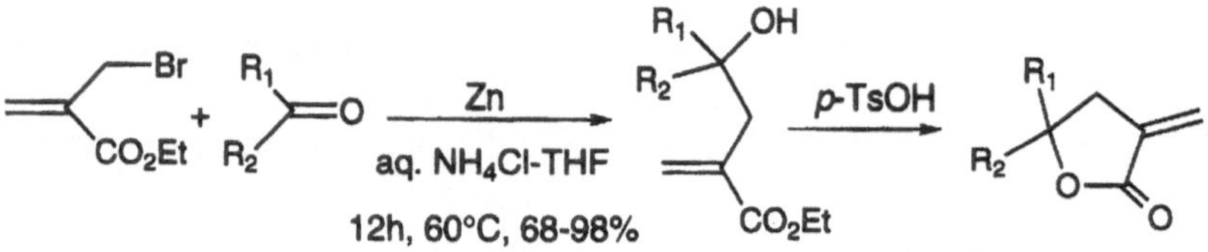

Wilson and Guazzaroni (1989) devised a modification of the Luche reaction by using a solid organic phase, such as reverse-phase C-18 silica gel, instead of tetrahydrofuran as the cosolvent. The solid organic phase can be reused and disposal of the solvent after the reaction is environmentally safe. Unfortunately, the syn/anti-selectivity was not improved regarding the Luche conditions. No chelation control was observed with β-alkoxyaldehydes.

More recently, a promising new technique based on the combined use of pulsed sonochemistry and electrochemistry was applied for the production of submicron crystalline metal powders. The first results showed that the zinc powder elaborated by such a method was about three times more reactive than the commercial one in an aqueous Barbier-type reaction excluding any organic cosolvent (Durant et al., 1995).

Zinc-mediated allylation in the Luche conditions (commercial zinc powder in a saturated aqueous NH_4 Cl/THF mixture) was applied in a five-step synthesis of (+)-muscarine, wherein an aldehyde derived from ethyl-lactate was treated with allyl bromide. No chelation control was observed with the α-alkoxyaldehyde, as well (Chan and Li, 1992a).

Better diastereoselectivities were obtained when α-ketoamides of proline benzyl ester were treated with allyl halides in the presence of zinc dust and pyridinium *p*-toluenesulfonate (PPTS) at 4°C in a THF/water mixture. The chiral auxiliary could be removed from the purified amides, yielding α-hydroxyketones with excellent enantioselectivities (Waldmann, 1990), as depicted in Scheme 4.2.

Treating 1-chloro-3-iodopropene and aldehydes or ketones with zinc powder in an aqueous medium led to the corresponding chlorhydrin:

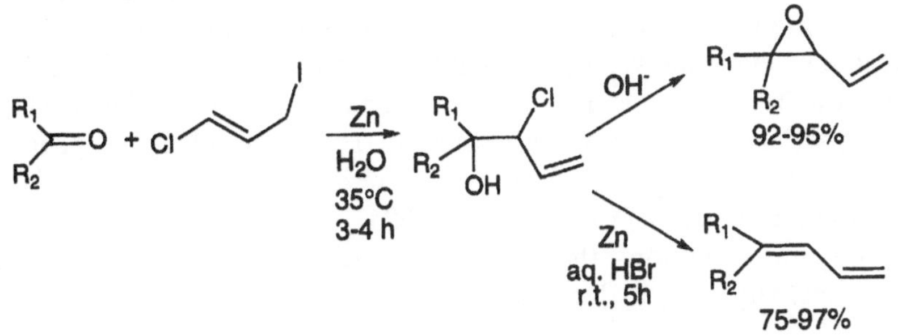

This is a convenient intermediate for the preparation of (*E*)-buta-1,3-dienes or vinyloxiranes (Chan and Li, 1990).

Another bifunctional compound, 2-(chloromethyl)-3-iodo-1-propene, was allowed to react in the same conditions with aldehydes or ketones, providing, after basic treatment, methylenetetrahydrofurans in high yield. The allylation is regioselective: conjugated carbonyl compounds reacted in a 1,2-fashion. It is also chemoselective, as evidenced by a competitive experiment between 5-nonanone and benzaldehyde (Li and Chan, 1991a):

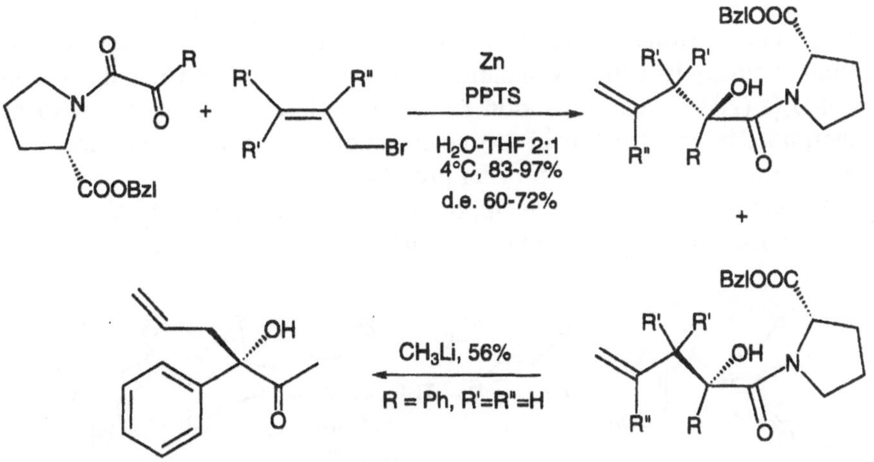

Scheme 4.2

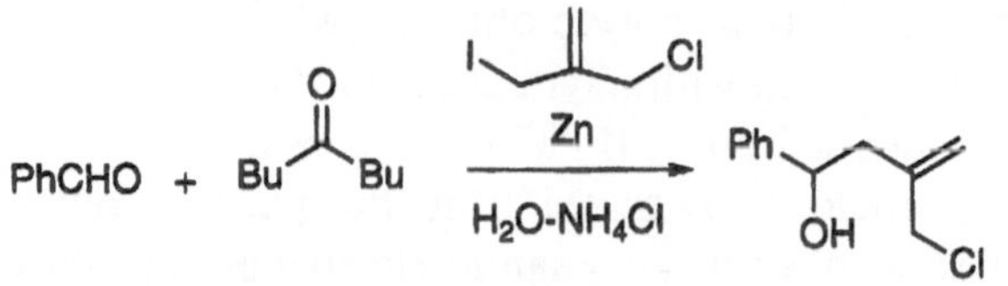

Similarly, 2,3-dichloropropene has been found to react smoothly with carbonyl compounds to give 3-chloro-homoallylic alcohols in excellent yields. Even, there, the presence of water was critical for the success of coupling, which has been achieved in a two-phase system of water and toluene containing a small amount of acetic acid (Oda *et al.*, 1992).

Reactions of cinnamyl chloride, unlike cinnamyl bromide, with aldehydes in the Luche conditions led to diastereomeric γ products, whereas the reaction with ketones gave mixtures of γ and α products. Phenylpropenes and dicinnamyls were detected, indicating the presence of cinnamyl radical intermediates in the reaction (Sjöholm *et al.*, 1994).

In the same conditions, the allylation of γ-aldoesters gave, with complete chemoselectivity, the γ-hydroxyesters:

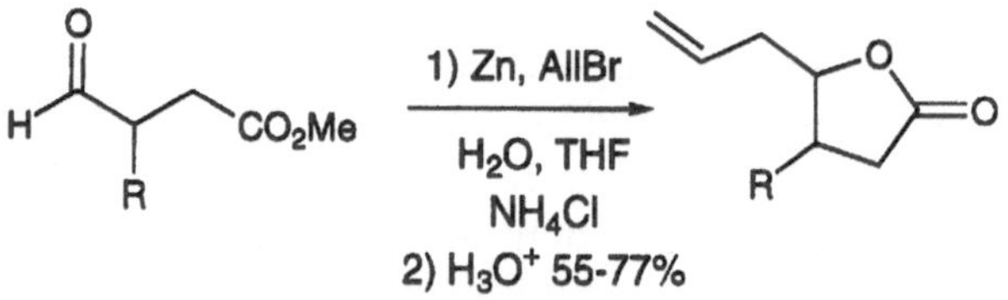

These were converted to γ-allyl-substituted γ-butyrolactones in a one-pot reaction (Kunz and Reissig, 1989). Both β- and γ-ketoesters underwent the same reaction. The *syn/anti*-diastereoselectivity was only moderate, except in the reaction between benzoyl acid esters with cinnamyl chloride (Ahonen and Sjöholm, 1995).

Facing these difficulties concerning the *syn/anti*-selectivity in carbonyl allylations, Marton *et al.*, (1996a) extensively studied the zinc-mediated allylation of aldehydes with allyl halides in cosolvent/H$_2$O (NH$_4$Cl) and in cosolvent/H$_2$O (NH$_4$Cl)/haloorganotin media. The stereochemistry seems to be determined by the structure of the radical anions, which were presumed to be formed through an electron-transfer process.

Of great interest is the use of Oppolzer's camphorsultam as chiral auxiliary in a zinc-mediated allylation of oximes in the Luche conditions:

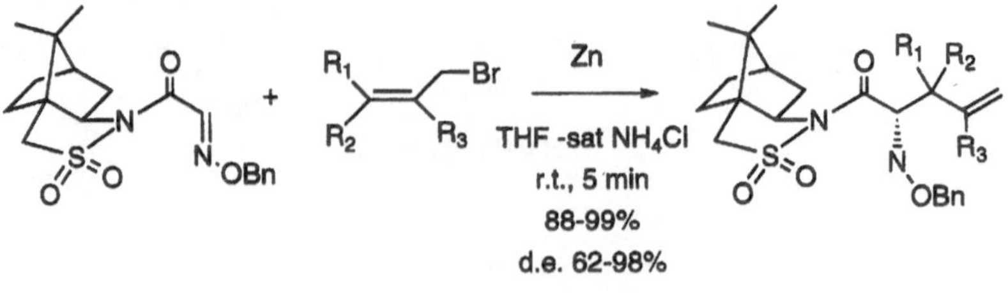

After reductive cleavage of the N–O bond in the presence of $Mo(CO)_6$, and removal of the sultam auxiliary with LiOH in $THF–H_2O$ solution, various allylglycines were obtained with enantioselectivities ranging from 62 to 98% (Hanessian and Yang, 1996).

4.2.3   Allylation of carbonyl compounds mediated by indium

Compared to other metals, indium has proved attractive, owing to three major properties: (i) indium is unaffected by boiling water; (ii) it does not form oxides readily in the air; and (iii) its first ionization potential is low (5.79 eV). Li and Chan (1991b) took advantage of these properties to undertake the first allylation of aldehydes and ketones mediated by indium in water, without the help of an additive, such as a salt or a cosolvent, as well as the help of ultrasound.

The interest of indium, compared to other metals, has been emphasized by Chan et al. (1994) in the allylation of carbonyl compounds containing acid-labile groups. Only indium as the metal gave the desired product in good yield:

metal	yield (%)
Zn	0
Sn	10
In	70

As the zero-valent indium species is regenerated by either zinc or aluminum, a catalytic amount of $InCl_3$, together with either zinc or aluminum, could be utilized in the allylation of carbonyl compounds in tetrahydrofuran–water mixture. The presence of water was found to be effective in suppressing undesired side-reactions (Araki et al., 1992).

Likewise, indium trichloride as the catalyst in the presence of the stoichiometric amount of tin could replace indium as the promotor in the aqueous allylation of trifluoroacetaldehyde (Loh and Li, 1996).

The problem of regeneration of the indium powder after its use in an allylation process, which is crucial from economical and ecological points of view, was resolved by simple electrochemical deposition of the metal on an aluminum cathode in a conventional electrolysis unit (Prenner et al., 1994).

The metallic indium-mediated allylation was applied with great success in carbohydrate chemistry in order to prepare higher-carbon sugars from water-soluble unprotected substrates. Thus, the naturally occurring deaminated sialic acid, 3-deoxy-D-glycero-D-galacto-nonulosonic acid (KDN), was prepared in three steps from D-mannose (Chan and Li, 1992b). The key step involving metallic indium and methyl α-bromomethylacrylate in water proceeded with good threo-selectivity (Scheme 4.3).

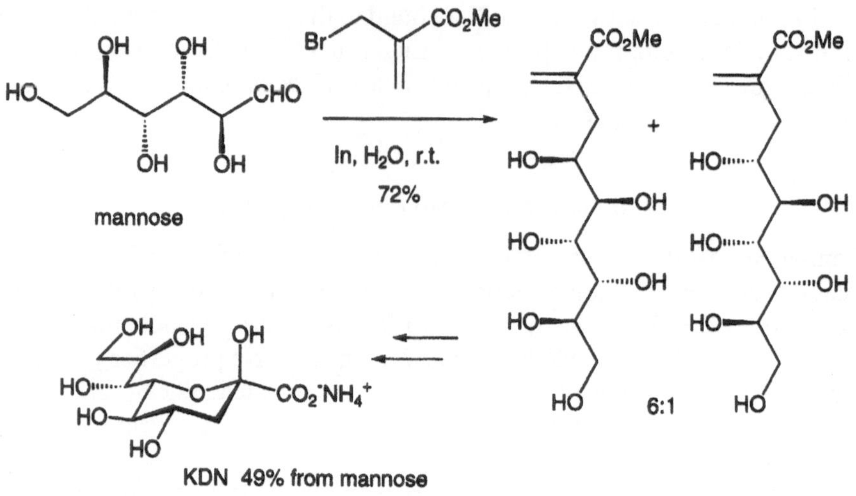

Scheme 4.3

Compared to tin-mediated allylation in ethanol–water mixture, the indium procedure is superior in terms of reactivity and selectivity. Indium-mediated allylation of pentoses and hexoses produced fewer by-products and was more diastereoselective. The reactivity and the *threo*-selectivity, compatible with a chelation-controlled reaction, were even improved when preformed allyl-dichloroindium was allowed to react with ribose in an ethanol–water mixture (Kim *et al.*, 1993). As an application of the indium methodology, a diastereoselective (*threo/erythro* = 4/1) allylation of N-acetyl-β-D- mannosamine (Scheme 4.4) with ethyl α-bromomethylacrylate and indium in a mixture of ethanol and 0.1 N aqueous hydrochloric acid could be achieved, allowing a straightforward synthesis of N-acetylneuraminic acid, the sialic acid that

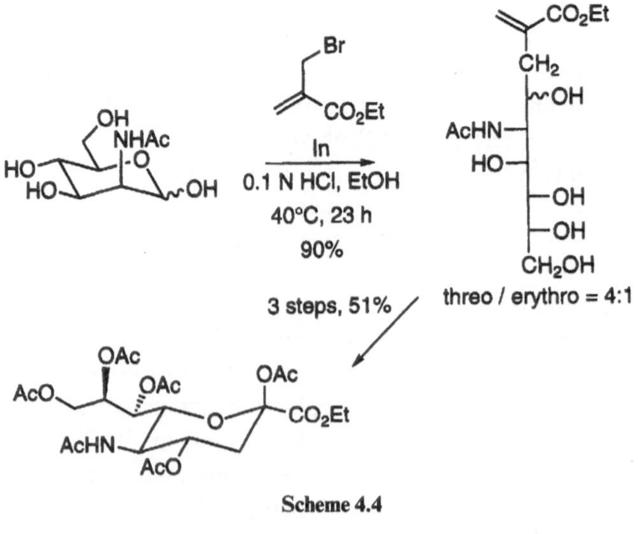

Scheme 4.4

plays a central role in molecular recognition on the surface of animal cells (Gordon and Whitesides, 1993). The beneficial influence of the acidity in the efficiency of this type of reaction had been previously pointed out (Kim *et al.*, 1993). The procedure was recently applied with success in the synthesis of several C(5) analogs of *N*-acetylneuraminic acids (Choi *et al.*, 1996).

Similarly, indium mediated the coupling of α-bromomethylacrylic acid with carbonyl compounds in aqueous media, yielding the corresponding γ-hydroxy-α-methylenecarboxylic acids, which can often be purified by crystallization. This reaction was applied for preparing KDN from mannose, and *N*-acetylneuraminic acid from *N*-acetylmannosamine. The *threo/erythro* ratio of the products arising from the chelation-controlled allylation was found to be 5/1 and 3/1, respectively (Chan and Li, 1995).

An inversion of diastereoselectivity occurred when using acetonide-protected aldoses (Binder *et al.*, 1994). This observation was exploited for a rapid synthesis (Scheme 4.5) of 3-deoxy-D-manno-2-octulosonic acid (KDO), the key constituent of the lipopolysaccharides of Gram-negative bacteria (Gao *et al.*, 1994).

All the stereochemical aspects of the indium-promoted allylation of α-oxygenated aldehydes were carefully analyzed in recent extensive studies (Paquette and Lobben 1996; Paquette and Mitzel, 1995, 1996a, b). Among the observations of these authors, noteworthy are the following: (i) the rate acceleration observed in water, compared to tetrahydrofuran, could be heightened by initial acidification; (ii) the pH dropped significantly as the reactions progressed; (iii) tetraethylammonium bromide enhanced both the reactivity and the selectivity (*syn*) of the reaction; and (iv) there was a consistent correlation between the reactivity of the aldehydes and their selectivity, i.e. α-hydroxyaldehydes were the most reactive and induced the highest selectivity (*syn*), which is indeed a consequence of chelated intermediates. Among the β-oxygenated aldehydes, only the β-hydroxyaldehydes gave good

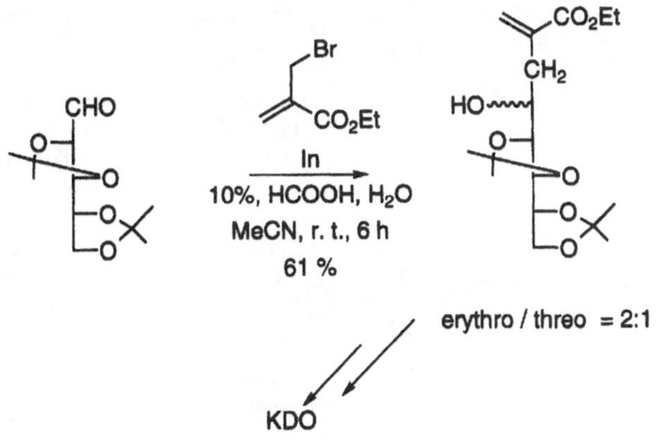

erythro / threo = 2:1

KDO

Scheme 4.5

diastereoselectivities (*anti*), which could also be due to a chelation control in water (Paquette and Mitzel, 1995, 1996a). Experiments with more rigid systems, such as 2-methylcyclohexanone, showed that the aqueous Barbier-type allylation reaction was still controlled by chelation. This chelation in aqueous medium via allylindium species is really efficient, so that π-facial diastereoselection reaches a maximum when the system is conformationally rigid (Paquette and Lobben, 1996).

Otherwise, the coupling of γ-substituted allyl bromides with aldehydes could induce some diastereoselectivity. The stereochemical course of the reaction, as well as its regiochemical aspect (Scheme 4.6), were carefully investigated by Isaac and Chan (1995a). The reaction was γ-regioselective in all cases, except when the γ-substituent was a silyl or a *t*-butyl group, which then induced α-isomers ((*E*) and (*Z*) mixture). The γ-regioselectivity is a general rule, even with substituents that could have induced an extra conjugation by α-addition. This result was recently confirmed with γ-cyanoallyl bromides, which led to γ-cyano-β-ethylenic secondary alcohols when reacting with aldehydes in the presence of indium in water (Sarrazin and Mauzé, 1996). Unfortunately, the indium-mediated allylation is moderately *anti*-selective, as already mentioned in tin- and zinc-mediated allylations. The *anti/syn* ratio increased with the size of aldehyde substituent; it was independent of the (*E*)/(*Z*) configuration of the double bond of the allyl bromide moiety (Chan and Isaac, 1996). An improvement of the *anti*-selectivity was observed in some experiments when stoichiometric tin was premixed with catalytic indium trichloride. An allylindium intermediate was then supposed to be the reactive species (Li and Loh, 1996). The rate and the diastereoselectivity of allylindium is enhanced in the presence of ytterbium trifluoromethanesulfonate in DMF–H_2O (Wang *et al.*, 1995) or lanthanum trifluoromethanesulfonate in water (Diana *et al.*, 1996).

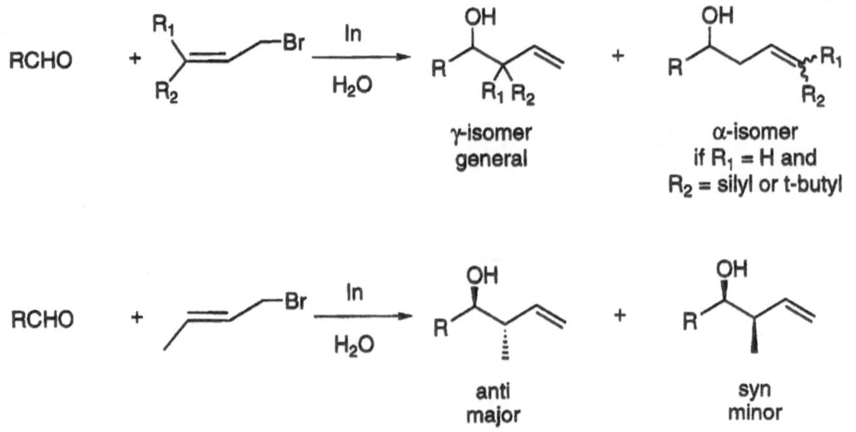

Scheme 4.6

Of great interest is the recent investigation of the stereochemical outcome of the addition of γ-substituted allylindium to α-oxyaldehydes. The diastereoselectivity related to the relative configuration of the α-oxy substituent toward the newly formed hydroxyl group is high in water for the addition of crotylindium to α-hydroxyaldehydes. An inversion of selectivity occurs when the α-oxy substituent of the aldehyde is the bulky t-butyl-dimethylsilyl group; with 3-bromoallylindium the selectivity reaches a maximum (syn/anti = 1/9–10). By contrast, the diastereoselectivity related to the relative configuration of the γ-substituent of the allyl moiety toward the newly formed hydroxyl group remains low in any solvent, but always in favor of the anti isomer (Paquette and Mitzel, 1996b).

The aqueous indium-mediated allylation of carbonyl compounds could also proceed intramolecularly, as described in carbocyclizations yielding five-, six- and seven-membered rings, which further cyclized to give fused α-methylene-γ-lactones:

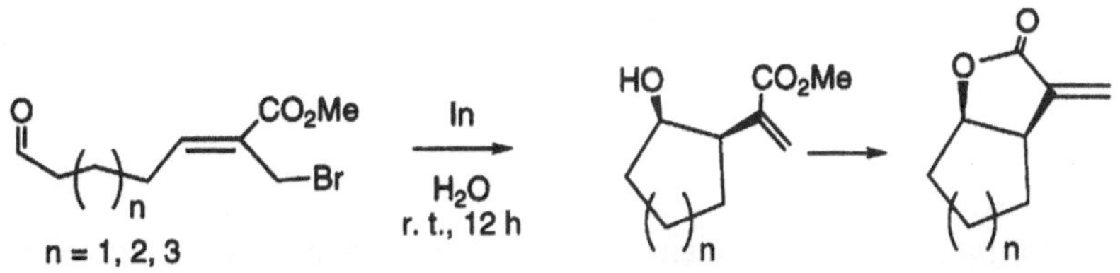

The ring junction stereochemistry in the fused lactones was found to be cis in all cases (Bryan and Chan, 1996). A two-atom ring expansion based on the indium-mediated Barbier-type allylation was recently reported by Li et al. (1996). This novel carbocycle enlargement was applied to five-, six-, seven-, eight- and twelve-membered ring compounds:

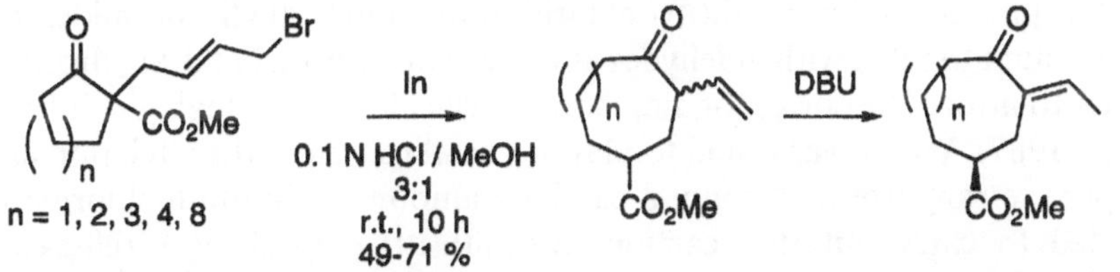

Another intramolecular cyclization involved 3-chloro-2-chloromethyl-propene as trimethylenemethane zwitterion equivalent (Lu and Li, 1996):

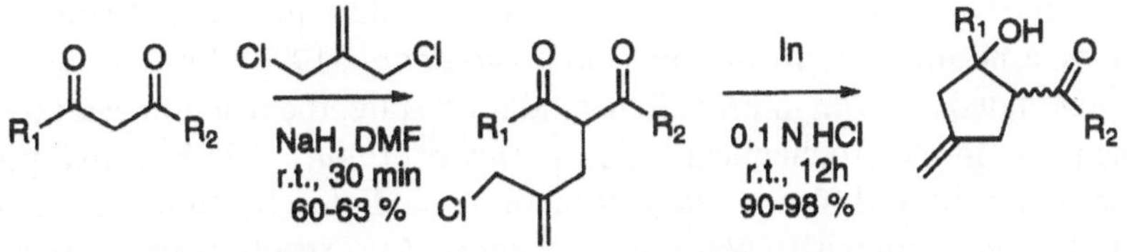

A trimethylenemethane dianion equivalent could be generated in water from 3-bromo-2-bromomethylpropene and indium. When reacting with aldehydes, this dianion equivalent gave the corresponding diols (Li, 1995):

$$RCHO \quad + \quad Br\diagup\diagdown\diagup Br \xrightarrow[\substack{0.1\ N\ HCl\ /\ MeOH \\ 4:1 \\ r.t.,\ 1\text{-}3\ h \\ 53\text{-}75\ \%}]{In} \quad R\diagup\diagdown\diagup\diagdown R$$

Interesting was the use of 1,3-dibromopropene in the presence of indium as *gem*-allyl dianion equivalent. Unfortunately, the coupling of this synthon with aldehydes gave a complicated mixture of four products:

$$RCHO \quad + \quad Br\diagup\diagdown\diagup Br \xrightarrow[\substack{H_2O \\ r.t.,\ 10\text{-}16\ h}]{In}$$

HO, OH
R R
major

+

OH OH
R R

+

OH
R

+

R

The conjugated diene, which was readily obtained in SnCl$_2$ Barbier-type allylation in dimethylformamide (Augé, 1985) or Zn-mediated allylation in water (Chan and Li, 1990), was detected as a by-product in the indium variant (Chen and Li, 1996).

4.2.4   Allylation reactions mediated by other metals

In the presence of bismuth(III) chloride–aluminum, allylic bromides have been found to react with aldehydes at room temperature in tetrahydrofuran–water to afford the corresponding homoallylic alcohols in high yields (Wada *et al.*, 1987). Water was found to play a crucial role since the allylation failed in pure tetrahydrofuran. Only a catalytic amount of bismuth chloride was needed to carry out the reaction. Bismuth(III) chloride was reduced by aluminum to zero-valent bismuth, which could insert into the carbon–bromine bond of the allylic bromide to afford an allylbismuth inter-mediate as the reactive species. The allylation reaction could occur with the couple Bi(0)–Al(0) in tetrahydrofuran–water only in the presence of a catalytic amount of hydrobromic acid (Wada *et al.*, 1990). Since bismuth(0) was postulated to be an intermediate oxidation state, the reaction was accom-plished via an electrochemical redox pathway (Figure 4.1) in a two-phase system (Minato and Tsuji, 1988). Reactions mediated by Bi(0) as the only promotor were sluggish (Wada *et al.*, 1990). An exception was, however, reported with the coupling between *p*-nitrobenzaldehyde and allyl iodide in water (Chan and Isaac, 1996).

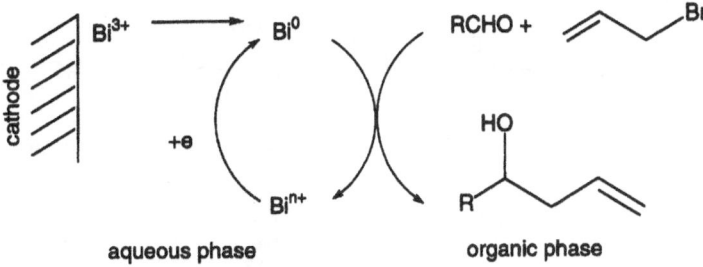

Figure 4.1 The electrochemical redox pathway for the allylation reaction mediated by bismuth in a two-phase system.

The reaction could be extended to the alkylation of immonium cations generated from secondary amines and aldehydes. The allylation of 1-(aminoalkyl)benzotriazoles proceeded smoothly only in the presence of water. More interestingly, alkylation agents used include benzyl and methyl halides hitherto unknown for an aqueous Barbier reaction (Katritzky *et al.*, 1992).

Recently, Sarangi *et al.* (1995) discovered that the Barbier-type allylation of carbonyl compounds could be mediated by zero-valent copper. In the experimental procedure copper(II) chloride in its hydrated form was mixed with magnesium powder. This reaction did not proceed with anhydrous $CuCl_2$, but addition of water to this anhydrous salt in the presence of magnesium allowed the allylation of aldehydes and ketones with allylic bromides. In the absence of halides, the hydrated reagent ($CuCl_2 \cdot 2H_2O/Mg$) turned out to be efficient in the reduction of aldehydes. However, the water of crystallization could not suffice and addition of water promoted the reduction process (Sarangi *et al.*, 1995).

4.2.5 Propargylation and allenylation of aldehydes

Water was supposed to limit rearrangements between allenyltributylstannane and dibutylstannyl dichloride while these reagents were allowed to react with aldehydes. Nevertheless, complete α-allenic stereoconvergence was not observed. Homopropargylic alcohols could be obtained as well, especially with formaldehyde or α,β-unsaturated aldehydes (Boaretto *et al.*, 1985b).

Straightforward propargylation and allenylation of aldehydes could be undertaken when using propargyl bromides and metallic tin in aqueous acidic media (Wu *et al.*, 1990). Unfortunately, no regioselectivity was observed. Unlike this reaction, indium-mediated coupling of aldehydes with propargyl bromides occurred regioselectively in aqueous media to give either homoprop-2-ynyl or allenylic alcohols depending on the γ-substituent of the prop-2-ynyl bromide (Isaac and Chan, 1995b):

X	allenic	acetylenic
H	12	88
Ph	95	5
Me	100	0
TMS	67	33

On the other hand, zinc-mediated coupling of aldehydes to propargyl bromides in the Luche conditions gave the corresponding homopropargylic alcohols in moderate to high yields together with small amounts of α-allenic alcohols (Yavari and Riazi-Kermani, 1995).

4.2.6   Mechanistic aspects

The salient features of the aqueous Barbier-type alkylation reaction that could enlighten its mechanism are the following. (i) An aldehyde can be allylated selectively in the presence of a ketone. (ii) The reaction is compatible with the presence of free hydroxyl or carboxyl groups. (iii) Allylic bromide is much more reactive than allylic chloride. (iv) The allylation occurs in the 1,2-fashion when an α,β-unsaturated aldehyde is the substrate. (v) The γ-position of the allyl metal species is the nucleophilic site, so that the allylation is γ-regioselective, except in rare cases (*vide infra*). (vi) The diastereoselectivity of the allylation is moderate to fairly good. Regardless of the stereochemistry of the double bond of the γ-substituted allylic bromide, the reaction is *anti*-selective, the best results being obtained with hindered aldehydes in the presence of indium. With α-hydroxylated aldehydes the diastereoselectivity is good, always in favor of the *syn* product as issued from a chelate intermediate. (vii) The reaction can be made enantioselective with an appropriate chiral auxiliary.

In organic solvents, it was demonstrated that the Barbier reaction did not necessarily involve the formation of an organometallic species. In some cases there is a radical pathway in which the anion radical resulting from single electron transfer from the metal to the halogenated compound is trapped by the ketone or the ketyl radical on the surface of the metal (Molle and Baner, 1982).

In aqueous zinc-promoted allylation, wherein the allylzinc species is considered unlikely, no evidence of the formation of a radical arising from allylic bromide could be obtained, but the radical intermediate formed could be associated with the metal surface (Wilson and Guazzaroni, 1989). The formation of a ketyl radical cannot be excluded either, as some ketones dimerized to the pinacol in the absence of the halide. But the process is slow and limited (Einhorn and Luche, 1987). The formation of a ketyl radical from saturated carbonyl compounds is endothermic, whereas from conjugated carbonyl compounds it is exothermic (Moyano *et al.*, 1990).

The initiation of the zinc-mediated reaction could then be attributed to the formation of an allylic radical anion of the type [CH$_2$=CH–CH$_2$–X]$\cdot^-$ on the metal surface. This radical surface could then react with the carbonyl group to give an alkoxide radical, which could add an electron and form the alcohols (Einhorn and Luche, 1987). This process takes place with an increase of the pH value of the reaction mixture (Sjöholm *et al.*, 1994). Likewise, Wurtz coupling products were detected with cinnamyl chloride, which was supposed to stem from an allylic radical anion intermediate. Moreover, in one experiment without aldehyde, Sjöholm *et al.* (1994) observed that all cinnamyl chloride was transformed into dicinnamyl.

Figure 4.2 is an illustration of the two possible mechanisms involving single electron transfer (SET) as suggested by Chan *et al.* (1994); however, path B must probably be precluded when working with saturated carbonyl compounds (Moyano *et al.*, 1990).

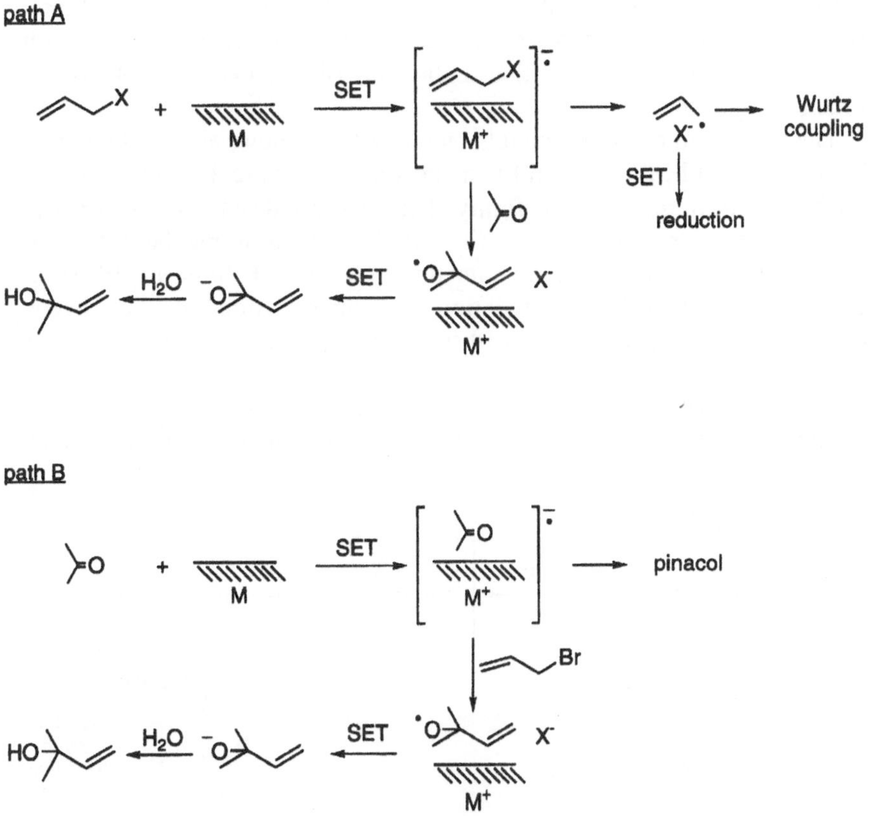

path A

path B

Figure 4.2 The two possible mechanisms involving single electron transfer proposed by Chan *et al.* (1994).

The existence of organotin and organoindium species in water might counter these types of mechanisms in tin- and indium-mediated allylations. In fact, the reaction of benzyl chloride with metallic tin in boiling water provided tribenzyltin chloride (Sisido *et al.*, 1961). Likewise, the preparation of mixed allyl and benzylorganostannanes is currently accomplished in aqueous media (Carofiglio *et al.*, 1992; Marton *et al.*, 1996b). Besides, allyldibutyltin chloride, allylbutyltin dichloride (Boaretto *et al.*, 1985a) and allyltributylstannanes (Grieco and Bahsas, 1987) were used successfully as allylating agents in water.

Moreover, aqueous reactions between the preformed organotin or organoindium reagents and sugars are faster than the heterogeneous aqueous reactions (Kim *et al.*, 1993). On the basis of similarity between the homogeneous and heterogeneous reactions, the authors postulated that the tin- and indium-mediated allylation of aldehydes proceeds through organotin or organoindium intermediates. On the other hand, indium-mediated and, to a smaller extent, tin-mediated reactions produce fewer by-products than zinc-mediated reactions.

Likewise, Li and Loh (1996) observed that premixing indium trichloride, tin and allylic bromide increased the rate and the diastereoselectivity of allylations, which could be another indication of the existence of transient organoindium species.

The formation of organometallic intermediates, however, can be initiated by an allylic radical anion on the metal surface (Figure 4.3). Noteworthy is the high value of the electron affinity of allylic bromide and, to a lesser extent, chloride (Moyano *et al.*, 1990). As a matter of fact, in tin-mediated reactions, Wurtz coupling products were sometimes detected (Kim *et al.*, 1993), but, unlike the zinc-mediated allylation, no dimerization of the carbonyl compounds was observed in the absence of the halide (Einhorn and Luche, 1987).

At this stage of the discussion, let us remember the stereochemical aspects of tin- and indium-mediated allylation. Though the indium-mediated

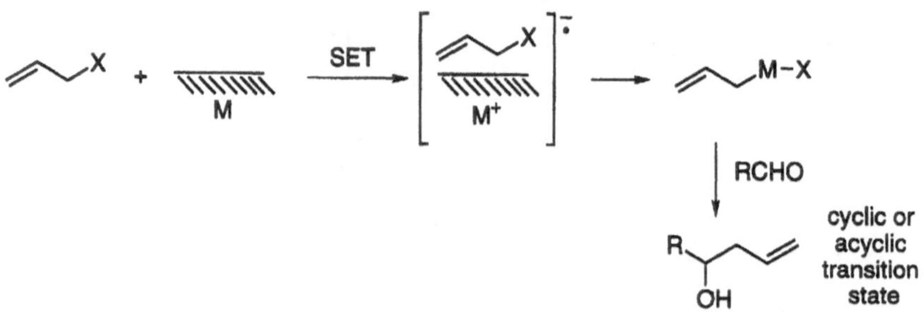

Figure 4.3 The formation of organometallic intermediates initiated by an allylic radical anion on a metal surface.

reactions are more diastereoselective, the stereochemical course of tin- and indium-mediated reactions is somehow the same (Kim *et al.*, 1993). Paquette and Mitzel (1996a) clearly demonstrated that organometallic chelate species are true intermediates in the indium-mediated allylation of α- and β-hydroxy-aldehydes in water:

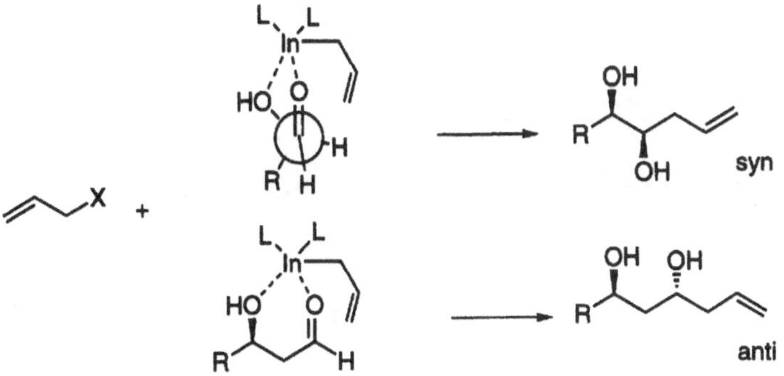

In summary, the initial formation of an allylic radical anion on the metal surface is the most likely event, which would explain the success of indium, as its first ionization potential is particularly low ($E_i = 5.79\,\text{eV}$). In tin- and indium-mediated reactions the second step should be the insertion of the metal cation into the carbon–bromine (chlorine) bond to afford organometallic intermediates, which are stable enough to be produced, but also highly reactive toward carbonyl compounds in aqueous media.

4.3   Conjugate 1,4-additions

4.3.1   *Organometallic additions*

Alkyl halides in the presence of a zinc–copper couple, as a mixture of zinc dust and copper(I) iodide, reacted smoothly with α-enones and α-enals in aqueous media (Scheme 4.7). Sonication enhanced the efficiency of the process, leading to 1,4 adduct in good yields (Pétrier *et al.*, 1986). Such a conjugate addition was later extended to various electron-deficient alkenes, including α,β-unsaturated esters, amides, nitriles (Dupuy *et al.*, 1991) or phosphine oxides (Pietrusiewicz and Zablocka, 1988). It appeared that the reactivity of the halide (RX) followed the order: tertiary > secondary >> primary and iodide > bromide >> chloride. The preferred solvent system was aqueous ethanol, but the parameter of highest importance was the solvent composition (Luche and Allavena, 1988a).

Initial formation of a radical anion of the type [RX]•⁻ seems highly probable as in the Barbier reaction. The radical ion absorbed on the metal surface, more or less readily cleaved in a radical, is trapped by the α,β-unsaturated carbonyl compound, affording an α-keto radical, which is further rapidly

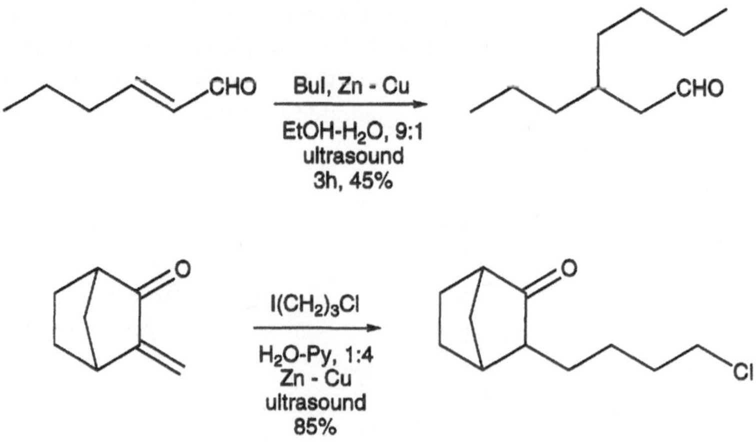

Scheme 4.7

reduced at the metal surface to the enolate, then protonated in a fast step (Luche and Allavena, 1988b).

Such a process also occurred when epoxyalkylhalides, except 2,3- and 3,4-derivatives, were mixed with electron-deficient alkenes in aqueous ethanol. In the case of 3,4-epoxyhalides, Sarandeses *et al.* (1992) observed a rapid cyclization of the radical intermediate, providing cyclopropylmethanols, which might be a useful preparation of such synthons.

4.3.2   Michael reactions

The Michael reaction is one of the most useful reactions for creating a carbon–carbon bond. The Michael reaction, which is promoted under pressure (Matsumoto, 1981), should be accelerated in water owing to the hydrophobic effect as has been postulated (Lubineau *et al.*, 1994a; Lubineau, 1986). As a matter of fact, the formation of the triketone depicted below:

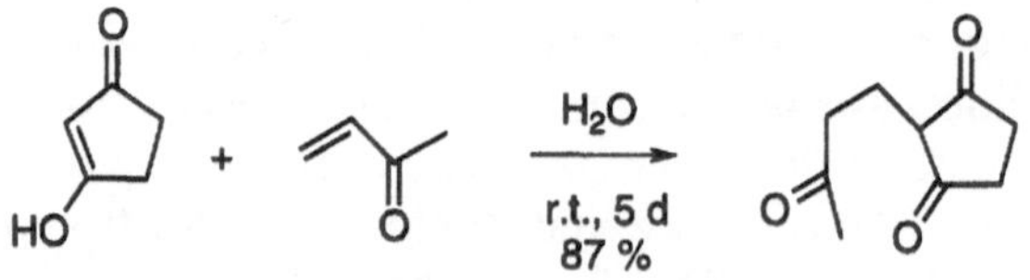

could be realized at room temperature in water, without any catalyst such as acid or base, which appreciably increased yield and purity of the product (Eder *et al.*, 1971; Hajos and Parrish, 1974). Likewise, 2-methyl-1,3-cyclopentadione was condensed with acrolein to give quantitatively a Michael adduct (Lavallée and Deslongchamps, 1988). An inorganic acidic catalyst was needed, however, to perform the Michael reaction between ascorbic acid and cyclic enones (Sussangkarn *et al.*, 1988):

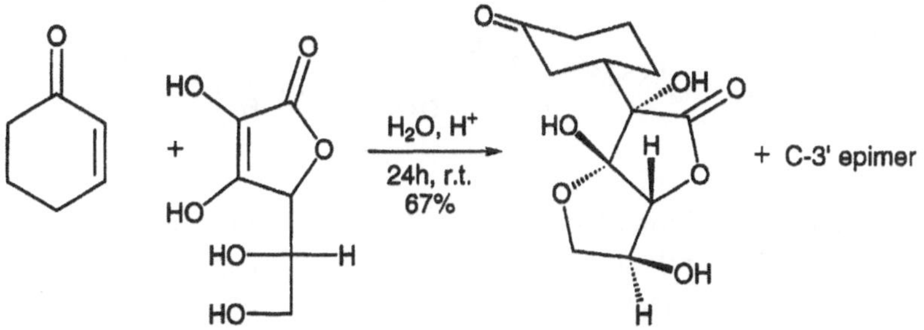

Ytterbium triflate, as a water-tolerant Lewis acid, catalyzed the Michael additions of various β-ketoesters toward α-enals (Keller and Feringer, 1996).

Water-soluble phosphines, such as 3,3′,3″-phosphanetriyltris(benzenesulfonic acid) (TPPTS) or *m*-monosulfonate, give quantitatively the water-soluble phosphonium salts when reacted with α,β-unsaturated acids in water (Larpent and Patin, 1988). They also react with activated alkynes, affording vinylphosphonium salts, vinylphosphine oxides or alkenes depending on the pH of the aqueous solution (Larpent *et al.*, 1990).

More recently, a huge acceleration of the reaction of nitroalkanes with buten-2-one was mentioned when going from nonpolar organic solvents to water (Lubineau and Augé, 1992). Indeed, the uncatalyzed Michael reaction, considered as impossible, proceeds nicely under neutral conditions when water is used as solvent, without any catalyst. It has been confirmed that the mixture of nitroalkanes and butene-2-one was unreactive under neat conditions or in aprotic solvents. The reaction depicted below is more rapid and selective in water than in methanol:

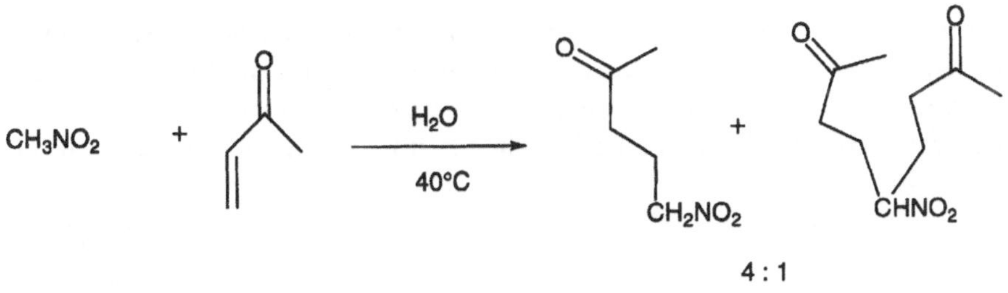

Solubility, thermodynamic and kinetic acidities of nitroalkanes are of great importance in the course of the reaction, but the hydrophobic effect (Lubineau *et al.*, 1994a) could be involved, since additives such as glucose or saccharose accelerate the reaction even more (Lubineau and Augé, 1992). Amphiphilic molecules can also influence the hydrophobic interactions. In fact, cetyltrimethylammonium chloride (CTACl) as cationic surfactant promoted the Michael reaction of various nitroalkanes with conjugated enones in dilute aqueous solutions of sodium hydroxide (Ballini and Bosica, 1996):

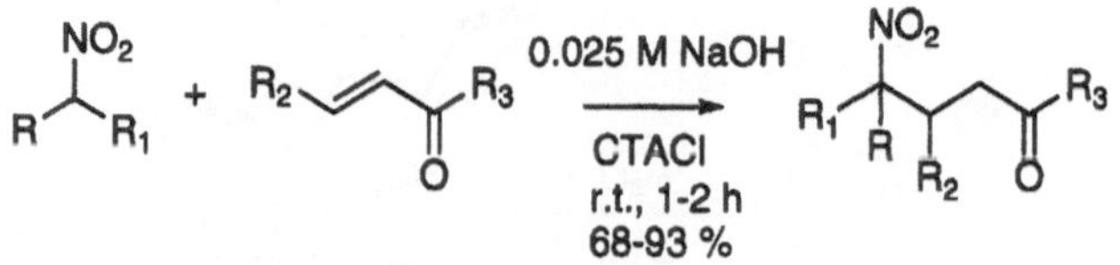

The conjugate addition of amines to α,β-ethylenic substrates was carefully investigated and several modes of activation, including high pressure, catalysis and hydrophobic effect, were compared (Jenner, 1996). Owing to a highly negative value for the activation volume ($\Delta V^{\ddagger} = -65\,cm^3\,mol^{-1}$ for one reaction), the conjugate addition of secondary amines to α,β-unsaturated esters, amides and nitriles is accelerated under pressure. The effect of water as solvent on the reaction rate is even greater when amines react with nitriles. The lack of apparent reactivity of α,β-unsaturated esters toward amines comes from the reverse reaction, which is particularly accelerated in water (Jenner, 1996).

A related reaction, known as the Baylis–Hillman reaction, was found to be greatly accelerated in water compared to the usual organic solvents. The first step of the reaction is the conjugate addition of 1,4-diazabicyclo[2.2.2]octane (DABCO) to acrylonitrile (Scheme 4.8), which is fast in water. The second step, which is rate-determining, is accelerated via a process in which the hydrophobic effect could be involved. Other structured solvents also enhanced the Baylis–Hillman reaction, but to a smaller extent (Augé et al., 1994).

4.4   Cross-aldol and Reformatsky-type reactions

The aldolisation reaction is one of the most popular carbon–carbon bond formation reactions as it involves carbonylated chemicals that are widely available from numerous industrial processes. In its simpler type, it is performed directly from the carbonylated substrates under acidic or basic

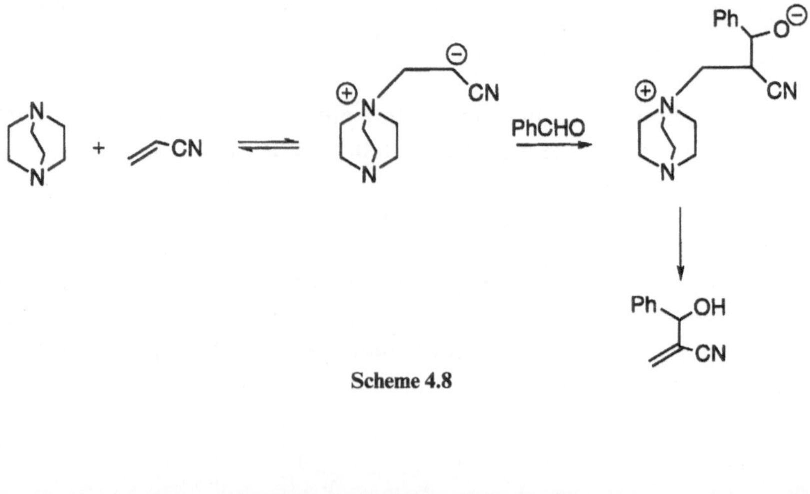

Scheme 4.8

catalysis, although the reaction might lead easily to dehydrated products as well as uncontrolled polymerization. Moreover, it is often not compatible with a cross-coupling process between two different carbonylated molecules. Usually, the coupling involves an enolizable ketone and either a ketone or an aldehyde for which enolization is slow or impossible, as for example the benzaldehyde derivatives:

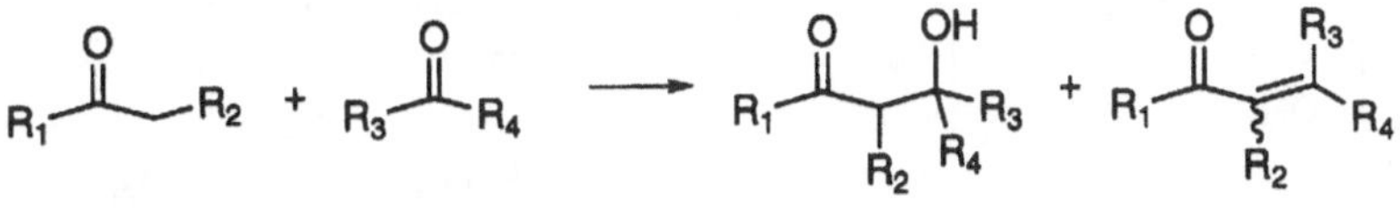

The hydroxyketones that are provided by the aldolisation process can exist in two diastereoisomeric forms (*syn* and *anti*) based on the transition-state geometry.

In an intramolecular case, it was shown that, in an aqueous medium, the nature of the acidic or basic catalyst had a dramatic effect on the outcome of the aldolisation (Denmark and Lee, 1992). Acid-induced aldol condensation of ketoaldehyde **1** provided the *syn* hydroxyketone **2**, while the *anti* isomer **3** arose from base-catalyzed reactions:

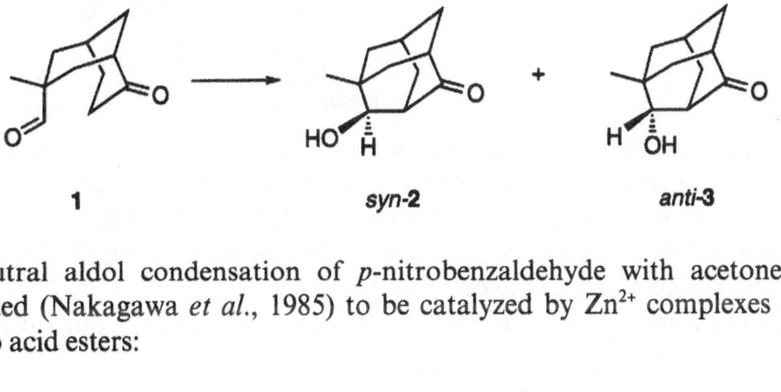

1	*syn*-**2**	*anti*-**3**

Neutral aldol condensation of *p*-nitrobenzaldehyde with acetone was reported (Nakagawa *et al.*, 1985) to be catalyzed by Zn^{2+} complexes of α-amino acid esters:

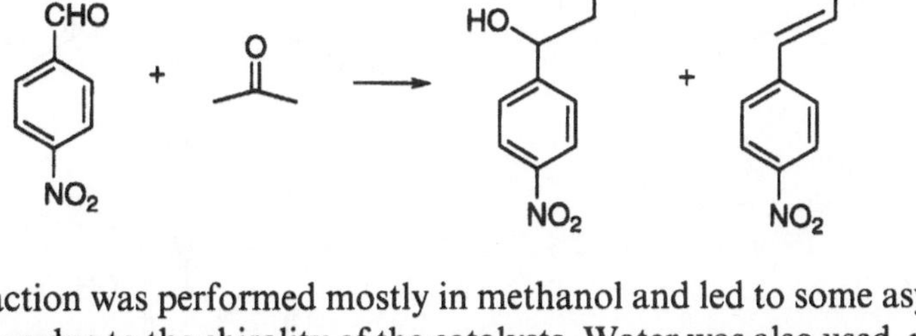

The reaction was performed mostly in methanol and led to some asymmetric induction due to the chirality of the catalysts. Water was also used, providing quantitative aldol condensation, although no optical activity could be obtained in this case (Buonora *et al.*, 1995). The influence of pH on this system was measured in order to minimize undesired dehydration of aldol products. Interesting effects of β-cyclodextrin were observed for the same reaction (Watanabe *et al.*, 1985; Zhang and Xu, 1989).

Unprotected ketoses could provide one-carbon extended ketoses by reaction with formaldehyde (Shigemasa *et al.*, 1994). Higher yields of aldol-type products were obtained in water compared to methanol:

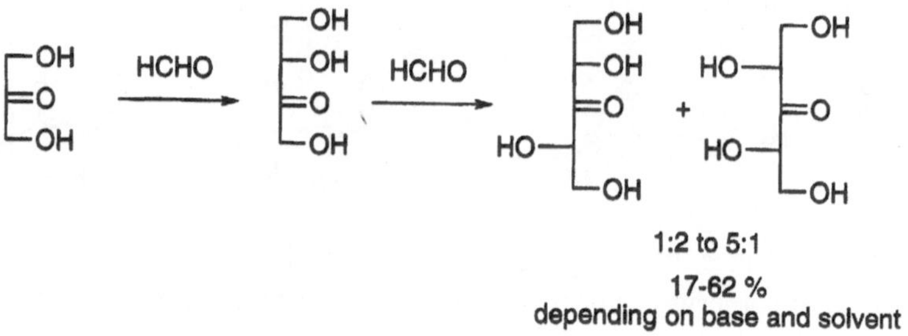

1:2 to 5:1
17-62 %
depending on base and solvent

Similar improvements due to the solvent were observed in the case of the reaction of phenolic enolates with aldehydes (Saimoto *et al.*, 1996):

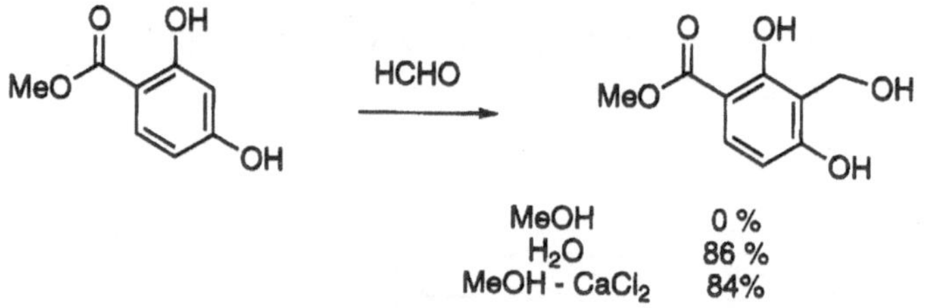

MeOH   0 %
H₂O   86 %
MeOH - CaCl₂ 84%

Using heterogeneous conditions and in the presence of anionic or cationic surfactants (Fringuelli *et al.*, 1994), a series of carbonyl compounds was shown to undergo condensation with excellent yields. Micellar catalysis afforded mainly the dehydrated products. The method was applied to an efficient one-pot synthesis of flavonols (Scheme 4.9).

Very important progress toward an efficient control of the outcome of the

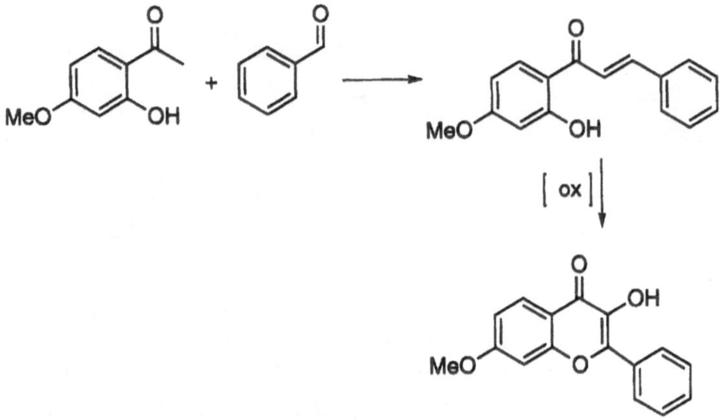

Scheme 4.9

aldolisation process was made by Mukaiyama *et al.* (1974), who discovered that a clear cross-coupling reaction could be achieved from a silyl enol ether by reaction with an aldehyde in the presence of a Lewis acid (stoichiometric titanium tetrachloride). Later, it was shown that, under high pressure, the same reaction could be performed without acidic promoter (Yamamoto *et al.*, 1983). In this case, the major product is the *syn* hydroxyketone, unlike for the TiCl$_4$ promoted reaction, which leads mostly to the *anti* addition product. The *syn* and *anti* stereochemistry is the result of two possible transition-state geometries having different activation volumes ($\Delta V^{\ddagger}_{syn} < \Delta V^{\ddagger}_{anti}$). This explains why increasing the external pressure leads to an increase in the *syn/anti* ratio.

The use of water as a solvent for such a reaction was therefore indicated. Indeed, water, through its high cohesive energy density, can enhance the rate of some chemical reactions, as they are under high pressure (Li, 1993; Lubineau *et al.*, 1994a). The first example of aqueous Mukaiyama cross-coupling was reported by Lubineau (1986). The reaction of the silyl enol ether of cyclohexanone with benzaldehyde was shown to proceed without any cata-lyst and at atmospheric pressure, with the same stereoselectivity as under high pressure (Scheme 4.10). The method was extended to other carbonyl compounds, such as formaldehyde, substituted benzaldehydes, and α,β-unsaturated aldehydes and ketones (Lubineau and Meyer, 1988). The unique side-reaction, i.e. the hydrolysis of the silyl enol ether, can be circumvented if a more reactive aldehyde is used, such as *p*-nitrobenzaldehyde. Ultrasound was shown to promote especially efficient phase mixing. With α,β-unsatu-rated ketones, only 1,4-addition occurred, because of the lack of reactivity of ketones as an electrophilic partner of the coupling. But in the case of α,β-

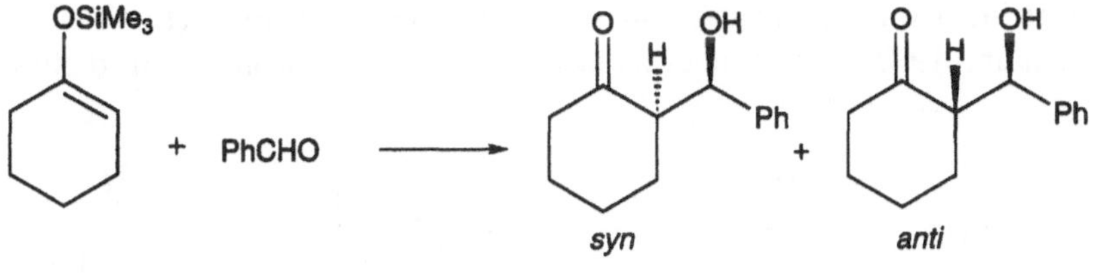

solvent	temp.	time	cond.	yield	syn : anti
CH$_2$Cl$_2$	20°C	2 hrs	1 eq TiCl$_4$	82	25:75
CH$_2$Cl$_2$	60°C	9 days	10000 atm	90	75:25
H$_2$O	20°C	5 days	stirring	23	85:15
H$_2$O - THF (1:1)	20°C	5 days	stirring	45	74:26
id	55°C	2 days	ultrasound	76	74:26

Scheme 4.10

unsaturated aldehydes such as acrolein, a mixture of 1,2- and 1,4-adducts was obtained:

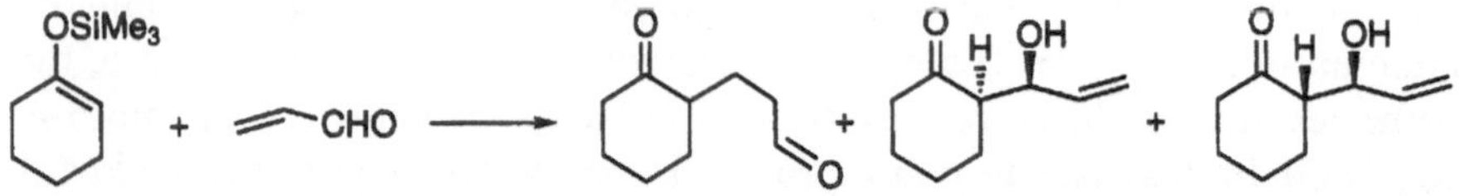

Although the yields were moderate, the neutrality of the medium allowed the isolation of the 1,2-addition products, which are extremely liable to dehydration.

Lanthanide triflates, which are water-tolerant Lewis acids, were shown to greatly improve the yields, and therefore the scope of the aqueous Mukaiyama reaction (Kobayashi, 1991; Kobayashi and Hachiya, 1992). This is detailed elsewhere in this book. In keeping with the same concept, indium chloride in catalytic amounts promoted efficient cross-coupling (Loh *et al.*, 1996b), as did the use of tris(pentafluorophenyl)boron (Ishihara *et al.*, 1993).

Aqueous Reformatsky-type cross-coupling reactions have also been reported (Chan *et al.*, 1990, 1994). Metallic zinc, tin or indium promote the aldol-type condensation of α-halocarbonyl compounds and aldehydes in aqueous media (Scheme 4.11). Likewise, α-haloesters (α-haloacids) were allowed to react with aromatic aldehydes to provide β-hydroxyesters (β-hydroxyacids) in moderate yields (Chan *et al.*, 1994).

Lanthanum trifluoromethanesulfonate was shown to promote an indium-mediated vinylogous Reformatsky coupling in water (Diana *et al.*, 1996):

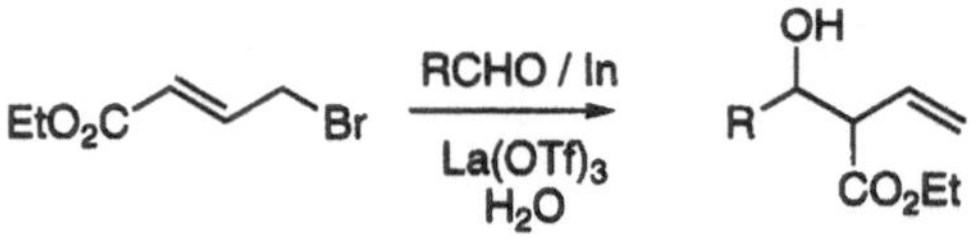

Starting from halogenoketones, it was recently shown (Lubineau and Bouchain, 1997) that the cycloaddition of α,α'-dibromo- (or dichloro-)

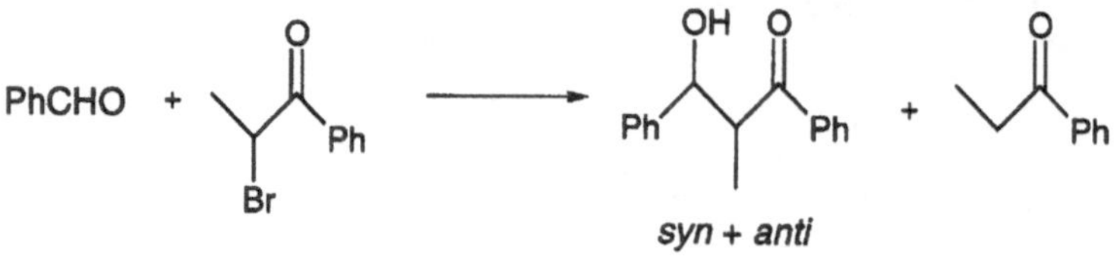

metal	yield	syn : anti
Zn	82%	71 : 29
Sn	67%	47 : 53
In	85%	92 : 8

Scheme 4.11

ketones with furan (or cyclopentadiene) formerly described by Noyori using $Fe_2(CO)_9$ in benzene (Takaya *et al.*, 1978), gave very good yields when the reaction was conducted in pure water using commercial iron powder. Related to this reaction is the condensation of monobromo- (or chloro-) ketone, which adds to furan (or cyclopentadiene) in water in the presence of triethyl-amine to give the corresponding cycloadducts in near-quantitative yields (Scheme 4.12). In both cases, the 2-oxyallyl cation, the formation of which is favored in water, was considered as the reactive intermediate.

4.5 Pinacol coupling reactions

General reviews on the coupling of carbonyl compounds providing pinacols have been published by Kahn and Rieke (1988), Pons and Santelli (1988), McMurry (1989), Fürstner (1993) and Wirth (1996). Examples of aqueous pinacol coupling reactions are relatively scarce. The use of aqueous titanium trichloride has been widely investigated by Clerici and Porta (Scheme 4.13). They reported such conditions for benzoyl cyanide reductive dimerization (Clerici *et al.*, 1981), which occurs in acidic media, whereas coupling of aromatic aldehydes and ketones requires basic conditions (Clerici and Porta, 1982a). Changing the solvent from acetic acid to acetone allowed the isola-tion of cross-coupling pinacol products (Clerici and Porta 1982b). This latter example was extended to acetyl pyridines, which provided pyridyl glycols (Clerici and Porta, 1983). Methyl phenylglyoxylate can generate a radical species that can add to the carbonyl carbon of a ketone, thus affording carboxylated pinacols (Clerici *et al.*, 1986). Starting from α,β-dicarbonyl compounds, α,β-dihydroxyketones were obtained in good to excellent yields (Clerici and Porta, 1989). In all these cases, water was brought into the system through the use of commercial titanium trichloride solution.

Water was used as the sole solvent of the reaction in the case of the

X	Y	Z	promoter	isolated yields
Br	Br	O	Fe	74 %
Br	Br	C	Fe	66 %
Cl	H	O	NEt$_3$	88 %
Cl	H	C	NEt$_3$	76 %

Scheme 4.12

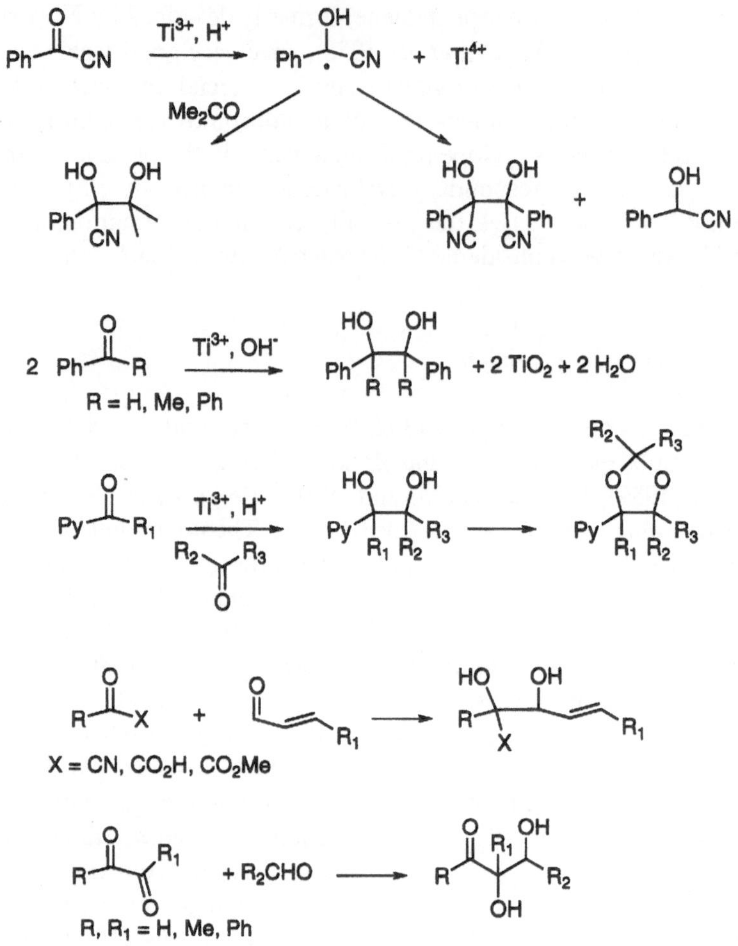

Scheme 4.13

coupling of an α,β-unsaturated ketone with acetone, promoted by the zinc–copper couple and under ultrasonic irradiation (Delair and Luche, 1989):

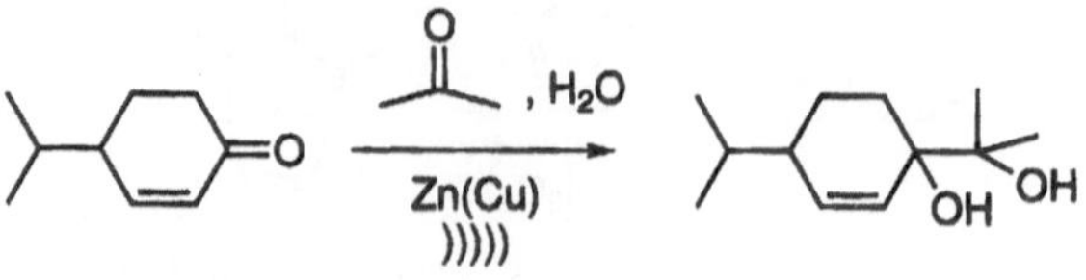

The zinc–zinc chloride system allowed the coupling of aromatic aldehydes and ketones to be achieved, to produce α-glycols in aqueous solution (Tanaka *et al.*, 1990). Although competitive reduction of aldehydes is observed when increasing the water content of the solvent, good yields of glycols were obtained in the case of aromatic ketones:

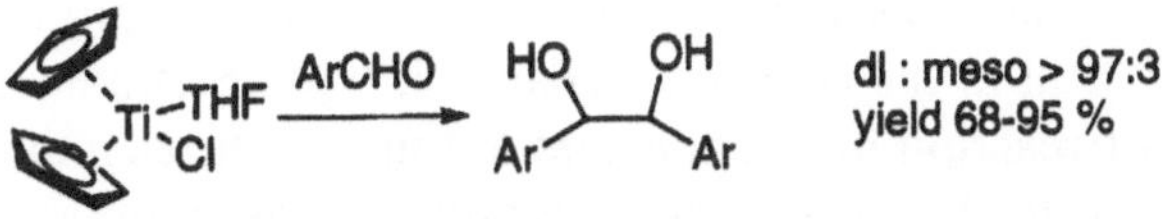

A stereoselective pinacol coupling was obtained by reaction of benzaldehyde or other benzylic aldehydes with $[Cp_2Ti^{III}(H_2O)]^+$, which is stable in water in the absence of oxygen (Barden and Schwartz, 1996):

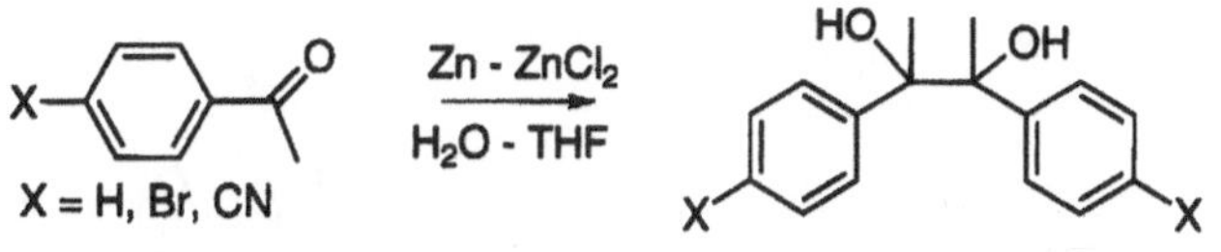

4.6 Miscellaneous reactions

The reaction of amines with aqueous formaldehyde provides iminium salts, which are used in various processes. In the presence of cyclopentadiene, the aza Diels–Alder reaction can take place to yield the azabicycloheptene skeleton (Larsen and Grieco, 1985):

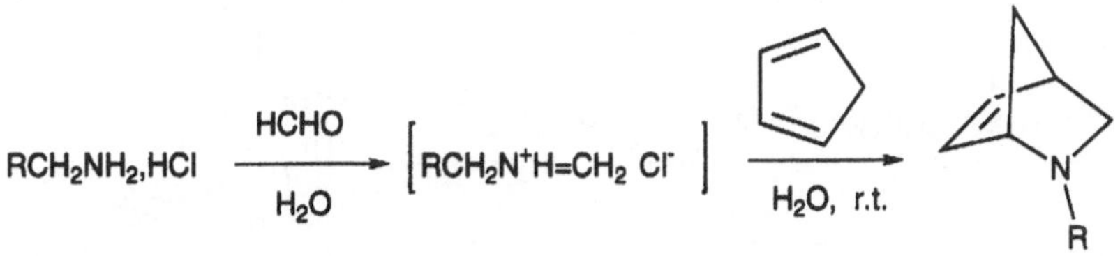

This is discussed further elsewhere in this book. Allylsilanes were also shown to add to iminium salts under Mannich-like conditions (Larsen et al., 1986). Piperidines were thus prepared via an aminomethano desilylation–cyclization process (Scheme 4.14).

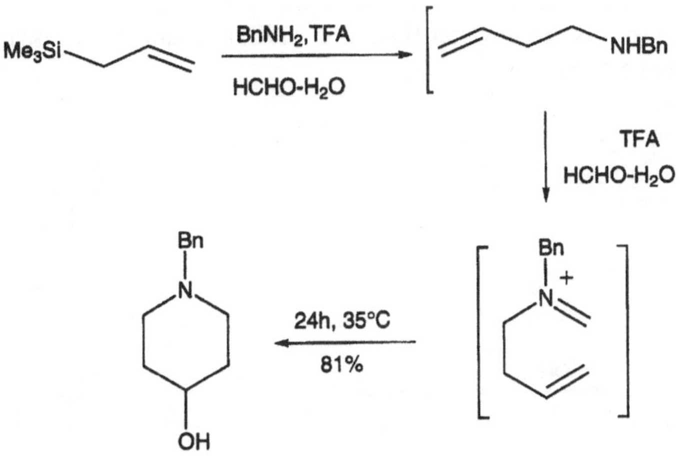

Scheme 4.14

This methodology was further extended to allylstannanes (Grieco and Bahsas, 1987). Still in the context of Mannich-type reactions, it was shown that the rate of reaction of phenols and ketones with secondary amines in the presence of formaldehyde was greatly increased in aqueous media compared with alcoholic or hydrocarbon solvents (Tychopoulos and Tyman, 1986). More recently, Kobayashi and Ishitani (1995) reported the catalysis of the reaction of vinyl ethers with iminium salts by lanthanide triflates:

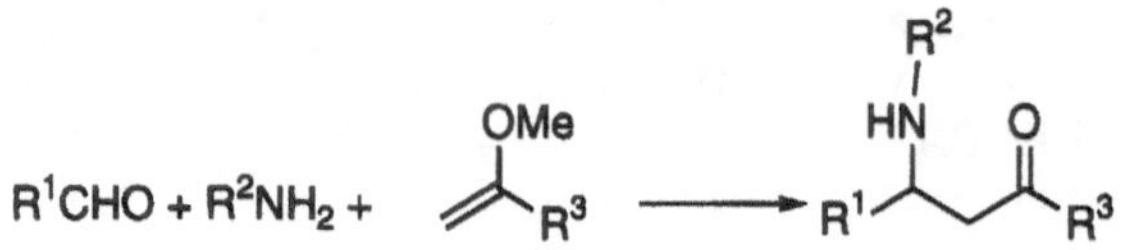

Another reaction type involving iminium salts led to pyrrolidines via the trapping of azomethine ylides generated from sarcosine and aqueous formaldehyde (Lubineau *et al.*, 1995). The efficiency of [3+2] cycloaddition was related to the water content. Indeed, adding THF to the reaction medium decreased the rate of the Michael addition, which competed with the desired pyrrolidine synthesis:

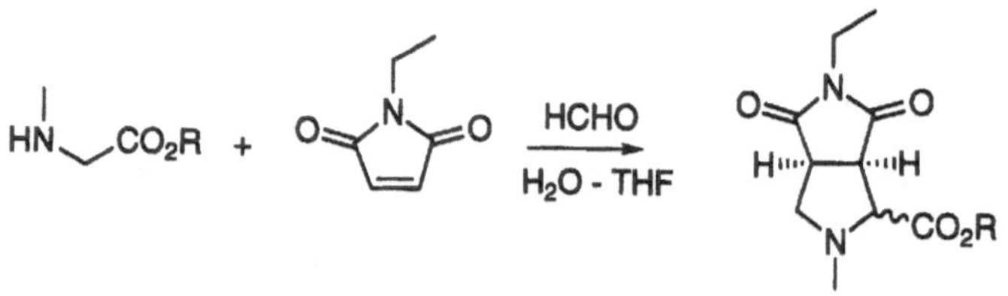

In addition to the iminium-salt-based chemistry and to cross-coupling processes, aqueous formaldehyde can also be used as the carbonylated substrate in the Prins reaction, an intermolecular ene-type acid-catalyzed reaction with alkenes providing 1,3-dioxanes (Adams and Bhatnagar, 1977).

Photosensitized reactions of tertiary amines with α,β-unsaturated esters were shown to proceed in water (Das *et al.*, 1996), leading to *N*-alkylpyrrolidones (Scheme 4.15).

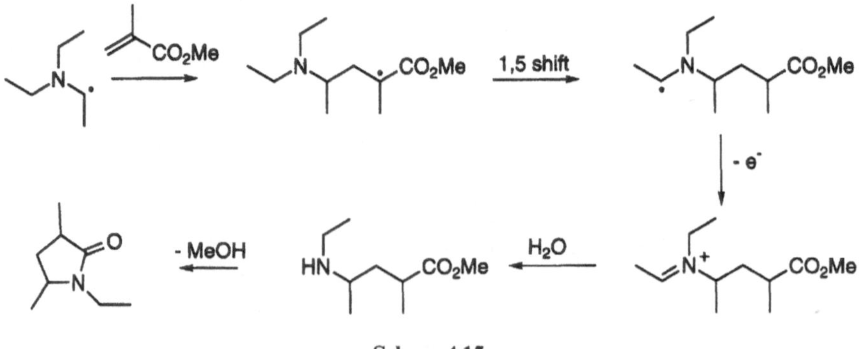

Scheme 4.15

The effects of inorganic salts on the rate of the aqueous cyanide-catalyzed benzoin condensation were discussed in terms of stabilization or destabilization of the water structure (Kool and Breslow, 1988):

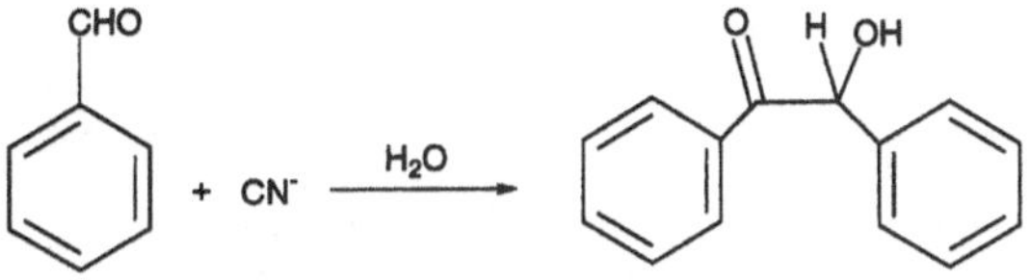

Of course, the carbonyl group is involved in the reactivity of unsaturated ketones or aldehydes toward dienes in cycloaddition pathways. Aqueous hetero Diels–Alder reactions of activated carbonyl compounds have also been reported, and are extensively described elsewhere in this book.

The Wittig–Horner olefination can also be achieved in aqueous heterogeneous media (Rambaud *et al.*, 1984). Thus (*E*)-vinylphosphonates were prepared in good yields from aldehydes and tetraethylmethylene diphosphonate in boiling water. Ethyl (diethylphosphono)acetates are alkylated in the same conditions providing in one pot α-acrylic acid esters (Kirschleger and Queignec, 1986):

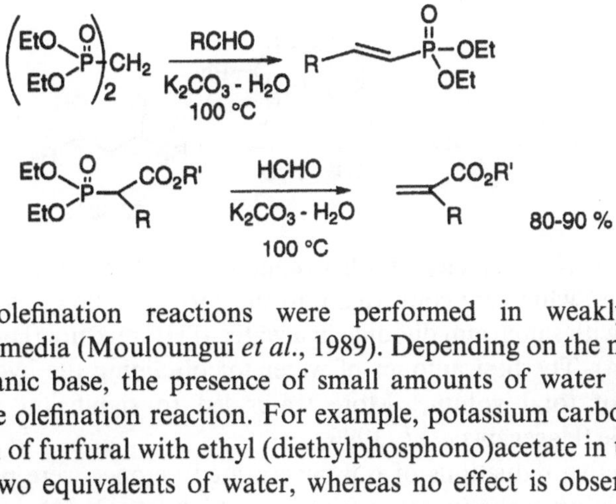

Similar olefination reactions were performed in weakly hydrated solid/liquid media (Mouloungui *et al.*, 1989). Depending on the nature of the solid inorganic base, the presence of small amounts of water was able to promote the olefination reaction. For example, potassium carbonate allows the reaction of furfural with ethyl (diethylphosphono)acetate in the presence of one or two equivalents of water, whereas no effect is observed for the cesium carbonate reaction.

To finish this chapter, the following examples deal with other reactions than carbon–carbon bond formation, but still concern the chemistry of the carbonyl group in water.

Baeyer–Villiger oxidation of ketones was shown to provide high yields of esters or lactones when performed in water (Fringuelli *et al.*, 1989).

The nucleophilic attack of the carbonyl group of an amide under hydrolyzing conditions is a very nice example of medium-assisted reaction (Menger and Fei, 1994). Long-chain ammonium amides could be hydrolyzed efficiently in the presence of fatty acid sodium salts; such a proximity effect was already described in the 1960s (Knowles and Parsons, 1967).

As a new illustration of this phenomenon, the esterification of sucrose with octanoyl chloride was shown to proceed in an aqueous medium:

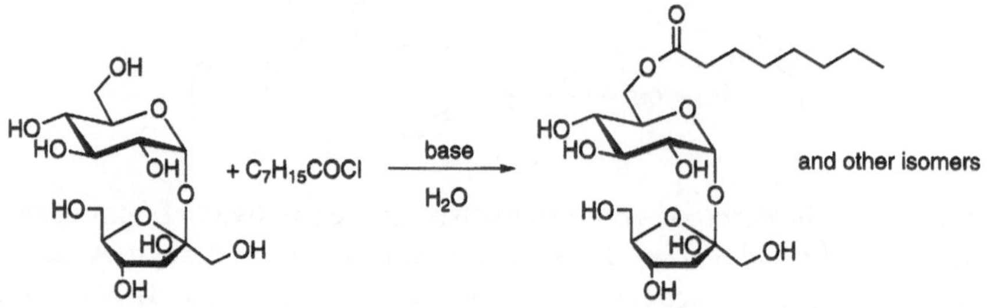

The polysubstitution was shown to be driven by the hydrophobic effect (Thévenet *et al.*, 1997). These findings are consistent with the known strengthening of the water structure due to large amounts of sucrose in solution (Lubineau *et al.*, 1993, 1994b).

In a reduction reaction, the presence of water led to significant changes in the product distribution. Irradiation of aroyl epoxyketones in aqueous THF led to good yields of β-hydroxyketones:

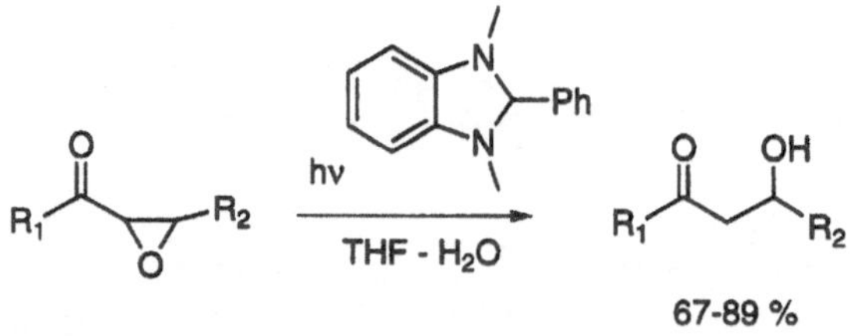

67-89 %

In Scheme 4.16 is depicted another reduction reaction, involving carbohydrate lactones, which are converted into the corresponding 2-deoxylactones by reaction with samarium diiodide in a water–THF mixture (Hanessian and Girard, 1994). The best amount of water for obtaining the hydroxyketone was 1–2% of total volume. More water led to significant amounts of benzaldehyde (Hasegawa *et al.*, 1996).

Regioselective reductions of α,β-unsaturated ketones were performed in water containing carbohydrate-derived amphiphiles. Cyclohexenones were converted to allylic alcohols, whereas cyclopentenones gave substantial amounts of 1,4-reduction (Denis *et al.*, 1996). Such a use of concentrated aqueous solutions of carbohydrates as the reaction medium has been described by Lubineau (1993, 1994b) in the context of many different chemical transformations including Diels–Alder, aldol and Michael reactions, as well as reductions of α,β-unsaturated ketones, all dealing with the chemistry of the carbonyl group either as the reaction site or as the activation site (Scheme 4.17).

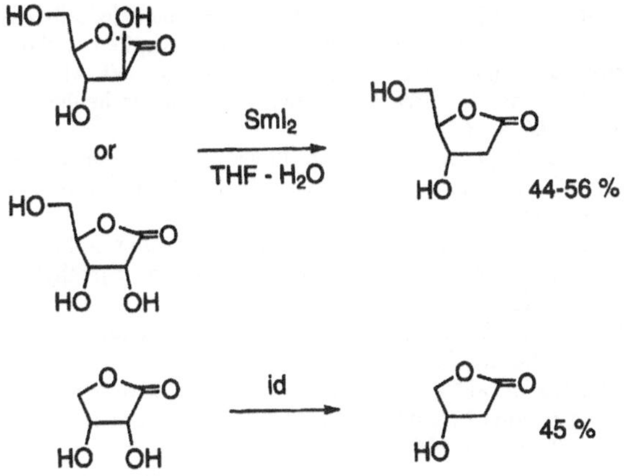

Scheme 4.16

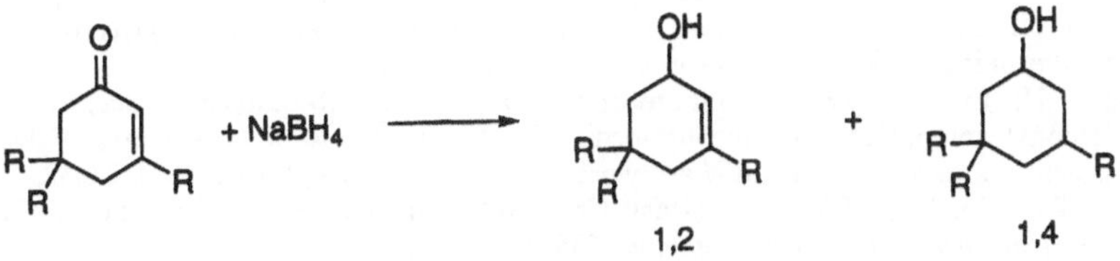

R	additive	1,2 : 1,4
H	2.5 M α-MeGlu	1:1
H	2.5M sucrose	2.4:1
Me	2.5 M sucrose	6.7:1
Me	2.5 M α-MeGlu	11.5:1

Scheme 4.17

References

Adams, D.R.; Bhatnagar, S.P. (1977) The Prins reaction, *Synthesis*, 661–72.

Ahonen, M.; Sjöholm, R. (1995) Zn-mediated chemoselective allylation of keto esters with allylic halides in aqueous medium, *Chem. Lett.*, 341–2.

Araki, S.; Jin, S.-J.; Idou, Y. (1992) Allylation of carbonyl compounds with catalytic amount of indium, *Bull. Chem. Soc. Jpn*, **65**, 1736–8.

Augé, J. (1985) Divalent tin-mediated synthesis: a very simple, stereoselective route to terminal conjugated (E) dienes and trienes, *Tetrahedron Lett.*, **26**, 753–6.

Augé, J.; Lubin, N.; Lubineau, A. (1994) Acceleration in water of the Baylis–Hillman reaction, *Tetrahedron Lett.*, **35**, 7947–8.

Ballini, R.; Bosica, G. (1996) The Michael reaction of nitroalkanes with conjugated enones in aqueous media, *Tetrahedron Lett.*, **37**, 8027–30.

Barbier, P. (1898) Synthèse du dimethylheptenol, *C.R. Acad. Sci. Paris*, **128**, 110–11.

Barden, M.C.; Schwartz, J. (1996) Stereoselective pinacol coupling in aqueous media *J. Am. Chem. Soc.*, **118**, 5484–5.

Binder, W. H.; Prenner, R.H.; Schmid, W. (1994) Indium-mediated allylation of aldehydes: a convenient route to 2-deoxy and 2, 6-dideoxy carbohydrates, *Tetrahedron*, **50**, 749–58.

Boaretto, A.; Marton, D.; Tagliavini, G. (1985a) Allylstannation. VI. Allylation and allenylation of aldehydes and ketones by allyl- and allenyl-tin chlorides in the presence of water, *J. Organomet. Chem.*, **286**, 9–16.

Boaretto, A.; Marton, D.; Tagliavini, G. (1985b) Preparation of α-allenic and β-acetylenic alcohols by treatment of a mixture of $Bu_3SnC\equiv C=CH_2$ and RCHO with Bu_3SnCl_2 and water, *J. Organomet. Chem.*, **297**, 149–53.

Bryan, V.J.; Chan, T. H. (1996) Indium mediated intramolecular carbocyclization in aqueous media. A facile and stereoselective synthesis of fused α-methylene-γ-butyrolactones, *Tetrahedron Lett.*, **37**, 534 1–2.

Buonora, P.T.; Rosauer, K. G.; Dai, L. (1995) Control of the aqueous aldol addition under Claisen–Schmidt conditions, *Tetrahedron Lett.*, **36**, 4009–12.

Carofiglio, T.; Marton, D.; Tagliavini, G. (1992) New simple route to allylstannanes by zinc-mediated coupling of allyl bromides with Bu_3SnCl or Bu_2SnCl_2 in an H_2O (NH_4Cl)/THF medium, *Organometallics*, **11**, 2961–3.

Chan, T. H.; Isaac, M.B. (1996) Organometallic-type reactions in aqueous media mediated by indium: application to the synthesis of carbohydrates, *Pure Appl. Chem.*, **68**, 919–24.

Chan, T.H.; Li, C.-J. (1990) Organic reactions in aqueous medium. Conversion of carbonyl compounds to 1,3-butadienes or vinyloxiranes, *Organometallics*, **9**, 2649–50.

Chan, T.H.; Li, C.-J. (1992a) A concise synthesis of (+)-muscarine, *Can. J. Chem.*, **70**, 2726–9.

Chan, T. H.; Li, C.-J. (1992b) A concise chemical synthesis of (+)-3-deoxy-D-glycero-D-galacto-nonulosonic acid (KDN), *J. Chem. Soc., Chem. Commun.*, 747–8.

Chan, T.H.; Li, C.-J. (1995) Indium-mediated coupling of α-(bromomethyl)acrylic acid with carbonyl compounds in aqueous media. Concise synthesis of (+)-3-deoxy-D-glycero-D-galacto-nonulosonic acid and N-acetylneuraminic acid, *J. Org. Chem.*, **60**, 4228–32.

Chan, T.H.; Li, C.-J.; Wei, Z.Y. (1990) Cross aldol type condensation reaction in aqueous media, *J. Chem. Soc., Chem. Commun.*, 505–7.

Chan, T.H.; Li, C.-J.; Lee, M.C.; Wei, Z.Y. (1994) Organometallic-type reactions in aqueous media, a new challenge in organic synthesis. *Can. J. Chem.*, **72**, 1181–92.

Chen, D.-L.; Li, C.-J. (1996) A gem-allyl dianion synthon in water, *Tetrahedron Lett.*, **37**, 295–8.

Choi, S.-K.; Lee, S.; Whitesides, G.M. (1996) Synthesis of C-5 analogs of N-acetylneuraminic acid via indium-mediated allylation of N-substituted 2-amino-2-deoxymannoses, *J. Org. Chem.*, **61**, 8739–45.

Clerici, A., Porta, O. (1982a) Pinacolic coupling of aromatic aldehydes and ketones promoted by aqueous titanium trichloride in basic media, *Tetrahedron Lett.*, **23**, 3517–20.

Clerici, A., Porta, O. (1982b). A novel reaction type promoted by aqueous titanium trichloride. Synthesis of unsymmetrical 1,2-diols, *J. Org. Chem.*, **47**, 2852–6.

Clerici, A.; Porta, O. (1983) A convenient synthesis of substituted pyridylglycols promoted by aqueous titanium trichloride,, *Tetrahedron*, **39**, 1239–46.

Clerici, A.; Porta, O. (1989) Radical addition to the carbonyl carbon promoted by aqueous titanium trichloride: stereoselective synthesis of α,β-dihydroxy ketones, *J. Org. Chem.*, **54**, 3872–8.

Clerici, A.; Porta, O.; Riva, M. (1981) Mixed carbonyl coupling induced by aqueous titanium trichloride. Some comments on the mechanism, *Tetrahedron Lett.*, **22**, 1043–6.

Clerici, A.; Porta, O.; Zago, P. (1986) Radical addition to the carbonyl carbon of ketones promoted by aqueous titanium trichloride in acidic medium. One step synthesis of pinacols and lactones, *Tetrahedron*, **42**, 561–72.

Das, S.; Kumar, J.S.D.; Shivaramayya, K.; George, M.V. (1996) Photoelectron transfer catalysed reactions of amines with α,β-unsaturated esters and acrylonitrile using different sensitizers, *J. Photochem. Photobiol., A: Chem.*, **97**, 139–50.

Delair, P.; Luche, J.-L. (1989) A new sonochemical carbonyl cross-coupling reaction *J. Chem. Soc., Chem. Commun.*, 398–9.

Denis, C.; Laignel, B.; Plusquellec, D.; Le Marouille, J.Y.; Botrel, A. (1996) Highly regio- and stereoselective reductions of carbonyl compounds in aqueous glycosidic media, *Tetrahedron Lett.*, **37**, 53–6.

Denmark, S.E.; Lee, W. (1992) Investigations on transition state geometry of the aldol condensation in aqueous medium, *Tetrahedron Lett.*, **33**, 7729–32.

Diana, S.-C.H.; Sim, K.-Y.; Loh, T.-P. (1996) Lanthanum trifluoromethanesulfonate [La(OTf)₃] promoted indium-mediated coupling of ethyl 4-bromocrotonate with carbonyl compounds in water, *Synlett*, 263–4.

Dupuy, C.; Pétrier, C.; Sarandeses, L.A.; Luche, J.-L. (1991) Ultrasound in organic synthesis. Further studies on the conjugate additions to electron deficient olefins in aqueous media, *Synth. Commun*, **21**, 643–51.

Durant, A.; Delplanche, J.L.; Winand, R.; Reisse, J. (1995) A new procedure for the production of highly reactive metal powder by pulsed sonoelectrochemical reduction, *Tetrahedron Lett.*, **36**, 4257–60.

Eder, U.; Saner, G.; Wiechert, R. (1971) New type of asymmetric cyclization to optically active steroid CD partial structures, *Angew. Chem. Int. Ed. Engl.*, **10**, 496–7.

Einhorn, C.; Luche, J.-L. (1987) Selective allylation of carbonyl compounds in aqueous media *J. Organomet. Chem.*, **322**, 177–83.

Fringuelli, F.; Germani, R.; Pizzo, F.; Savelli, G. (1989) Baeyer–Villiger reaction in water, *Gazz. Chim. Ital.*, **119**, 249.

Fringuelli, F.; Pani, G.; Piermatti, O.; Pizzo, F. (1994) Condensation reactions in water of active methylene compounds with arylaldehydes. One-pot synthesis of flavonols, *Tetrahedron*, **50**, 11499–508.

Furlani, D.; Marton, D.; Tagliavini, G., Zordan, M. (1988) Hydrated σ-bonded organometallic cations in organic synthesis. I. Allyl-, crotyl-, l-methylallyl-, cyclohex-2-enyl-, and cinnamyl-stannation of carbonyl compounds in water, *J. Organomet. Chem.*, **341**, 345–56.

Fürstner, A. (1993) Chemistry of and with highly reactive metals, *Angew. Chem. Int. Ed. Engl.*, **32**, 164–89.

Gao, J., Härter, R.; Gordon, D M.; Whitesides, G.M. (1994) Synthesis of KDO using indium-mediated allylation of 2,3:4,5-di-*O*-isopropylidene-D-arabinose in aqueous media, *J. Org. Chem.*, **59**, 3714–15.

Gordon, D.M.; Whitesides, G.M. (1993) Indium-mediated allylation of unprotected carbohydrates in aqueous media: a short synthesis of sialic acid, *J. Org. Chem.*, **58**, 7937–8.

Grieco, P.A., Bahsas, A. (1987) Reactions of allylstannanes with *in situ* generated immonium salts in protic solvents: a facile aminomethano destannylation process, *J. Org. Chem.*, **52**, 1378–80.

Hajos, Z.; Parrish, D. R. (1974) Synthesis and conversion of 2-methyl-2-(3-oxobutyl)-1,3-cyclopentanedione to the isomeric racemic ketols of the [3.2.1] bicyclo octane and of the perhydroindan series, *J. Org. Chem.*, **39**, 1612–15.

Hanessian, S.; Girard, C. (1994) One step α-deoxygenation of unprotected aldonolactones using samarium diodide–THF/H₂O) system. A new synthesis of 2-deoxy-D-ribose, *Synlett*, 861– 2.

Hanessian, S.; Yang, R.-Y. (1996) The asymmetric synthesis of allylglycine and other unnatural α-amino acids via zinc-mediated allylation of oximes in aqueous media, *Tetrahedron Lett.*, **37**, 5273–6.

Hasegawa, E.; Kato, T.; Kitazume, T.; Yanagi, K.; Hasegawa, K.; Horaguchi, T. (1996) Photo induced electron transfer reactions of α,β-epoxy ketones with 2-phenyl-*N,N*-dimethylbenz-imidazoline (PDMBI): significant water effect on the reaction pathway, *Tetrahedron Lett.*, **37**, 7079–82.

Isaac, M.B. Chan, T.H. (1995a) Indium-mediated coupling of aldehydes with allyl bromides in aqueous media. The issue of regio- and diastereo-selectivity, *Tetrahedron Lett.*, **36**, 8957–60.

Isaac, M.B.; Chan, T.H. (1995b) Indium-mediated coupling of aldehydes with prop-2-ynyl bromides in aqueous media, *J. Chem. Soc.. Chem. Commun.*, 1003–4.

Ishihara, K.; Hananki, N.; Yamamoto, M. (1993) Tris(pentafluorophenyl) boron as a new efficient, air stable, and water tolerant catalyst in the aldol-type and Michael reactions, *Synlett*, 577–9.

Jenner, G. (1996) Comparative study of physical and chemical activation modes. The case of the synthesis of β-amino derivatives, *Tetrahedron*, **52**, 13557–68.

Kahn, B.E., Rieke, R.D. (1988) Carbonyl coupling reactions using transition metals, lanthanides and actinides, *Chem. Rev.*, **88**, 733–45.

Katritzky, A.R.; Shobana, N.; Harris, P.A. (1992) Bismuth(III) chloride–aluminium-promoted alkylations of immonium cations to amines in aqueous media. Unstabilized carbanion equivalents for use in the presence of water, *Organometallics*, **11**, 1381–4.

Keller, E.; Feringer, B. L. (1996) Ytterbium triflate catalyzed Michael additions of β-ketoesters in water, *Tetrahedron Lett.*, **37**, 1879–82.

Killinger, T.A.; Boughton, N.A.; Runge, T.A.; Wolinsky, J. (1977) Alcohols as solvent for the generation and reaction of allylic zinc halides with aldehydes and ketones, *J. Organomet. Chem.*, **124**, 131–4.

Kim, E.; Gordon, D. M.; Schmid, W.; Whitesides, G.M (1993) Tin- and indium-mediated allylation in aqueous media: application to unprotected carbohydrates, *J. Org. Chem.*, **58**, 5500–7.

Kirschleger, B.; Queignec, R. (1986) Heterogeneous mediated alkylation of ethyl diethyl phosphonoacetate. A one-pot access to α-alkylated acrylic esters, *Synthesis*, 926–8.

Knowles, J.R.; Parsons, C.A. (1967) Proximity effects in bimolecular reactions in solutions, *J. Chem. Soc., Chem. Commun.*, 755–67.

Kobayashi, S. (1991) Lanthanide trifluoromethanesulfonates as stable Lewis acids in aqueous media. Yb(OTf)$_3$ catalyzed hydroxymethylation reaction of silyl enol ethers with commercial formaldehyde solution, *Chem. Lett.*, 2187–90.

Kobayashi, S. (1994) Rare earth metal trifluoromethanesulfonates as water-tolerant Lewis acid catalysts in organic synthesis, *Synlett*, 689–701.

Kobayashi, S.; Hachiya, I. (1992) The aldol reaction of silyl enol ethers with aldehydes in aqueous media, *Tetrahedron Lett.*, **33**, 1625–8.

Kobayashi, S.; Ishitani, H. (1995) A novel Mannich-type reaction in aqueous media. Lanthanide triflate-catalyzed condensation of aldehydes, amines and vinyl ethers for the synthesis of β-amino ketones, *J. Chem. Soc., Chem. Commun.*, 1379.

Kool, E.T.; Breslow, R. (1988) Dichotomous salt effects in the hydrophobic acceleration of the benzoin condensation, *J. Am. Chem. Soc.*, **110**, 1596–7.

Kunz, T.; Reissig, H. U. (1989) Radical additions in aqueous medium: direct synthesis of 5-allyl-substituted γ-lactones from allylic bromides/zinc and methyl γ-oxocarboxylates, *Liebiegs Ann. Chem.*, 891–3.

Larpent, C.; Patin, H. (1988) Nucleophilic addition of water-soluble phosphines on activated olefins, *Tetrahedron*, **44**, 6107–18.

Larpent, C.; Meignan, G.; Patin, H. (1990) Organic chemistry in water. Nucleophilic addition of water-soluble phosphines on activated alkynes: an efficient synthesis of new vinylphosphonium salts and of specifically derivated olefins, *Tetrahedron*, **46**, 6381–98.

Larsen, S.D.; Grieco, P.A. (1985) Aza Diels Alder reactions in aqueous solution: cyclocondensation of dienes with simple iminium salts generated under Mannich conditions *J. Am. Chem. Soc.*, **107**, 1768–9.

Larsen, S.D.; Grieco, P.A.; Fobare, W.F. (1986) Reactions of allyl silanes with simple iminium salts in water: a facile route to piperidine via an aminomethano desilylation–cyclization process, *J. Am. Chem. Soc.*, **108**, 3512–13.

Lavallée, J.-F.; Deslongchamps, P. (1988) One-step construction of a 13-α-methyl 14-α-hydroxy steroid via a new anionic polycyclization method, *Tetrahedron Lett.*, **29**, 6033–6.

Li, C.-J. (1993) Organic reactions in aqueous mediated – with a focus on carbon–carbon bond formation, *Chem. Rev.*, **93**, 2023-35.

Li, C.-J. (1995) Trimethylenemethane dianon equivalent in aqueous medium, *Tetrahedron Lett.*, **36**, 517-18.

Li, C.-J. (1996) Aqueous Barbier–Grignard type reaction: scope, mechanism and synthetic applications, *Tetrahedron*, **52**, 5643–68.

Li, C.-J.; Chan, T.-H. (1991a) Organometallic reactions in aqueous media. Convenient synthesis of methylenetetrahydrofurans, *Organometallics*, **10**, 2548–9.

Li, C.-J.; Chan, T.-H. (1991b) Organometallic reactions in aqueous media with indium, *Tetrahedron Lett.*, **32**, 7017–20.

Li. C.-J.: Chen. D.-L.; Lu, Y.-Q.; Haberman, J.X.; Mague, J.T. (1996) Novel carbocycle enlargement in aqueous medium, *J. Am Chem. Soc.*, **118**, 4216–17.

Li, X.-R.; Loh, T.-P. (1996) Indium trichlororide-promoted tin-mediated carbonyl allylation in water: high simple diastereo- and diastereofacial selectivity, *Tetrahedron Asymmetry*, **7**, 1535–8.

Loh, T.-P.; Li, X. R. (1996) A versatile and practical synthesis of α-trifluoromethylated alcohols from trifluoroacetaldehyde ethyl hemi-acetal in water, *J. Chem. Soc., Chem. Commun.*, 1929–30.

Loh, T.-P.; Pei, J.; Cao, G.-Q. (1996) Indium trichloride catalyzed Mukaiyama aldol reaction in water, *J. Chem. Soc., Chem. Commun.*, 1819–20.

Lu, Y.-Q.; Li, C.-J. (1996) Novel [3+2] annulation via a trimethylenemethane zwitterion equivalent in water, *Tetrahedron Lett.*, **37**, 471–4.

Lubineau, A. (1986) Water-promoted organic reactions: aldol reaction under neutral conditions, *J. Org. Chem.*, **51**, 2142–4.

Lubineau, A.; Augé, J. (1992) Water-promoted organic reactions. Michael addition of nitroalkanes to methylvinylketone under neutral conditions, *Tetrahedron Lett.*, **33**, 8073–4.

Lubineau, A.; Bouchain, G. (1997) Water-promoted reactivities: generation of oxyallyl intermediates and their [4+3] cycloadditions with furan and cyclopentadiene. Facile access to bridged cycloheptenones, *Tetrahedron Lett.*, in press.

Lubineau, A.; Meyer, E. (1988) Water-promoted organic reactions. Aldol reaction of silyl enol ethers with carbonyl compounds under atmospheric pressure and neutral conditions, *Tetrahedron*, **44**, 6065–70.

Lubineau, A.; Augé, J.; Bienaymé, H.; Queneau, Y.; Scherrmann, M.-C. (1993) Aqueous sugar solutions as solvent in organic synthesis: new reactivity and selectivity, in *Carbohydrates as Organic Raw Materials II*, Descotes, G., Ed.; VCH, Weinheim, pp. 99–112.

Lubineau, A.; Augé, J.; Queneau, Y. (1994a) Water-promoted organic reactions, *Synthesis*, 741–60.

Lubineau, A.; Bienaymé, H.; Queneau, Y.; Scherrmann, M.-C. (1994b) Aqueous cycloadditions using glyco-organic substrates. Thermodynamics of the reaction, *New J. Chem.*, **18**, 279–85.

Lubineau, A.; Bouchain, G.; Queneau, Y. (1995) Reactivity of the carbonyl group in water. Generation of azomethine ylides from aqueous formaldehyde: Michael addition versus dipolar trapping, *J. Chem. Soc., Perkin Trans. 1*, 2433–7.

Luche, J.-L.; Allavena, C. (1988a) Ultrasound in organic synthesis. Optimisation of the conjugate additions to α,β-unsaturated carbonyl compounds in aqueous media, *Tetrahedron Lett.*, **29**, 5369–72.

Luche, J.-L.; Allavena, C. (1988b) Ultrasound in organic synthesis. Mechanistic aspects of the conjugate addition to α-enones in aqueous media, *Tetrahedron Lett.*, **29**, 5373–4.

Luche, J.-L.; Damiano, J.-C. (1980) Ultrasounds in organic syntheses. Effect on the formation of lithium organometallic reagents, *J. Am. Chem. Soc.*, **102**, 7926–7.

Mandai, T.; Nokami, J.; Yano, T.; Yoshinaga, Y.; Otera, J. (1984) Facile one-pot synthesis of bromo homoallyl alcohols and 1,3-keto acetates with allyltin intermediates, *J. Org. Chem.*, **49**, 172–4.

Marton, D.; Tagliavini, G.; Vanzan, N. (1989) Hydrated σ-bonded organometallic cations in organic synthesis. Allylstannation of aldehydes by crotyltin chlorides in acid media, *J. Organomet. Chem.*, **376**, 269–76.

Marton, D.; Stivanello, D.; Tagliavini, G. (1996a) Stereochemical study of the allylation of aldehydes with allyl halides in cosolvent/water(salt)/Zn/haloorganotin media, *J. Org. Chem.*, **61**, 2731–7.

Marton, D.; Russo, V.; Stivanello, D.; Tagliavini, G. (1996b) Preparation of benzylstannanes by zinc-mediated coupling of benzylbromides with organotin derivatives. Physicochemical characterization and crystal structures, *Organometallics*, **15**, 1645–50.

Masuyama, Y.; Takahara, J.P.; Kurusu, Y. (1989) Palladium-catalyzed carbonyl allylation by allylic alcohols with SnCl₂. A solvation-controlled diastereoselection, *Tetrahedron Lett.*, **30**, 3437–40.

Masuyama, Y.; Nimura, Y.; Kurusu, Y. (1991) Palladium-catalyzed carbonyl allylation by 2-(hydroxymethyl) acrylate derivatives: synthesis of α-methylene-γ-butyrolactones, *Tetrahedron Lett.*, **32**, 225–8.

Masuyama, Y.; Kishida, M.; Kurusu, Y. (1996) γ-*Syn*-selective carbonyl addition by l-bromo-2-butene with tin(II) iodide and tetrabutylammonium bromide, *Tetrahedron Lett.*, **37**, 7103–6.

Matsumoto, K. (1981) High pressure Michael addition catalyzed by fluoride ions, *Angew. Chem. Int. Ed. Engl.*, **20**, 770–1.

Mattes, H.; Benezra, C. (1985) Use of bromomethyl acrylic acid for the synthesis of α-methylene-γ-butyrolactones, *Tetrahedron Lett.*, **26**, 5697–8.

McMurry, J.P. (1989) Carbonyl-coupling reactions using low-valent titanium, *Chem. Rev.*, **89**, 1513–24.

Menger, F.M.; Fei, Z.X. (1994) Fast amide cleavage under mild conditions. An evolutionary approach to bioorganic catalysis, *Angew. Chem. Int. Ed. Engl.*, **33**, 346–8.

Minato, M.; Tsuji, J. (1988) Allylation of aldehydes in an aqueous two-phase system by electrochemically regenerated bismuth metal, *Chem. Lett.*, 2049–52.

Molle, G.; Bauer, P. (1982) The Barbier synthesis; a one-step Grignard reaction?, *J. Am. Chem. Soc.*, **104**, 3481–7.

Mouloungui, Z.; Delmas, M.; Gaset, A. (1989) The Wittig–Horner reaction in weakly hydrated solid/liquid media: structure and reactivity of carbanionic species formed from ethyl (diethyl phosphono)acetate by adsorption on solid inorganic bases, *J. Org. Chem.*, **54**, 3936–41.

Moyano, A.; Pericas, M. A.; Rieca, A.; Luche, J.-L. (1990) A theoretical study of the Barbier reaction, *Tetrahedron Lett.*, **31**, 7619–22.

Mukaiyama, T.; Banno, K.; Narasaka, T. (1974) New cross-aldol reactions. Reactions of silyl enol ethers with carbonyl compounds activated by titanium tetrachloride, *J. Am. Chem. Soc.*, **96**, 7503–9.

Mukaiyama, T.; Harada, T.; Shoda, S. (1980) An efficient method for the preparation of homo allylic alcohol derivatives by the reaction of allyl iodide with carbonyl compounds in the presence of stannous halide, *Chem. Lett.*, 1507–10.

Nakagawa, M.; Nakao, H.; Watanabe, K. I. (1985) Steric effects of chiral ligands in a new type of aldol condensations catalyzed by zinc(II) complexes of α-amino acid esters, *Chem. Lett.*, 391–4.

Nokami, J.; Otera, J.; Sudo, T.; Okawara, R. (1983) Allylation of aldehydes and ketones in the presence of water by allylic bromides, metallic tin, and aluminium, *Organometallics*, **2**, 191–3.

Nokami, J.; Wakabayashi, S.; Okawara, R. (1984) Intramolecular allylation of carbonyl compounds. A new method for five and six membered ring formation, *Chem. Lett.*, 869–70.

Nokami, J.; Tamaoka, T.; Ogawa, H.; Wakabayashi, S. (1986) Facile synthesis of 2-methylene-4-butyrolactones, *Chem. Lett.*, 541–4.

Oda, Y.; Matsuo, S.; Saiko, K. (1992) An efficient synthesis of 3-chlorohomoallyl alcohols. Zinc-promoted 2-chloroallylation of carbonyl compounds with 2,3-dichloropropene in an aqueous solvent system, *Tetrahedron Lett.*, **33**, 97–100.

Paquette, L.A.; Lobben, P.C. (1996) π-Facial diastereoselection in the 1,2-addition of allylmetal reagents to 2-methylcyclohexanone and tetrahydrofuranspiro-(2-cyclohexanone), *J. Am. Chem. Soc.*, **118**, 1917–30.

Paquette, L.A.; Mitzel, T.M. (1995) Chelation control associated with organometallic addition reactions in water. The high stereoselectivity offered by α- and β-hydroxyl substituents obviates the need for protecting groups, *Tetrahedron Lett.*, **36**, 6863–6.

Paquette, L.A.; Mitzel, T.M. (1996a) Addition of allylindium reagents to aldehydes substituted at C_α or C_β with heteroatomic functional groups. Analysis of the modulation in diastereoselectivity attainable in aqueous, organic, and mixed solvent systems, *J. Am. Chem. Soc.*, **118**, 1931–7.

Paquette, L.A.; Mitzel, T.M. (1996b) Comparative diastereoselectivity analysis of crotylindium and 3-bromoallylindium additions to α-oxy aldehydes in aqueous and nonaqueous solvent systems, *J. Org. Chem.*, **61**, 8799–804.

Pétrier, C.; Luche, J.-L. (1985) Allylzinc reagent additions in aqueous media, *J. Org. Chem.*, **50**, 910–12.

Pétrier, C.; Einhorn, J.; Luche, J.-L. (1985) Selective tin and zinc mediated allylations of carbonyl compounds in aqueous media, *Tetrahedron Lett.*, **26**, 1449–52.

Pétrier, C.; Dupuy, C.; Luche, J.-L. (1986) Conjugate additions to α,β-unsaturated carbonyl compounds in aqueous media, *Tetrahedron Lett.*, **27**, 3149–52.

Pietrusiewicz, K.M.; Zablocka, M. (1988) Optically active phosphine oxides. Conjugate addition to vinyl phosphine oxides in aqueous media, *Tetrahedron Lett.*, **29**, 937–40.

Pons, J.M.; Santelli, M. (1988) Reductions promoted by low valent transition metal complexes in organic synthesis, *Tetrahedron*, **44**, 4295-312.

Prenner, R.H.; Binder, W.H.; Schmid, W. (1994) Indium-assisted allylation in carbohydrate chemistry: a convenient route to D-glycero-D-galacto and D-glycero-L-galacto-heptose, *Liebigs Ann. Chem.*, 73–8.

Rambaud, M.; Del Vecchio, A.; Villieras, J. (1984) Wittig-Horner reaction in heterogeneous media: V. An efficient synthesis of alkene-phosphonates and α-hydroxymethyl-α-vinyl phosphonate in water in the presence of potassium carbonate, *Synth. Commun.*, **14**, 833–41.

Saimoto, H.; Yoshida, K.; Murakami, T.; Morimoto, M.; Sashiwa, M.; Shigemasa, Y. (1996) Effect of calcium reagents on aldol reactions of phenolic enolates with aldehydes in alcohol, *J. Org. Chem.*, **61**, 6768–9.

Sarandeses, L.A.; Mouriño, A.; Luche, J.-L. (1992) Sonochemistry of epoxyalkylhalides in the presence of a zinc–copper couple, *J. Chem. Soc., Chem. Commun.*, 798–9.

Sarangi, C.; Nayak, A.; Nanda, B.; Das, N.B. (1995) A novel Cu(II)–Mg-system for allylation and reduction of carbonyl compounds, *Tetrahedron Lett.*, 36, 7119–22.

Sarrazin, L.; Mauzé, B. (1996) An efficient indium-mediated synthesis of α-cyano-β-ethylenic secondary alcohols in water, *Synth. Commun.*, **26**, 3179–91.

Schmid, W.; Whitesides, G.M. (1991) Carbon–carbon bond formation in aqueous ethanol: diastereoselective transformation of unprotected carbohydrates to higher carbon sugars using allyl bromide and tin metal, *J. Am. Chem. Soc.*, 113, 6674–5.

Shigemasa, Y.; Yokohama, K.; Sashira, H.; Saimoto, H. (1994) Synthesis of threo- and erythro-3-pentulose by aldol type reaction in water, *Tetrahedron Lett.*, 35, 1263–6.

Sisido, K.; Takedo, Y.; Kimugawa, Z. (1961) Direct synthesis of organotin compounds. Di- and tribenzyltin chlorides, *J. Am. Chem. Soc.*, 83, 538–41.

Sjöholm, R.; Rairama, R.; Ahonen, M. (1994) Zinc mediated allylation of aldehydes and ketones with cinnamyl chloride in aqueous medium, *J. Chem. Soc., Chem. Commun.*, 1217–18.

Sussangkarn, K.; Fodor, G.; Karle, I.; George, C. (1988) Ascorbic acid as a Michael donor. Reaction with alicyclic enones, *Tetrahedron*, 44, 7047–54.

Takaya, H.; Makino, S.; Hayakawa, Y.; Noyori, R. (1978) Reactions of polybromo ketones with 1,3-dienes in the presence of iron carbonyls. New 3+4 → 7 cyclocoupling reactions forming 4-cycloheptenones, *J. Am. Chem. Soc.*, 100, 1765–77.

Talaga, P.; Schaeffer, M.; Benezra, C.; Stampf, J.-L. (1990) A new synthesis of γ-substituted α-methylene-γ-butyrolactones (2-methylene-4-alkanolides) using catalysis by SnCl$_2$/Amberlyst 15, *Synthesis*, 530.

Tanaka, K.; Kishigami, S.; Toda, F. (1990) A new method for coupling aromatic aldehydes and ketones to produce α-glycols using Zn–ZnCl$_2$ in aqueous solution and in the solid state, *J. Org. Chem.*, **55**, 2981–3.

Thévenet, S.; Descotes, G.; Bouchu, A.; Queneau, Y. (1997) Hydrophobic effect driven esterification of sucrose in aqueous medium, *J. Carbohydr. Chem.*, 16, 691–6.

Tychopoulos, V.; Tyman, J.H.P. (1986) Enhancement of the rate of Mannich reactions in aqueous media, *Synth. Commun.*, 16, 1401–9.

Uneyama, K.; Kamaki, N.; Moriya, A.; Torii, S. (1985) Sn(II)–Al-promoted allylation of aldehydes with allyl chloride in an aqueous solvent system, *J. Org. Chem.*, 50, 5396–9.

Uneyama, K.; Ueda, K.; Torii, S. (1986a) Reductive generation of active zero-valent tin in SnCl$_2$–Al system and its use for highly diastereoselective reaction of cinnamyl chloride and aldehydes, *Tetrahedron Lett.*, 27, 2395–6.

Uneyama, K.; Nanbu, H.; Torii, S. (1986b) An SnCl$_2$-promoted α-methylene-γ-butyrolactone synthesis in an aqueous medium, *Chem. Lett.*, 1201–2.

Wada, M.; Ohki, H.; Akiba, K. (1987) Bismuth(III) chloride–aluminium-promoted allylation of aldehydes to homoallylic alcohols in aqueous solvent, *J. Chem. Soc, Chem. Commun.*, 708–9.

Wada, M.; Ohki, H.; Akiba, K. (1990) A Grignard-type addition of allyl unit to aldehydes by using bismuth and bismuth salt, *Bull. Chem. Soc. Jpn.*, 63, 1738–47.

Waldmann, H. (1990) Proline benzyl ester as chiral auxiliary in Barbier-type reactions in aqueous solutions, *Synlett*, 627–8.

Wang, R.; Lim, C.-M.; Tan, C.-H.; Lim, B.-K.; Sim, K.Y.; Loh, T.-P. (1995) Ytterbium trifluoromethanesulfonate [Yb(OTf)$_3$] promoted indium mediated allylation reactions of carbonyl compounds in aqueous media, *Tetrahedron Asymmetry*, 6, 1825–8.

Watanabe, K.-I.; Yamada, Y.; Goto, K. (1985) New type aldol condensations catalyzed by metal(II) complexes of α-amino acid esters and that with cyclodextrin systems, *Bull. Chem. Soc. Jpn*, 58, 1401–6.

Wilson, S.R.; Guazzaroni, M.E. (1989) Synthesis of homoallylic alcohols in aqueous media, *J. Org. Chem.*, 54, 3087–91.

Wirth, T. (1996) 'New' reagents for the 'old' pinacol coupling reaction., *Angew. Chem. Int. Ed. Engl.*, 35, 61–3.

Wu, S.; Huang, B.; Gao, X. (1990) Reaction of propargyl bromide with aldehydes in the presence of metallic tin. Synthesis of homopropargylalcohols and homoallenylalcohols, *Synth. Commun.*, **20**, 1279–86.

Yanagisawa, A. Inoue, H.; Morodome, M.; Yamamoto, H. (1993) Highly chemoselective allylation of carbonyl compounds with tetraallyltin in acidic aqueous media, *J. Am. Chem. Soc.*, **115**, 10356–7.

Yamamoto, H.; Maruyama, K.; Matsumoto, K. (1983) Organometallic high-pressure reactions. 2. Aldol reaction of silyl enol ethers with aldehydes under neutral conditions, *J. Am. Chem. Soc.*, **105**, 6963–5.

Yavari, I.; Riazi-Kermani, F. (1995) Zinc-promoted Barbier-type reaction of propargyl bromide with aldehydes in aqueous media, *Synth. Commun.*, **25**, 2923–8.

Zhang, Y.; Xu, W.; (1989) The aldol condensation catalyzed by metal-(II)-β-cyclodextrin complexes, *Synth. Commun.*, **19**, 1291–6.

Zhou, J.; Lu, G.; Wu, S. (1992) A new approach for the synthesis of α-methylene-γ-butyrolactones from α-bromomethylacrylic acid (or esters), *Synth. Commun.*, **22**, 481–7.

5 Aqueous transition-metal catalysis

I.P. BELETSKAYA and A.V. CHEPRAKOV

5.1 Trends in aqueous transition-metal catalysis

The application of water in organic transition-metal catalysis was from its birth a spontaneous idea, and as such it is rather ill-defined. It is usual to justify research on the use of water in catalytic processes by environmental reasons. Certainly, the trend to get rid of organic solvents is quite common for modern chemical technology. More and more of both end-user products and industrial chemicals, which earlier were invariably associated with organic solvents, such as paints, varnishes, textile processing agents, etc., are now manufactured based on water. This means a decrease in the amount of uncontrolled release of toxic organic wastes into the environment, and a dramatic reduction of the hazards for both industrial personnel and consumers. It is natural to try to propagate this trend back to the reactors in which the chemicals themselves are being produced, in an attempt to make the whole chemical industry function as safely and efficiently as the most sophisticated chemical plant of Nature, the living cell.

Water is a unique solvent, having a rare combination of properties. However general a theory of solvation and solutions might be elaborated, there always remains an asymmetry in how different solvents fit into it. Water always succeeds in finding a separate place, being either the sole well-behaved liquid on which a given theory is based, or the sole but absolute exclusion, which makes a certain other theory just an unreliable hypothesis. Suffice it to mention that only water spawns the whole science of micellar and other self-aggregation effects (exclusions indeed exist, but are comparatively so few that their very existence helps to emphasize the unique nature of water) and inter-facial phenomena occurring due to the formation of structured aggregates, which at the high end define the shape of living matter.

However, in the first place the application of water in catalysis pursues several, although less ambitious, but rather practical, goals.

Water may be a reagent, being the cheapest source of either oxygen for the formation of oxygen-containing products (alcohols, acids, ketones, alde-hydes, etc.), or hydrogen, as in some reduction processes.

Water is an amphoteric compound, and thus it can lend all types of assis-tance, whichever is required in a given process, or even a combination of several. It may itself be a general acid or base catalyst, or, in the presence of other acids and bases, serve as an environment deploying proton or hydroxide ion for specific acid or base catalysis.

Water and aqueous solutions are preferred media for reactions involving complex natural compounds, many of which (e.g. carbohydrates, poly-peptides, nucleosides, etc.) are strongly hydrophilic. In order to make such compounds take part in reactions in common organic media, hydrophilic sites must first be hydrophobized by large protective groups, which make the compounds change their natural conformations and reactivity.

Water is a solvent that can be used to implement approaches combining the advantages of homogeneous and heterogeneous catalysis by distributing the reagents, products and catalysts between different phases, or pseudophases, or interphase regions (self-structured layers formed by amphiphilic compounds). Three distinct types of such approach can be distinguished based on phase transfer, phase separation and solubilization, though most real processes usually involve combinations of the above phenomena.

5.1.1   Water as reagent

Water is a moderately reactive nucleophile and, as such, is involved in several well-known catalytic cycles, in which it (or its conjugate base, hydroxide ion) usually attacks either carbonyl ligands or other ligands in nucleophilic cleavage or reductive elimination steps. Such processes may be encountered, for example, in hydroxycarbonylation reactions:

$$M\overset{R}{\underset{CO}{\diagdown}} \xrightarrow{H_2O \text{ or } OH^-} M\overset{R}{\underset{COOH}{\diagdown}} \longrightarrow M^{(n-2)} + RCOOH$$

or in the so-called water-gas shift reaction (WGSR):

$$M-CO \xrightarrow{H_2O \text{ or } OH^-} M-COOH \longrightarrow M-H + CO_2$$

which is a source of metal hydride complexes formed in the absence of molecular hydrogen. Further transformations triggered by WGSR occurring with or without the involvement of CO are extremely versatile.

The attack of water on unsaturated ligands can be traced in the famous Wacker-type olefin oxidations as well as in catalytic hydration. Again, both pathways may involve other reactive species (CO, other unsaturated molecules, etc.), thus generating a versatile chemistry:

As water is almost never added to reaction mixtures in stoichiometric amounts, such reactions are usually carried out in the presence of excess water, which at the same time plays the role of solvent or cosolvent.

5.1.2 Water as solvent

The solvent properties of water are unique, and are not reproduced in full scope by any other liquid known, though any one of its properties taken separately (polarity, polarizability, specific solvation due to hydrogen bonding, or basicity, etc.) is easily outperformed. It is generally accepted that at least part of the unique behaviour of water is accounted for by its perfectly structured character in the liquid state. This deserves the comment that the structure of water is quite fragile, being a property of only the bulk liquid, either pure or containing low concentrations of additives. Such structure cannot form in mixtures of water with other solvents, especially when water is the minor component. Thus, there is indeed a great difference between aqueous organic solvents and neat water, as only the latter bears the properties associated with the structure of water, e.g. those accounted for by micellar and other interfacial phenomena. Moreover, the properties of mixtures of water with other solvents do not usually vary incrementally with the concentration of water. The addition of small amounts of water to organic solvents (5–10%) causes a profound change of solvent properties, which are immediately reflected in the reactivity. Further addition of water usually has a much smaller effect, if any.

Water may indeed play a very important role in so-called phosphine-less palladium catalytic systems. Indeed, in many such reactions, palladium is introduced in Pd(II) form, while the catalytic cycles are driven by Pd(0). Therefore, there must be some process that accounts for primary generation of Pd(0) from PdII to trigger the then self-sustained catalytic cycle. It is well known that Pd(II) salts (PdCl$_2$, and especially Pd(OAc)$_2$) are quite strong oxidants, but their oxidative properties are strongly dependent on the solvent used. Such simple palladium salts possess polymeric structures with bridged ligands. The inertness of simple palladium complexes in catalytic processes in organic solvents may be associated with this structural feature, as: (i) palladium atoms are blocked by the relatively inert bridged ligands; and (ii) the oxidative properties of Pd(II) are much less pronounced. The addition of water helps to destroy at least partially these bridged polymeric structures, owing to both partial ligand exchange with water, and hydrogen bonding making the ligands more labile and liable to exchange reactions. Besides, in water-containing solvents, Pd(II) possesses a sufficiently high oxidative power, which may help to trigger catalytic cycles just by oxidizing any electron-rich species in solution, and thus one of the roles of phosphine ligands no longer becomes needed.

5.1.3   Reactions in homogeneous aqueous solutions

The skill of meticulously drying organic solvents was always regarded as a prime virtue of any organometallic chemist. Thus, the idea that some organometallic reactions may be conducted in wet solvents, or even in water–organic mixtures with deliberately added water, might look like a shocking revelation to the majority of the organometallic community. Everyone can easily find a couple or more papers in which the addition of water brings no positive results, serves no clear goals, and is a pure tribute to the new fashion. However, there are indeed good examples of an essential or even huge effect of water on some catalytic processes. It must be stressed that currently there is no explanation of the nature of such effects. One of the most evident approaches may be the increase of solvent polarity and the change of specific solvating effects leading, for example, to the facilitation of ligand exchange and cleavage of unwanted ligands blocking the coordination sphere of the transition-metal catalyst. Many catalytic processes lead to the co-formation of halide ions, the increase in concentration of which may dynamically change the ligand environment of catalytic species, and induce catalyst poisoning. In this case the high polarity and hydrogen-bonding capability of water may help to remove halide ions and to keep the catalytic cycle alive. Similar effects of revitalizing heterogeneous catalysts by washing the active surface poisoned by reaction by-products are well known.

5.1.4   Reactions in neat water

This approach may be regarded as a particular variety of the above technique. But concerning the environmental requirements and possible technological perspectives, such reactions can justify the high expectations and meet the new standards of environmental and occupational safety of the chemical industry. The recycling of aqueous organic solvents is always a more complex task than the recycling of pure organic liquids. The addition of water cannot help to eliminate the hazards associated with organic solvents. Thus, the addition of water to organic solvents can be justifiable only if the positive effects (increase of selectivity or catalytic efficiency) are so high that the gains outweigh the increased cost of reaction mixture workup and solvent utilization.

Reactions in neat water can be feasible if and only if at least the major reagents are soluble in water. In the case when the efficiency of the catalytic process is sufficiently high, the process may run even if the solubility of the major reagent(s) is moderate or even quite poor. In such a case, the reagent that is poorly soluble in water shall form a separate phase serving as a feed-stock for the reaction, which actually takes place in the aqueous phase. This technique is widely used, and owing to its practical importance it is described below in the paragraph on reactions in biphasic systems under phase-separation conditions.

In general, reactions in which at least one of the reacting compounds is insoluble in water can be either homogenized by the addition of as much as possible of a suitable water-miscible cosolvent, or carried out in heterogeneous systems, using, as was noted above, (i) phase-transfer, (ii) phase-separation, or (iii) solubilization techniques.

(i) Phase transfer requires that the contact between the reagents, normally occupying different phases, be furnished by phase-transfer agents. Normally, phase transfer is applied to ionic reagents, forming amphiphilic ion pairs with the respective phase-transfer agents, and thus transferred from the aqueous phase to the organic phase (normal liquid–liquid phase transfer). There are, however, cases in which hydrophobic compounds are transferred to the aqueous phase by a highly hydrophilic phase-transfer agent, e.g. cyclodextrin forming a host–guest complex (reverse phase transfer). The operation of water-soluble catalysts in heterogeneous systems often implies that phase transfer can occur within the catalytic cycle, when hydrophilic and hydrophobic species are combined to form an amphiphilic intermediate. The latter case is extremely interesting, though it appears that it has never been explicitly investigated.

(ii) Phase separation is not the description of a distinct phenomenon (as phase transfer certainly is), but rather a utilitarian approach to the realization of homogeneous processes in a heterogeneous mode, thus allowing resolution of the major drawback of homogeneous catalysis, i.e. the problem of separation of catalyst from products. Despite an overwhelming number of advantages, homogeneous catalysis is still losing the battle to heterogeneous processes in industry. Indeed, the application of transition-metal catalysis in chemical technology is severely retarded because of the high cost of the catalysts, the most powerful of which are, by a whim of Nature, derivatives of the precious platinum-group metals. The problem is actually not the current cost itself, but rather the scarcity of natural sources of these elements. High demand for any such catalyst by industry may very soon result in the depletion of known sources, and thus to dramatic increase of cost, making the perspectives for such industry very uncertain.

Heterogenization of homogeneous catalyst can be done in a number of ways. The catalyst can be immobilized on a solid organic or inorganic support, by physical absorption or chemical bonding. Immobilization is a well-established method, having a brilliant history of outstanding achievements. Still, it has a major drawback, severely undermining its use in industrial catalysis. Immobilized catalysts often have quite meager efficiency and productivity due to a trivial reason – the catalytic centers are localized in space and have no self-mobility, and thus the rate of reactions is limited by diffusion control.

Aqueous catalysis brought forward an ingenious and amazingly simple solution: to combine the reaction itself and the workup of the reaction mixture, commonly done by shaking it with water in a separatory funnel. The reagents and the catalyst are contained in different liquid phases from the beginning. At the end, the catalyst and the products remain in different phases. The layer containing the catalyst can be reused, or at least processed to isolate the precious metal catalyst. An ideal description of such a liquid–liquid phase separation method leaves only a single question: 'How can the reaction occur if the reagents cannot meet?' The answer to this question is at the same time the end of the ideal scheme. The reactions in such systems can take place if and only if the reacting species can penetrate into the other phase, either because their own solubility is quite considerable, or because an implicit phase transfer is involved. Thus, actual phase-separation methods are always a compromise between separation (always associated with leaching of the catalyst into the organic phase) and reaction rate.

(iii) The third technique mentioned above is solubilization, or the enhancement of solubility due to the interaction with molecular aggregates. At first glance there is no big difference between solubilization and phase transfer, as both lead to the increase of apparent solubility, though by entirely different mechanisms. Phase transfer operates on a molecular level through the bonding between two molecules forming an amphiphilic complex. Solubilization requires the formation of molecular aggregates (micelles, microemulsions, swollen micelles, bilayers, vesicles), which can be treated as new phases or pseudo-phases. A common feature of all such aggregates is the formation of interfacial layers, within which the aqueous and organic phases interpenetrate each other, and which have intermediate properties. It is natural to assume that the reactions between hydrophilic and lipophilic reagents can occur within these interfacial layers. Thus, solubilization can be considered as a phenomenon that is opposite or complementary to phase transfer, as the former is based on the localization of reacting species in the interfacial layer, while the latter depends on the transport of reagents across interfacial boundaries. In a technical sense, solubilization, which eliminates the distinct separation of immiscible phases, is opposed by phase-separation techniques: the advantages of the former are the deficiencies of the latter, and vice versa.

The simplest realization of solubilization phenomena is the well-known micellar catalysis, and the most complex is the operation of biological membranes [1, 2]. Practical application of solubilization depends on the development of systems that allow processing on the preparative scale. Simple micelles can rarely satisfy this requirement, as the solubilizing ability of such systems is quite low. High solubilizing ability is a feature of well-balanced surfactant systems, such as microemulsions and related media, establishing a perfect interpenetration of immiscible reagents.

5.1.5   Water-soluble phosphine ligands

Among the derivatives of all the transition metals, palladium and rhodium complexes are undoubtedly the most powerful and versatile catalysts. As both classical catalysts – PdCl$_2$(PPh$_3$)$_2$ and Wilkinson complex RhCl(PPh$_3$)$_3$ – are strongly hydrophobic, the idea to study similar complexes with hydrophilic phosphines for catalytic processes in water-based media is quite natural. The hunt for new hydrophilic phosphines during about two decades brought dozens of examples of such ligands. As often happens, none of them seems able to match the popularity and performance of the first and most straightforward discovery, the sulfonated triphenylphosphine ligands – the so-called TPPMS (sodium salt of monosulfonated triphenylphosphine) and TPPTS (trisodium salt of trisulfonated triphenylphosphine).

Early results on the preparation and application of hydrophilic phosphines were reviewed in 1986 [3]. Though to date all conceivable approaches to hydrophilization of phosphine ligands had been tried, sulfonated ligands still occupy a privileged place, being not only the most numerous but undoubtedly the best-studied group of phosphines.

There are two basic approaches to sulfonated phosphines: (i) the sulfonation of known phosphines; and (ii) the assembly of a new ligand from already sulfonated and non-sulfonated parts.

Sulfonation of phosphines bearing phenyl rings attached directly to the phosphorus atom can be carried out only under quite harsh conditions by prolonged reaction of the initial ligand with oleum. The phosphorus atoms of phosphine are protonated by strong acids, and thus the aromatic ring is strongly deactivated toward substitution. Sulfonation is regioselective, and sulfo groups enter exclusively at the *meta* position to the phosphorus atom, though, as the addition of each new sulfo group has a small influence on the properties of other phenyl rings, sulfonation results in the formation of all possible molecules, with the number of sulfo groups varying from one to the total number of phenyl rings. The degree of sulfonation can be controlled only in two ways: (i) by introducing only one sulfo group in the reaction with less sulfonating agent than is needed by stoichiometry and subsequent easy separation of water-soluble sulfonated ligand from hydrophobic nonreacted precursor; and (ii) by exhaustive sulfonation by allowing excess of sulfonating agent under as severe conditions as can be endured by the ligand. The resulting ligand can be purified either by simple fractional crystallization, or by procedures already developed for the purification and separation of other important types of organic polysulfonates (dyes, tensides, etc.). It must be stressed that in most cases there is no definite evidence that the ligand thus produced is actually an individual molecule with strict stoichiometry. Purification of polysulfonated derivatives is an extremely sophisticated problem, as the undersulfonated derivatives have similar properties, and the compounds include variable amounts of water of crystallization, are insoluble in most organic solvents and are too readily soluble in aqueous media to allow for accurate recrystallization. As a result, a certain degree of care must be reserved when comparing results obtained with sulfonated ligands from different sources.

Several other factors, which are uncommon in organic chemistry and conventional methods of homogeneous catalysis, must be taken into consideration when interpreting the results obtained with sulfonated phosphines (as well as with some other water-soluble phosphines). The first is the effect of ionic strength. The presence of deliberately added ions in aqueous solution may affect the state of ionic ligands, their conformations and the complexation equilibria, which in turn may affect the reactivity and selectivity of catalytic systems [4]. There may also be a more subtle effect as the ligand itself may be the source of ionic strength effects in cases when phosphine ligand is added to reaction mixtures in large quantities. In some cross-coupling reactions, the loadings of palladium catalyst are 10–20 mol% or even higher, while the ligand may be added in two- to five-fold excess with respect to metal catalyst precursor, and each ligand may bear three or four ionic groups. In such systems the concentration of ionic species may be equal to or more than the concentration of substrate, which would be equivalent to the addition of an excess of electrolyte to the reaction mixture. A high concentration of electrolyte must have a strong influence on the conformations of some substrates, the partition of organic components between aqueous and organic layers in biphasic systems, etc. As far as we are aware, ionic effects coming from the ligand have never been explicitly noted or investigated.

The other important factor is the amphiphilicity of ligands and catalysts. Indeed, most water-soluble ligands and complexes consist of hydrophilic and hydrophobic parts, and thus must possess the abilities of partitioning between aqueous and organic phases, of aggregation at interfaces and of self-aggregation. Amphiphilicity has numerous implications, many of which are quite difficult to foresee. Below is only a brief list of the most evident ones:

(i) Amphiphilic compounds behave differently depending on whether the concentration of amphiphile is lower or higher than the specific concentration at which self-aggregation occurs. Aggregation should, for example, account for increased leaching of hydrophilic complex into the organic phase.

(ii) Aggregation depends on the presence of other types of species in solution. Charged species, and especially multicharged counter-ions, alter the properties of interfacial layers, and the solubilization ability of aggregates, which again should affect the complexation equilibria, and the rates and selectivity of catalytic reactions [5, 6].

(iii) The presence of amphiphilic compounds alters the properties of phases, e.g. by lowering the interfacial tension, which in turn results in the formation of emulsions. Emulsification introduces a lot of problems for the separation of phases in biphasic methods, and may eliminate altogether the advantages of phase-separation techniques.

The example in Figure 5.1 is illustrative of the new features introduced by aqueous chemistry with water-soluble ligands. The crystal structure of Pd(TPPMS)$_3$ complex is quite peculiar [7]. The molecules of complex are

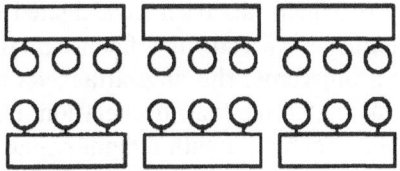

Figure 5.1 Crystal structure of the Pd(TPPMS)₃ complex. Circles represent ionic sulfonate groups and rectangles the remaining hydrophobic part of the molecule.

packed in a perfectly ordered layered way, where circles correspond to sulfonate ionic groups and the rectangle denotes the hydrophobic remainder of the molecule. Such structures are well known, being exactly what is called a bilayer [2], the main structural unit of a vast number of types of self-organized matter from the simplest vesicles to biological membranes. Bilayers are spontaneously formed by amphiphilic compounds, the aggregates of which have negative – or zero – curvature interfaces, because the area of ionic headgroups is roughly the same as or less than the footprint of the hydrophobic part being projected onto the interface (in contrast to simple amphiphiles like long-chain carboxylates, sulfates, sulfonates, etc., which form spherical micellar aggregates). Thus, it is evident that the nucleation and growth of a Pd(TPPMS)₃ crystal passes through spontaneous aggregation to a bilayered structure, which, like all other aggregates of amphiphilic compounds, is capable of solubilization of both water and hydrophilic species, and hydrophobic reagents. Again, as far as we know, this problem has never been explicitly investigated, though this phenomenon may have a strong influence on the behavior of reagents both in biphasic systems (bilayered structure formed at the interface functions as a gate between immiscible phases, thus facilitating reactions) and in homogeneous solutions, in which in some cases the concentration of catalyst is so high that it may form aggregates solubilizing or encapsulating other reagents, and thus either facilitate or retard the reactions.

Sulfonation can be applied to prepare hydrophilic derivatives of those chiral phosphines (in which chirality is defined by the saturated hydrocarbon skeleton) which can endure the harsh conditions of this reaction. Thus, sulfonation of the following chiral phosphines can be done by 20% oleum at 0°C to give mixtures of di-, tri- and tetrasulfonated products in moderate yields:

PPh₂ PPh₂	Ph₂P⠀⠀PPh₂	Ph₂P⠀⠀PPh₂	Ph₂P⠀⠀PPh₂
(S,S)-cyclobutanediop	(S,S)-BDPP	(S,S)-chiraphos	(R)-prophos

Chiraphos and prophos ligands are structurally similar to DPPE, the sulfonation chemistry of which has been investigated in detail. It must be noted that sulfonation must have a strong influence on the ligating abilities of these ligands, as sulfo groups enter the rings attached directly to phosphorus atoms, with a possible outcome that dechelation becomes easier. High enantiomeric purity can be achieved with ligands capable of keeping a chiral environment of reagents at the metal atom; the facilitation of any disorganized movements of the ligands would have a negative effect on the efficiency of asymmetric induction. In this respect sulfonated ligands of this type can be expected to give results inferior to those obtained with the best hydrophobic chiral ligands.

Bis(diphenylphosphino)ethane (DPPE) was the only one of the widely used series of chelating 1,ω-bis(diphenylphosphino)alkanes that deserved a special study on sulfonation. The results may be regarded as largely discouraging. The sulfonation of DPPE by 20% oleum at 0°C leads to a complex mixture of products, including not only sulfonated phosphines but also phosphine oxides. The tetrasulfonated ligand DPPETS can be separated in a reasonably pure form by recrystallization from aqueous MeOH in about 30% yield. The ligand was shown to form water-soluble complexes with Rh, Ni, Pt and Pd. However, the results obtained for hydroformylation with the rhodium complex of this ligand were inferior even in comparison with the standard and much less expensive TPPTS [8].

Phosphines bearing electron-withdrawing substituents in the benzene ring possess enhanced π-acidities, which are believed to account for better selectivity in hydroformylation reactions. However, direct sulfonation of such strongly deactivated ligands is a serious challenge. There is only one report dealing with the sulfonation of such a ligand – tris(4-fluorophenyl)phosphine (TFPP). Sulfonation by a huge excess of 25% oleum during 17 days gave only disulfonated phosphine with a negligible amount of trisulfonate. The mixture thus obtained was claimed to be a better ligand in rhodium-catalyzed hydroformylation in a biphasic system, giving better regioselectivity than TPPTS or Ph_3P [9].

Direct sulfonation of ligands in which phenyl rings are not attached directly to the phosphorus atom, but are separated by saturated hydrocarbon chains or other groups, can be run under much milder and controllable conditions to afford exhaustively sulfonated ligands with all phenyl rings bearing sulfo groups at *para* positions. This approach was applied to a large number of interesting ligands to give some highly stimulating results, which is not surprising as in this approach the sulfonated phosphine ligand can be regarded as a truly bifunctional molecule. The part that accounts for the hydrophilicity and the nearest phosphorus atom environment are separated by relatively inert links, and have a negligible interaction. In this respect this method is close to the convergent approach in which a hydrophilic ligand is built by modification of a well-known phosphine by attaching hydrophilic moieties of any conceivable nature.

A series of highly basic hydrophilic phosphines was obtained by sulfonation of readily available tris(ω-phenylalkyl)phosphines $P[(CH_2)_x(C_6H_5)]_3$, where $x = 1, 2, 3$ and 6:

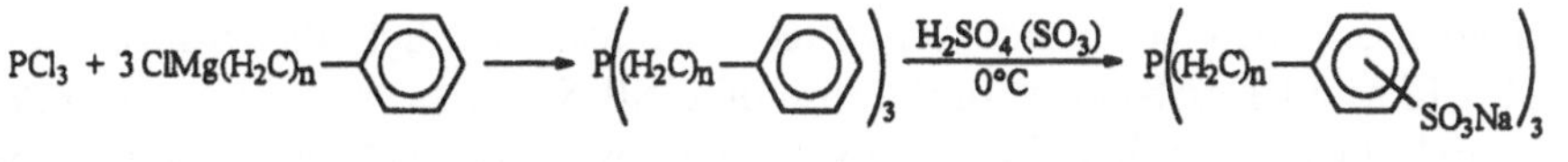

$$PCl_3 + 3\,ClMg(H_2C)_n\!-\!\bigcirc \longrightarrow P\!\left((H_2C)_n\!-\!\bigcirc\right)_3 \xrightarrow[0°C]{H_2SO_4\,(SO_3)} P\!\left((H_2C)_n\!-\!\bigcirc_{SO_3Na}\right)_3$$

Owing to the deactivation of the benzene ring by protonated phosphorus, even separated by methylene groups, it is necessary to use 20% oleum for $n = 1$ or 2. For longer hydrocarbon spacers, no retardation is observed and the phosphines are quantitatively sulfonated by sulfuric acid at 0°C as easily as alkylbenzenes. Sulfo groups enter both *para* and *ortho* positions, and, though the partial rate factor for *para* substitution is larger, the *ortho,para,para* isomer is the major product of sulfonation just because of the two-fold statistical correction for *ortho* positions. However, it is the *para,para,para* isomer that is the most easily obtained in pure form by simple recrystallization of the crude product. The ligands thus obtained can be regarded as a hydrophilized tributylphosphine. For $n = 3$ and $n = 6$ the basicity of these ligands is almost the same as the basicity of PBu_3, while the steric bulk is only slightly larger, which allows these phosphines to be considered as possible ligands for cobalt-catalyzed hydroformylation [10].

The sulfonation of bis(8-phenyloctyl)menthylphosphine gave the corresponding sulfonated chiral ligand [11]:

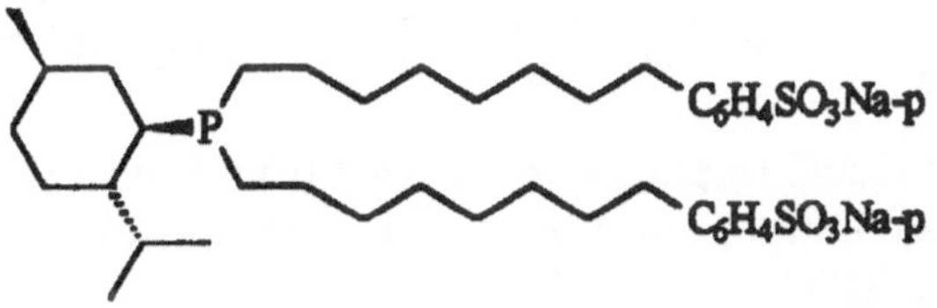

In this case the sulfonated groups are located far from the phosphorus atom, but monodentate ligands with chirality carried by the menthyl residue are notoriously inefficient in asymmetric syntheses.

Both of the most famous *trans*-chelating phosphines, BISBI [12] and the similar 2,2'-bis(diphenylphosphinomethyl)-1-1'-binaphthalene [13], can be sulfonated:

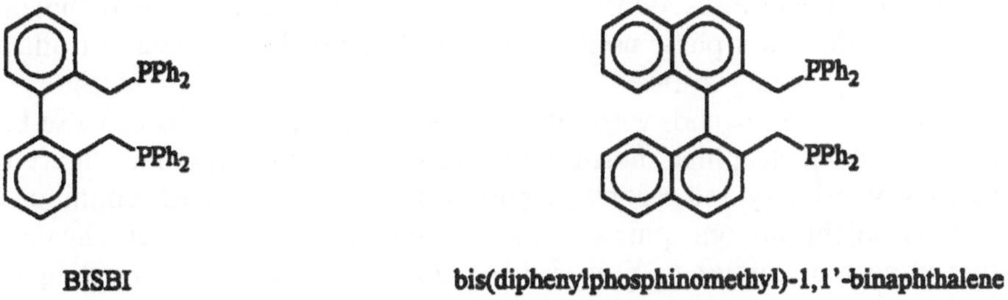

BISBI   **bis(diphenylphosphinomethyl)-1,1'-binaphthalene**

The sulfonation must be performed with oleum containing up to 65% SO_3, as the deactivating effect of positive phosphorus is conducted through CH_2 links to the aromatic rings not directly bonded to the phosphorus atom. Under such conditions, sulfonation gives mixtures of polysulfonated products, which were not separated but used as ligands (respectively, BISBIS and BINAS) as is. In the case of BISBIS, it was shown that even the phenyl rings of diphenylphosphino groups were partially sulfonated. The resulting ligands showed very high performance in rhodium-catalyzed hydroformylation in biphasic media. The binaphthyl derivative ligand is similar to the famous chiral ligand BINAP, which can also be sulfonated and serves to perform asymmetric hydrogenation in aqueous systems (*vide infra*).

A number of interesting ligands, derivatives of 1-phosphanorbornene and 1-phosphanorbornadiene, were obtained as Diels–Alder adducts of 2*H*-phospholes and various dienophiles. When such adducts contain phenyl substituents, sulfonation of such compounds can be performed without the destruction of the cage structure to give highly hydrophilic ligands. Thus, sulfonation of the adduct of dimethylphenylphosphole and diphenylacetylene gave a pure trisulfonated ligand NORBOS with very high activity in hydroformylation in biphasic media, more reactive than BISBIS, and inferior only to BINAS [13].

The tripodal ligand sulphos was obtained from benzyltris(chloromethyl)-methane by sulfonation, and subsequent reaction with the diphenylphosphide anion in DMSO:

This ligand forms electroneutral zwitterionic rhodium complexes, which are not soluble in water but are very soluble in alcohol–water mixtures [14], which is required for reactions in biphasic systems with aqueous methanol or ethanol as the lower phase used to enhance the solubility of olefins or other moderately polar substrates.

Pure sulfonated ligands with sulfo groups in the aromatic ring can also be produced by nucleophilic substitution. The substitution of aromatic fluorine by PH_3 or primary and secondary phosphines under superbasic conditions makes available not only pure sulfonated tertiary phosphines but also ligands that altogether cannot be prepared by sulfonation, such as, for example, secondary phosphines or highly sulfonated phosphines with two sulfo groups per aryl ring:

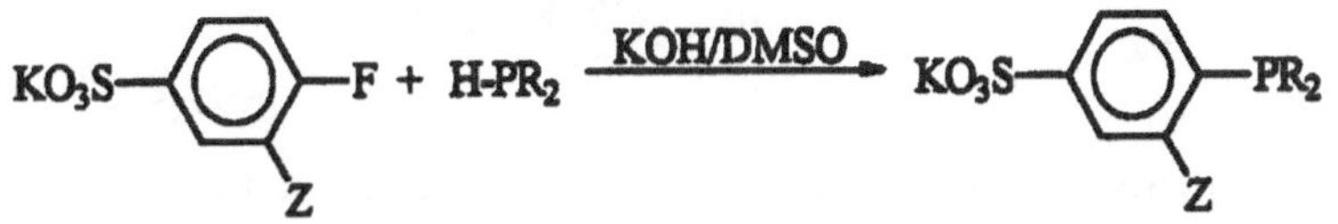

where $Z = H$ or SO_3K.

The introduction of sulfo groups in *para* positions had a much smaller influence on the properties of the ligand than *meta*-sulfo groups in phosphines obtained by direct sulfonation. X-ray structural analysis of the pure potassium salt of tris(*p*-sulfophenyl)phosphine (crystallized as P(*p*-$C_6H_4SO_3K)_3 \cdot KCl \cdot 0.5H_2O$) prepared by this method showed that the geometry of this ligand is almost identical to that of unsubstituted Ph_3P [15].

Besides sulfo groups, a lot of other hydrophilic moieties were tried for modification of phosphine ligands. In some cases the modification pursued several goals, e.g. to impart not only hydrophilicity but also phase-transfer properties, as in the case of polyether- or crown ether-substituted phosphines [16].

A general approach to the preparation of various hydrophilic phosphines was developed to involve free-radical addition of diphenylphosphine, phenylphosphine or bis(phenylphosphino)ethylene to unsaturated molecules bearing carboxylic groups or other substituents that can be transformed into groups rendering the molecule water-soluble [17, 18]. This method allows one to attach the diphenylphosphine group to a vast variety of hydrophilic moieties, including polyols, polyethyleneglycols and their monoethers, and sugars, by first attaching the allyl group, with subsequent free-radical addition of Ph_2PH or $PhPH_2$ induced either by conventional free-radical initiators such as AIBN or by U.V. irradiation.

Another highly hydrophilic ligand bearing two carboxylate groups can be obtained from the readily available dichloromaleic anhydride:

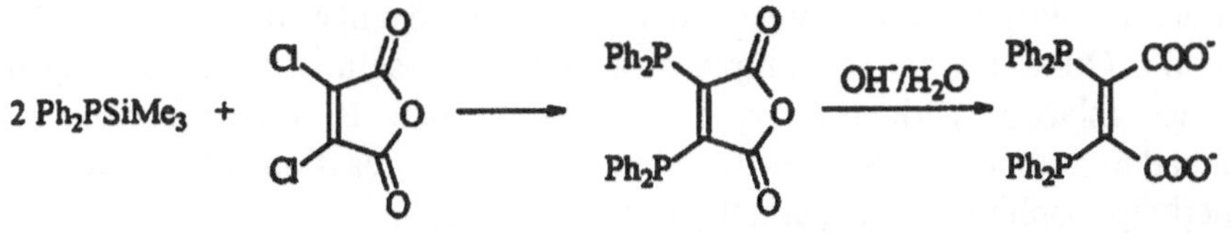

The reactivity of the double bond in this molecule toward water would, however, interfere with its application as a ligand for aqueous transition-metal catalysis [19].

Phosphanorbornene and norbornadiene ligands with phosphonate or phosphonamide groups can be prepared by the addition of substituted phosphole to ethynylphosphonates and ethynylphosphonamides, which gave both 5- and 6-substituted derivatives:

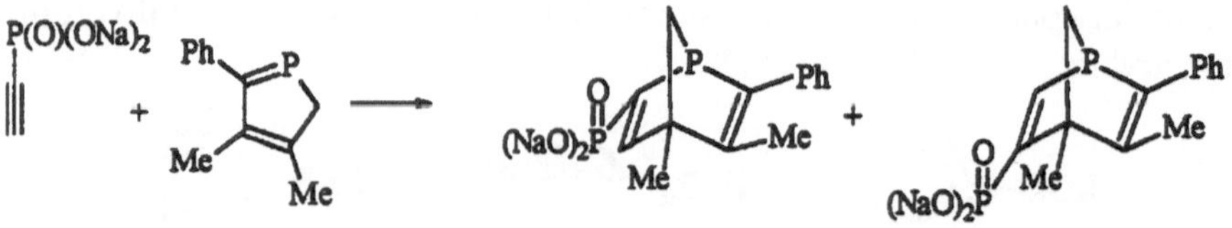

A minor product of this Diels–Alder addition turned out to be a highly hydrophilic ligand, giving good results in rhodium-catalyzed hydrogenation of the model substrate (Z)-α-(N-acetamido)cinnamic acid. On the other hand, the major 6-substituted adducts were much less efficient due to additional chelation of the rhodium atom by the phosphonate oxygen [20].

Another ligand of this series obtained as a hydrolyzed adduct with maleic anhydride showed very high solubility in water and therefore negligible leaching into organic solvents, unmatched by many sulfonated ligands [21]:

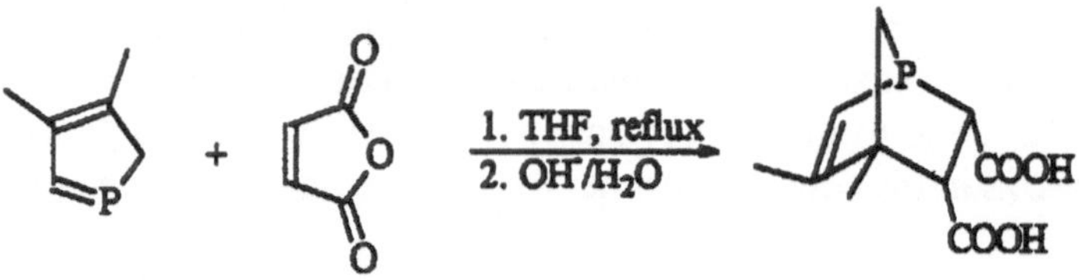

Other examples of recent work on water-soluble phosphine ligands include TPPMP (triphenylphosphine monophosphonate), as the first example of a triarylphosphine ligand bearing an anionic phosphonate group [22]; and a series of derivatives of triphenylphosphine with one, two, or three phenyls bearing anionic carboxylate groups [23].

Yet another approach to the hydrophilization of phosphine ligands is the introduction of one or more CH_2OH groups by reaction of P–H compounds with formaldehyde. Data on the possible use of such ligands in catalysis are quite controversial. A recent example of one such ligand, 1,2-bis[bis-(hydroxymethyl)phosphino]ethane can be found in reference [24]. The reaction of the parent phosphine with CH_2O gives the simplest of such ligands, $P(CH_2OH)_3$. Recently this molecule has found a new application in aqueous catalysis by opening a convenient route to a new and very appealing ligand. Quite a notable example of a new water-soluble ligand having no direct relation with the types already studied is 1,3,5-triaza-7-phosphaadamantane (PTA), easily obtained by a reaction of trihydroxymethylphosphine with hexamethylenetetramine [25]:

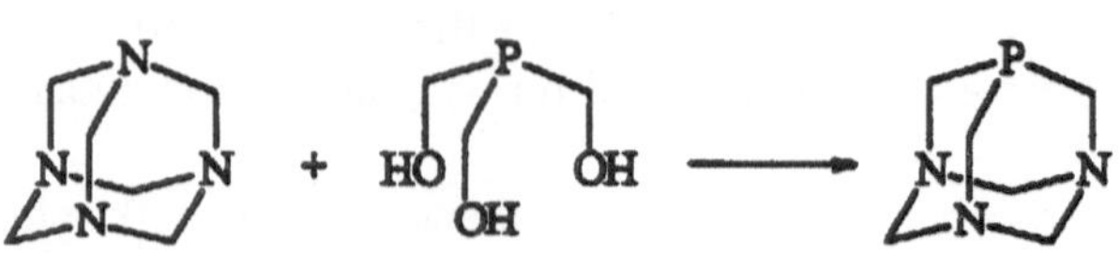

The ligand itself and its complexes (e.g. with Ru, Rh, etc.) are readily soluble in water. Unlike sulfonated phosphine ligands, PTA is stable towards oxidation by air, but nonetheless, being a trialkylphosphine, it possesses an enhanced basicity to form more stable complexes with metals. Besides, owing to much smaller steric bulk than that of the standard sulfonated phosphines TPPMS or TPPTS, more ligands may fit into the coordination shell to form complexes, which are structurally similar to those formed by normal triaryl- or trialkylphosphines. This property may be decisive in controlling the selectivity of catalytic processes, in which early or too easy de-ligation often triggers a variety of side-reaction pathways giving products other than those expected.

Phosphines containing amino groups or nitrogen-containing heterocyclic residues were suggested to be used for the design of a new class of amphiphilic catalysts with changeable hydrophile–lipophile balance. The basic idea of such an approach is that the catalyst must be reasonably soluble in organic solvents to carry out the reactions in homogeneous solutions, where there are no solubility restrictions severely limiting the efficiency of processes in biphasic systems. Protonation of nitrogen atoms can be done only after the reaction to extract the protonated complex into water for separation and recycling.

Amphiphilic phosphines with pendent amino groups were obtained, for example, by stepwise alkylation of PH_3 with or without modification by long alkyl chains either by quaternization of dimethylamino groups or by free-radical addition of primary or secondary phosphines to long-chain olefins [26, 27]:

$$PH_3 \xrightarrow[\text{KOH, DMSO}]{\text{ClCH}_2\text{CH}_2\text{NMe}_2} H_2P\!\!\diagup\!\!\diagdown\!\!\text{NMe}_2$$

$$\xrightarrow[\text{KOH, DMSO}]{\text{ClCH}_2\text{CH}_2\text{NMe}_2} HP(\text{CH}_2\text{CH}_2\text{NMe}_2)_2 \xrightarrow[\text{2. ClCH}_2\text{CH}_2\text{NMe}_2]{\text{1. } n\text{-BuLi}} P(\text{CH}_2\text{CH}_2\text{NMe}_2)_3$$

The same approach was applied to obtain primary phosphines with longer chains R_2N–$(CH_2)_m$–PH_2 ($R = Me$, n-Bu, etc.; $m = 2, 3, 6, 10, 11$). The phosphines can be selectively quaternized at the N atom using phase-transfer alkylation to afford cationic primary phosphines, $[R'R_2N$–$(CH_2)_m$–$PH_2]^+I^-$, which are stable to air, and possess strong amphiphilic properties with hydrophile–lipophile balance dependent on the chain length. Longer-chain compounds have a strong affinity to hydrophobic phases, but may be extracted back to water by acidification and protonation of the phosphorus atom [28]. A similar class of amphiphilic phosphines containing quaternized phosphorus, $Ph_2P(CH_2)_nPMe_3^+$ ($n = 2, 3, 6, 10$), was studied as ligands for iron complexes [29].

A number of ligands with controllable amphiphilicity can be obtained by substituting some or all of the benzene rings by pyridyl or 4-aminophenyl residues in common ligands, e.g. in triphenylphosphine itself [30, 31]. This approach was applied in a recent attempt to obtain a series of new ligands combining the outstanding performance of the *trans*-chelating ligand BISBI with controllable amphiphilicity [32]:

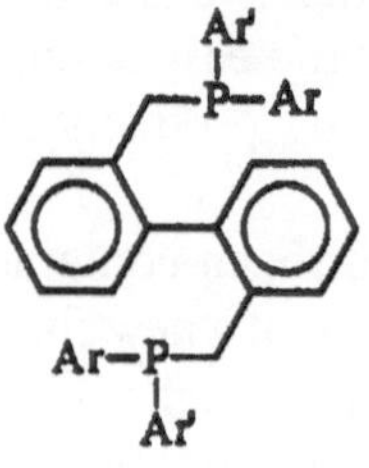

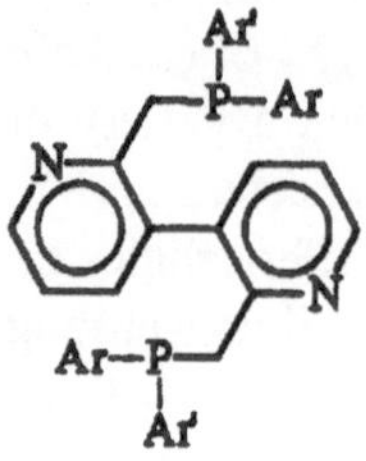

| Ar, Ar' = Ph | BISBI | Ar, Ar' = Ph | L2 |

| Ar = Ph; Ar' = 3-Py | L1 | Ar = Ph; Ar' = 3-Py | L3 |

Ar = Ph; Ar' =4-Et$_2$NCH$_2$C$_6$H$_4$ L4

The ligands thus produced can be used for rhodium-catalyzed hydro-formylation in toluene solution, and can be recovered by extraction with aqueous acids. However, only the ligand **L4** bearing diethylaminomethyl groups could be reused for re-extraction into the fresh organic phase. All other ligands (**L1, L2, L3**) with pyridyl fragments gave complexes that decomposed during this procedure.

Phosphines can be tailored to hydrophilic polymers. Two such poly-ligands were prepared. The first was obtained by attaching the diphenylphos-phinobenzyl group to polyacrylic acid to form a linear polymer with high hydrophilicity due to the residual carboxylic groups (the ratio of free carboxylic groups to modified ones was kept as 5:1):

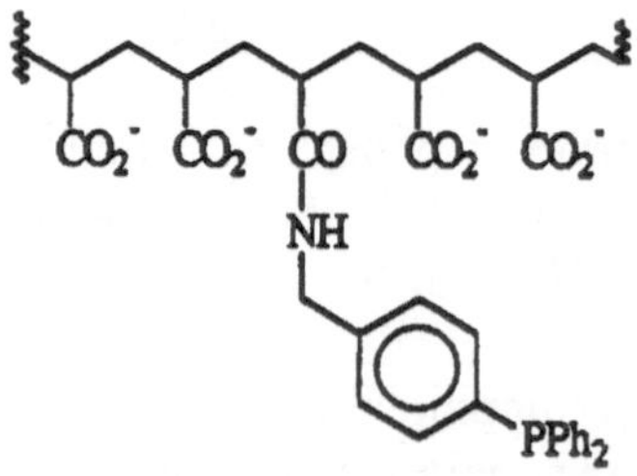

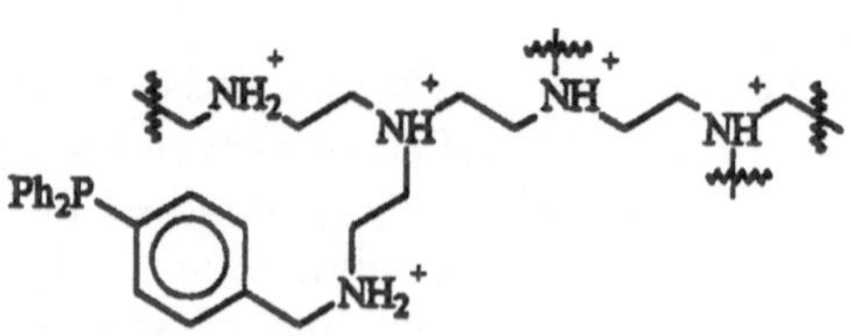

In the second poly-ligand, the same phosphorus-containing group was attached to commercial polyethyleneimine, which is a poorly characterized crosslinked polymer containing variable amounts of tertiary, secondary and primary amino groups. [The structure given here is not the same as given in the original reference [33], but is more compatible with the material produced and sold under the name of polyethyleneimine.] The method chosen for modification – the formation of imine with (diphenylphosphino)benzalde-hyde and subsequent reduction with $NaBH_4$ – allowed the phosphorus-containing group to be attached only to primary amino groups. As the resulting poly-ligand was made hydrophilic by protonation, it possessed a high acidity. The catalytic potential of both ligands was not described, except their ability to form cobalt complexes with $Co_2(CO)_8$, which can be used in, for example, hydroformylation [33].

In the following sections we shall discuss research on transition-metal catalysis in which water played a significant role. Almost all of the reactions treated below deal with palladium, rhodium or ruthenium catalysis. Examples of catalytic reactions involving water may be found for almost any of the transition metals, and a comprehensive survey cannot fit into the format of this book. Also, we have avoided the discussion of reactions involving liquid–liquid phase-transfer catalysis (PTC). However, we believe that the chemistry discussed below is quite representative of the approaches, problems and trends of the reviewed area.

5.2 Palladium-catalyzed cross-coupling reactions

5.2.1 Cross-coupling with organoboron compounds

Cross-coupling reactions of boronic acids and other organoboron compounds with organic halides (the Suzuki reaction) are now considered as one of the most important synthetic methods for $C(sp^2)$–$C(sp^2)$ bond forma-tion. Unlike other cross-coupling reactions between organic halides and organometallic compounds occurring due to transmetallation, the reaction with boronic acids and other organoboron compounds is specific as it requires the presence of bases. The role of base is most probably to form an electron-rich intermediate with the tetracoordinated boron atom, which is more reactive in the reaction with Pd(II) complexes than is the initial tricoor-dinated organoboron compound. Additionally, as organoboron compounds are often quite stable to protolytic decomposition by water, it is evident that the Suzuki reaction is the most inviting case for the application of water. It is indeed so, though in earlier research the use of water was to a certain extent unintentional. In the traditional version, this method often involved water, which was used to dissolve the inorganic base before its introduction into the reaction mixture, so that the reaction was actually run either in aqueous organic solvent (such as dimethoxyethane, ethanol, diglyme, etc.) or in a biphasic system (PhH–water, etc.), though there are a considerable number of examples of the non-aqueous technique using, for example, sodium ethylate in ethanol.

In the classical method often used in most early papers, the reaction was carried out in refluxing benzene in contact with an aqueous solution of the base. This technique is highly interesting as it actually employs phase transfer. Indeed, it is well known that such bases as hydroxide or carbonate are so highly hydrophilic that it is impossible to transfer these ions to the nonpolar organic phase. The hydroxide ion is believed not to cross the phase boundary, and the biphasic reactions involving OH⁻ ion are commonly believed to take place at the interface. A very high concentration of alkali is usually used to overcome the limitations imposed by concentration and mass-transfer factors. In the Suzuki reaction, the actual mechanism of the action of base may be different (Figure 5.2). Indeed, boronic acids in the presence of strong bases must form the respective boronate salts, which consist of a hydrophilic ionic boronate group and an organic and usually hydrophobic moiety, and thus are definitely amphiphilic. Thus, boronates themselves must function as the phase-transfer agent, while the borate and halide ions formed in the transmetallation step are continuously removed from the organic layer into the aqueous phase.

While in most papers utilizing this technique the content of water in the system was quite small, there were several examples published in which the content of water in the solvent was much higher. For example, the preparation of hydrophilic polyphenylene by Suzuki reaction in the presence of water-soluble phosphine ligands was carried out in aqueous DMF with up to 70% of water (v/v) [34] (Scheme 5.1). Sulfonated triphenylphosphine ligands were used to carry out the Suzuki reaction in aqueous solvents (H_2O–MeCN, 50:50; H_2O–MeOH–PhH, 70:15:15; H_2O–EtOH, 60:40; H_2O–MeOH, 70:30) [7] in the presence of inorganic base (Na_2CO_3). The reaction with the water-soluble salt of p-bromophenylacetic acid was run in neat water, though with a very high amount of palladium catalyst.

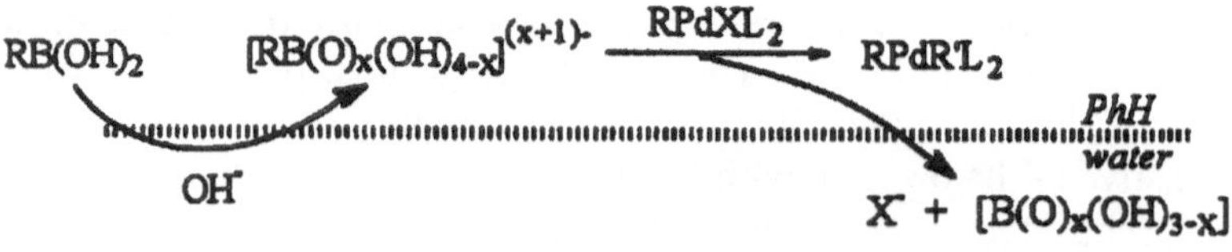

Figure 5.2 The mechanism of the Suzuki reaction.

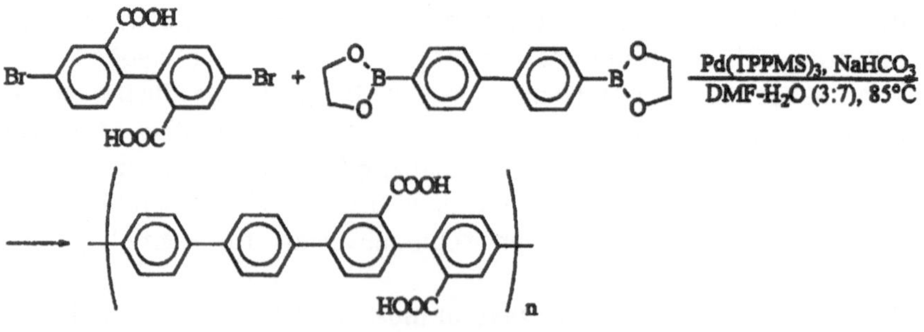

Scheme 5.1

The use of anhydrous organic solvents in the Suzuki reaction brings no noticeable advantages. On the other hand, in this case it becomes necessary either to use organic bases like Et_3N or to add phase-transfer salts and carry out the reaction in solid–liquid PTC mode. In many cases, such as, for example, in the reactions of alkenylboranes with bromoalkenes and bromoalkynes leading to dienes or enynes [35], no essential difference was noted between organic base in organic solvent versus inorganic base in aqueous solvent. An extensive study of the influence of various factors, including solvent effects, the nature of the base, etc., was carried out for the cross-coupling with B-alkyl-9-borabicyclo[3.3.1]nonanes catalyzed by $PdCl_2(DPPF)$ or $Pd(PPh_3)_4$:

$$RX + Alk-B\diagdown \xrightarrow{PdCl_2(DPPF), base} Alk-R$$

The reactions were carried out in anhydrous solvents such as THF and DMF, in aqueous mixtures such as $THF–H_2O$ (5:1), and in biphasic systems such as $PhH–H_2O$, etc., with a wide selection of bases, such as NaOH, TlOH, MeONa, K_2CO_3, K_3PO_4, etc., with invariable success giving high yields of target products [36]. Again, little difference between aqueous and non-aqueous techniques could be noticed.

There are, however, some cases in which the aqueous technique can lead to the formation of side-products, as e.g. in arylation of 3-bromochromone by sterically hindered mesitylboronic acid, which is complicated by protonolysis [37] when the reaction is carried out in aqueous solvent in the presence of Na_2CO_3. This reaction can be made selective if it is run without water with Tl_2CO_3 as a solid base:

$$+ ArB(OH)_2 \xrightarrow[PhH]{PdL_4, Tl_2CO_3}$$

Similarly, in the reaction of methyl 5-bromonicotinate with sterically hindered o-tolylboronic acid, the use of the biphasic technique with Na_2CO_3 as a base gave poor results in comparison with the reaction in anhydrous DMF in the presence of Et_3N [38]:

$$MeOOC \quad \text{Br} + \text{B(OH)}_2 \text{Me} \xrightarrow{Pd(PPh_3)_4}$$

Na_2CO_3, PhH - H_2O	12%
Et_3N, DMF	81%

The use of aqueous TlOH often allows one to lower the temperature, even in the case of cross-coupling with sterically hindered reagents. Thus, the reactions of aryl iodides with mesitylboronic acid can be performed at room temperature in dimethylacetamide with 10% aqueous TlOH giving a high yield of cross-coupling product [39]:

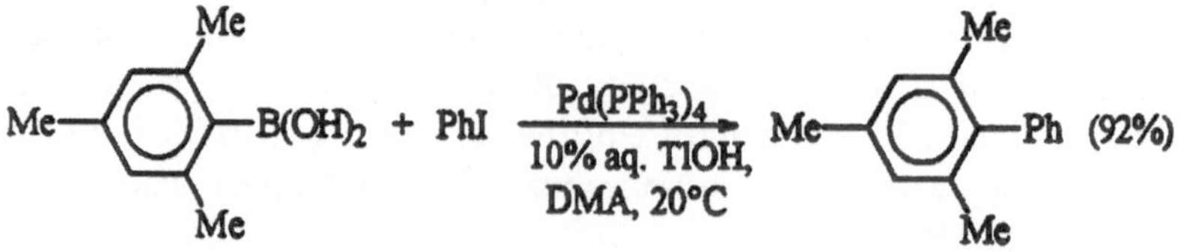

Dramatic acceleration by aqueous TlOH was also described for cross-coupling between vinylboronic acids and vinyl halides giving dienes. Thus, under comparable conditions the reaction in the presence of TlOH was 1000 times faster than the reaction in the presence of KOH. Such influence of the thallium cation was ascribed to assistance in the removal of the halide ligand from the intermediate $RPdXL_2$ formed at the oxidative addition step of the catalytic cycle due to the formation of insoluble thallium halide, which may enhance the reactivity of this intermediate at the transmetallation step. Alternatively, it may be argued that thallium boronate can bind to the $RPdXL_2$ intermediate due to Tl-halide coordination to facilitate ligand exchange:

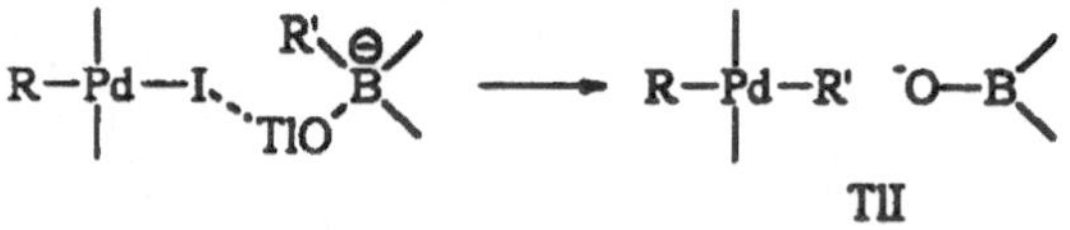

It should be noted that cross-coupling in the presence of phosphine complexes of palladium usually requires high amounts of catalyst, with initial loadings of 25–30 mol% not being uncommon. An entirely new approach to the Suzuki reaction is phosphine-free palladium catalysis. The use of palladium catalysts without the addition of phosphine ligands for cross-coupling with organoboron compounds in aqueous media opened a new chapter in the story of this powerful synthetic method. This approach allows catalyst efficiency to be dramatically increased, and the reaction to be performed under milder conditions.

The reaction of arylboronic acids with water-soluble organic halides can be performed at room temperature in the presence of simple palladium salts ($PdCl_2$, $Pd(OAC)_2$, etc.), and inorganic bases (NaOH, Na_2CO_3, K_2CO_3, K_3PO_4, etc.):

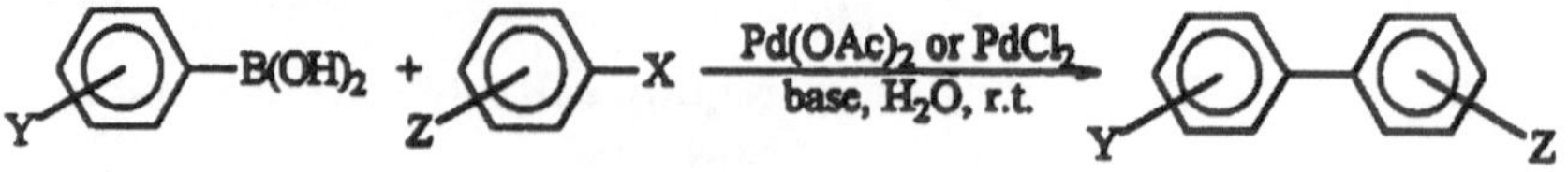

where X = Br, I; Y = H, p-F, p-(4-n-amylcyclohexyl); Z = m-,p-OH, o-,m-,p-COOH. The yields of products are in most cases near to quantitative [40].

The reaction was successfully applied to the preparation of 5-arylsalicylic acids:

where Y = H, F.

Esters of arylboronic acids can be used in place of free acids. The reactivity of the catalytic system is so high that both aryl iodides and bromides react with comparable ease, though the substrates used bear strong electron-donor groups, and thus must be less reactive in oxidative addition. In many respects this process poses a lot of questions. It is not clear what is the primary reducing agent for Pd(II), which is so efficient even at room temperature. It seems that the only reasonable candidate for this role is boronic acid itself, or rather the electron-rich anionic boronates formed in water in the presence of hydroxide ions. Anyway, it is clear that there are no good ligands for Pd(0) in the system, which may form stable complexes and reduce the reactivity of the catalyst. Thus all Pd(0) is efficiently trapped by oxidative addition to aryl halide, and deactivation due to clustering of non-ligated palladium species and further nucleation leading to growth of inactive large metal particles do not occur.

The reaction with water-insoluble aryl halides can also be run in neat water, though in this case the use of aqueous DMF brings better results, as e.g. for cross-coupling with 3-bromopyridines. However, in this case elevated temperatures are required to achieve high yields:

where Z = NO_2, NH_2, EtO.

The addition of phosphine ligands often leads to a substantial drop of catalytic efficiency, and even to the formation of by-products [41, 42]. In the case when arylboronic acid is sterically hindered, the reaction in organic solvents in the presence of Ph_3P may lead to the transfer of the phenyl group from phosphine.

Inhibition of the Suzuki reaction by phosphine ligands was also noted when catalyst precursors were either $Pd_2(DBA)_3 \cdot PhH$ or $[(\eta^3\text{-allyl})PdCl]_2$. In the case when phosphines were absent from the reaction mixtures, higher turnover numbers were observed at milder conditions. The reactions were also accelerated at higher pH values, which is consistent with the hypothesis that the reactive form of organoboron compound is the anionic tetrahedral complex [43].

Tetraarylborate salts, and in the first place readily available sodium tetraphenylborate, can be used in cross-coupling with aromatic halides. All four aryl groups are transferred if the reaction is carried out in aqueous media in the presence of inorganic bases, in sharp contrast with the process in anhydrous systems, in which it is usually possible to use only one aryl group.

Water-soluble aryl halides react readily in neat water, giving the respective biaryl in high yields at room temperature. Water-insoluble aryl halides can also be made to react, though in this case the addition of an organic cosolvent like DMF or acetone helps to accelerate the process [44]:

$$Ph_4BNa \quad + \quad \overset{Z}{\underset{Z}{\bigcirc}}-X \quad \xrightarrow[\text{base, H}_2\text{O, r.t.}]{Pd(OAc)_2 \text{ or } PdCl_2} \quad Ph-\overset{}{\underset{Z}{\bigcirc}}$$

where $X = Br$, I; $Z = m\text{-OH}$, $p\text{-OH}$, $m\text{-COOH}$, $p\text{-COOH}$, $p\text{-CHO}$, $p\text{-NO}_2$, $m\text{-NO}_2$; and also $ArBr = 5$-bromosalicylic acid, 5-bromo-2-methylbenzothiazole, etc.

As the possibility to use all four phenyl groups depends on the presence of inorganic base in aqueous solution, in which all such bases may be represented by the hydroxide anion, the reactive organoboron species must be tetrahedral borate anions, The stepwise stripping of phenyl groups from boron converges at the fourth turn to the above described reaction of phenylboronic acid (the state and composition of anionic organoboron species should depend on water content and pH value for a given reaction mixture). This is shown in Scheme 5.2.

As the reactivity of aryl bromides is lower than that of aryl iodides, stepwise substitution of halogen can be achieved:

$$\overset{Br}{\underset{I}{\bigcirc}} + Ph_4BNa \quad \xrightarrow[\text{H}_2\text{O, r.t.}]{PdX_2, \text{ base}} \quad \overset{Br}{\underset{Ph}{\bigcirc}} \longrightarrow \overset{Ph}{\underset{Ph}{\bigcirc}}$$

The high reactivity of this catalytic system can be further illustrated by the

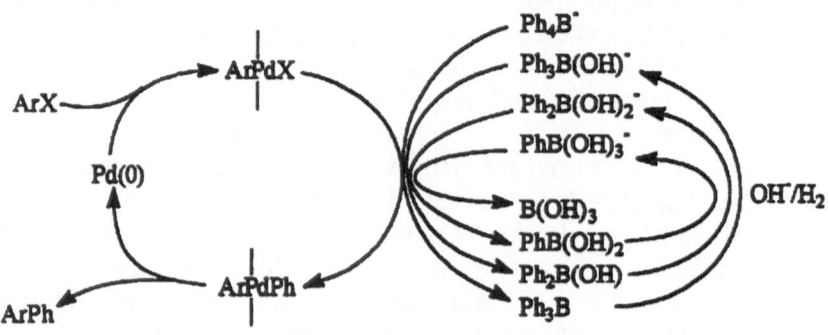

Scheme 5.2

following. Palladium black was found to catalyze the reaction of phenyl-boronic acid with water-soluble aryl iodides at room temperature:

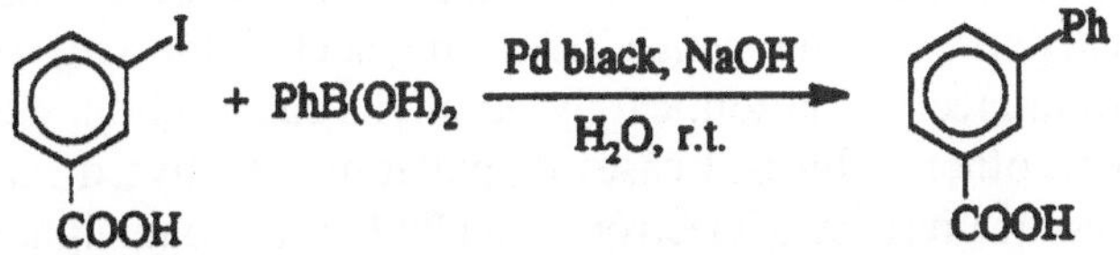

Reactions with sodium tetraphenylborate showed very high catalytic effi-ciency. So, for example, the reaction of m-bromobenzoic acid gives a quanti-tative yield of phenylated product even if the initial loading of $PdCl_2$ is as low as 0.0004 mol%, which corresponds to a turnover number of 250 000:

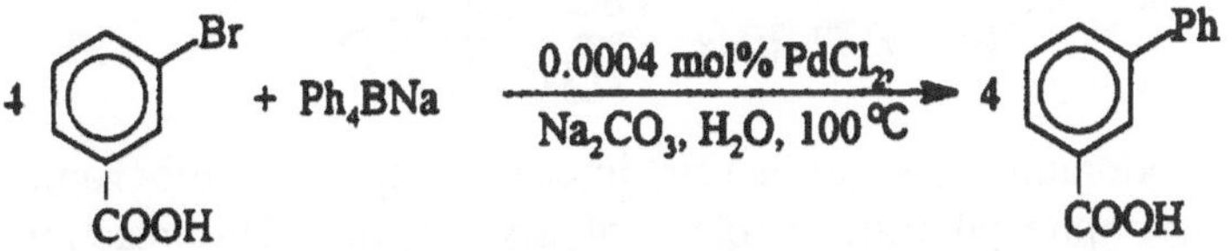

The efficiency of this catalytic system is so high that some water-soluble aryl chlorides, such as p-chlorobenzoic acid, can react, though only at reflux:

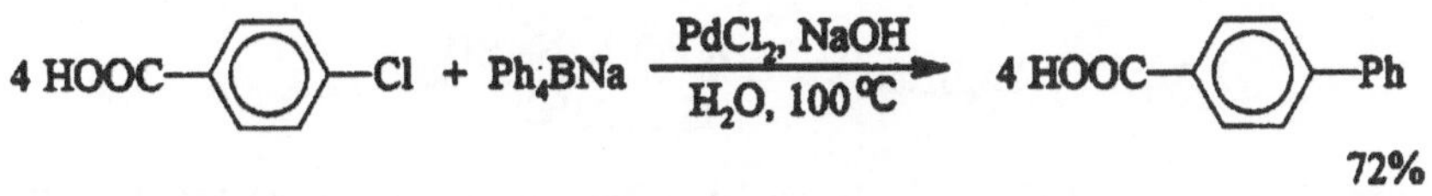

72%

The influence of various factors is very clearly seen for this reaction, as in the reactions with aryl iodides or bromides the reactivity is so high that all positive and negative effects are leveled and are practically not important. For example, here the reaction is faster at higher pH values, as the use of Na_2CO_3 instead of NaOH leads to a marked decrease of yield. Both hydrophobic (DPPF) and hydrophilic (TPPMS) phosphines inhibit the reaction. It is quite interesting that the yields of p-phenylbenzoic acid are higher and approach 95% in the presence of chromate ions. The role of the latter is possibly to prevent early deactivation of Pd(0) due to aggregation and formation of Pd black, as in this case the rate of oxidative addition may not be so high to ensure fast and quantitative trapping of the Pd(0) species [45].

Other water-soluble chloroarenes, like all three chlorophenols, 2,4-dichlorophenol and 3-nitro-4-chlorobenzoic acid, can also be made to react. The reaction with water-insoluble aryl chlorides, such as p-chloroanisole or p-chloroacetophenone, can be run in aqueous organic solvents, such as DMF or HMPA at 130–140°C in the presence of Na_2CO_3. In the absence of water in DMF, the yield of phenylation product is half of that obtained with a small amount of water (DMF:H_2O = 13:1) added to the solvent. Though the yields are usually moderate (40–60%, based on chloroarene), it is evident that the reactivity of this system is unrivaled by any other currently available palladium-catalyzed system based on oxidative addition to the carbon–halogen bond. This reaction thus serves as a certain feasibility test on the limitations of such reactions, among which there are such important processes as carbonylation, Heck arylation, catalyzed allylic substitution, etc. It may be noted that in other published cases of palladium catalyzed reactions with chloroarenes the presence of electron-rich alkyl- or cycloalkylphosphines was required [46].

The reaction can be extended to non-aromatic halides. For example, allyl bromide reacts with Ph$_4$BNa to give allylbenzene in 60% yield after 64 h at room temperature, and 76% after 3 h at reflux:

$$\diagdown\diagdown_{Br} + Ph_4BNa \xrightarrow[H_2O,\ 100\ ^\circ C]{PdCl_2,\ Na_2CO_3} \diagdown\diagdown_{Ph}$$

Diaryliodonium salts can be used in cross-coupling with boronic acids or sodium tetraphenylborate in place of aryl halides. This reaction can be performed in water in the presence of catalytic amounts of PdCl$_2$. At 80°C both the aryl groups of the iodonium salt are involved. The addition of base is required to enable the transfer of all four phenyl groups of Ph$_4$BNa [47]:

$$2\ Ar_2IX + Ph_4BNa \xrightarrow[H_2O,\ 80\ ^\circ C]{1\ mol\%\ PdCl_2,\ Na_2CO_3} 4\ Ar\text{-}Ph$$

where $X = HSO_4$, BF_4, CF_3COO; $Ar = Ph$, $p\text{-}FC_6H_4$, $m\text{-}O_2NC_6H_4$. The reaction must involve two distinct steps, and as the first aryl group of the iodonium salt is transferred, the formation of water-insoluble aryl iodide must occur. The fact that in this case both aryl groups are used may mean that aryl iodide reacts so fast that no accumulation and formation of heterophase occurs.

Phosphine-less catalysis can be applied as well for cross-coupling of aryl-boronic acids and arenediazonium salts, though in this case the application of aqueous media gave no distinct advantages over anhydrous organic solvents [48].

5.2.2   Cross-coupling with organotin compounds

Cross-coupling reactions of aryl (vinyl) halides or triflates with organotin compounds (the Stille reaction) can now be regarded as one of the classical methods of organic synthesis [49]. Again, the basic technique requires palladium complexes with phosphine ligands, and the reactions are carried out in anhydrous organic solvents like DMSO, HMPA, etc., often at elevated temperatures.

Later it was shown that this method also benefits from phosphine-less catalysis. The use of phosphine-less complexes, such as Pd(DBA)$_2$,

$PdCl_2(MeCN)_2$ and $[(\eta^3\text{-allyl})PdCl]_2$, or even simple palladium salts allows one to perform the reactions under milder conditions, at lower temperatures in less-polar solvents (acetone, THF, ether, even benzene). The elementary steps of the catalytic cycle – oxidative addition and especially transmetallation – are accelerated, which is quite important as the reactivity of organotin compounds is relatively low compared with other organometallic compounds used in cross-coupling reactions [50].

The addition of small amounts of water helps to increase the selectivity and yield in, for example, cross-coupling of vinyloxiranes with $PhSnMe_3$ by favoring allylic rearrangement over direct substitution [51]:

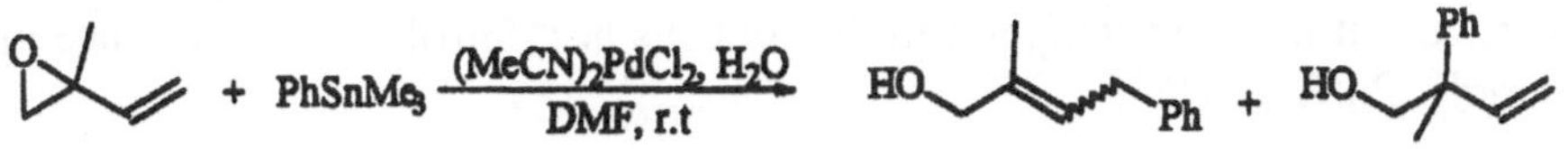

For instance, in the absence of water, the ratio of rearranged to non-rearranged product was 27:1, but increased to 65:1 on addition of 10 equivalents of water. Simultaneously, water helped to improve the $(E)/(Z)$ ratio of the rearranged product. The effect of water in this reaction was not unique: the increase of yield, regio- and stereoselectivity was also observed for the addition of other acidic or basic components like Et_3N or Bu_3SnOAc, but none of them was as efficient in simultaneously increasing the yield, regio- and stereoselectivity. This reaction was successfully used for the stereoselective synthesis of a sex attractant:

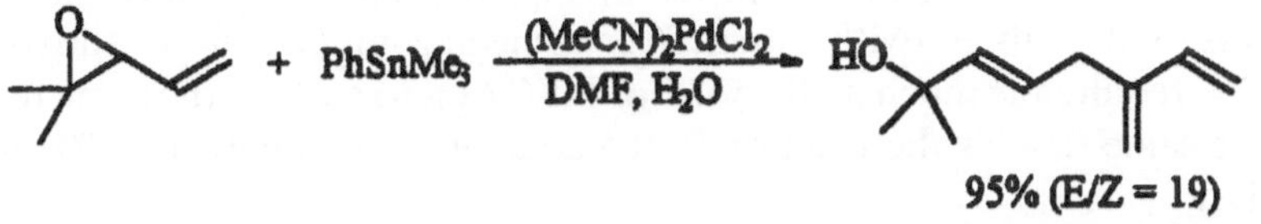

95% (E/Z = 19)

The cross-coupling of α-stannylated enol ether with 2,4-dimethoxy-5-iodopyrimidine in aqueous ethanol yielded the 5-acetyl derivative, though no data for comparison with anhydrous conditions, as well as other examples of this method, were published [52]:

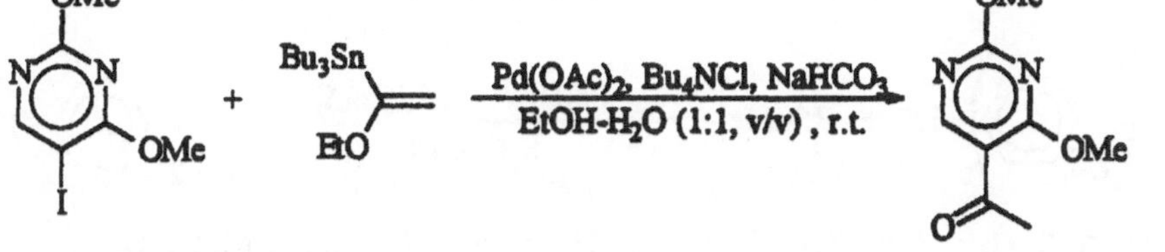

Arenediazonium salts can be used in cross-coupling with organotin compounds instead of aryl halides, thus allowing the use of readily available aromatic amines, and in most cases just to skip the preparation of aryl halides usually done by the Sandmeyer reaction. When first discovered, this reaction was accomplished in anhydrous solvents with isolated diazonium salts [53]. Later it was shown that it is possible to carry out the reaction in water with diazotized amine without prior isolation of the diazonium salt. The trick here is to use the acidic solution of the diazonium salt, as it is well known that the addition of alkali leads to the formation of covalent diazo compounds and, further on, diazoates, which are very unstable and would lead to extensive tarring. Nevertheless, the high reactivity of phosphine-less palladium catalysts enables one to work within the narrow limits set by the low thermal stability of diazonium salts, the necessity to keep pH < 7, and the moderate reactivity of organotin compounds toward transmetallation.

The reaction of various benzenediazonium salts with tetramethyltin can be carried out at room temperature in aqueous acetonitrile in the presence of 1 mol% $Pd(OAc)_2$ [54]:

$$Z\text{—}C_6H_4\text{—}N_2X + Me_4Sn \xrightarrow[\text{MeCN-H}_2\text{O (1:1, v/v), r.t.}]{1\text{ mol% Pd(OAc)}_2} Z\text{—}C_6H_4\text{—}CH_3$$

where $X = Cl$, HSO_4; $Z = o\text{-},p\text{-}NO_2$, $p\text{-}I$, $p\text{-}Br$, $p\text{-}MeO$, $p\text{-}Me$. The reaction is equally successful with both electron-donor and electron-withdrawing substituents in the diazonium salt molecule. Neither bromine nor even iodine atoms in the aromatic ring are substituted, which makes the method potentially suitable for stepwise introduction of different groups by repeated palladium-catalyzed transformations.

The reaction is essentially faster in aqueous solvents. Thus the same process, but realized with isolated benzenediazonium salt in anhydrous MeCN, requires as much as 10 mol% of $Pd(OAc)_2$ to achieve the same results for the same time as the reaction in the aqueous system with 1 mol% of the catalyst.

The only major problem limiting the scope of this method is competition with the protonolysis of the C–Sn bond in acidic solution, which is a serious problem for more reactive organotin compounds. The cross-coupling reaction can be performed also with diaryliodonium salts, which do not require acidification to survive. Diaryliodonium salts can be made to react readily with various organotin compounds in neat water [55]:

$$\left(Z\text{—}C_6H_4\right)_2 IX + RSnMe_3 \xrightarrow[\text{H}_2\text{O, 60°C.}]{Pd(OAc)_2 \text{ or } Pd(dba)_2} Z\text{—}C_6H_4\text{—}R$$

where $X = BF_4$, CF_3COO, HSO_4; $Z = H$, $m\text{-}NO_2$; $R = Me$, $m\text{-}MeC_6H_4$.

The main drawback of the reactions with organotin compounds described above is the utilization of only one of the four organic radicals, which leads to the formation of highly toxic waste R_3SnX, thus complicating the workup and being uneconomic. The best solution would be to use readily available monosubstituted organotin compounds $RSnX_3$, which among other advantages possess relatively low toxicity and are less volatile. However, straightforward application of such compounds is hardly feasible owing to the very low reactivity of the C–Sn bond. To overcome this obstacle, an approach that is roughly similar to that used for cross-coupling with organoboron compounds may be applied. The tin atom is capable of increasing the coordination number by binding basic ligands, which results in a buildup of negative charge on the tin atom and an increase of reactivity toward transmetallation. Thus, compounds $RSnCl_3$ are readily dissolved in aqueous alkali to give stannate salts with the anion $[RSn(OH)_{3+n}]^{n-}$, which was shown to be quite reactive in cross-coupling with various aryl halides. Water-soluble aryl iodides readily react in the presence of phosphine-less catalyst $PdCl_2$, while water-insoluble aryl iodides and all bromides required the presence of TPPTM ligand. Aryl iodides gave high yields of cross-coupling product independently of whether electron-donor or electron-withdrawing substituents were present in the aromatic ring of ArI, while aryl bromides gave high yields only when the substituent was electron-withdrawing [56]:

$$RSnX_3 \xrightarrow{OH^-/H_2O} [RSn(OH)_{3+n}]^{n-} \xrightarrow{ZC_6H_4Hlg,\ PdCl_2\ or\ PdCl_2(TPPMS)_2} Ar\text{-}R$$

where $X = Cl$, Br; $R = Ph$, Me, CH_2CH_2COOH; and $Z(Hal) = m\text{-}COOH$ (Br, I), $p\text{-}OH$ (I), $p\text{-}COMe$ (I), $p\text{-}Me$ (I), $o\text{-}NH_2$ (I), $p\text{-}OCH_2COOH$ (I), etc.

The reaction was quite successful and ran without unexpected complications even in the most unfavorable case of a donor substituent in the *ortho* position to iodine, as in *o*-iodoaniline, though in this case cyclized products were obtained in cross-coupling with, for example, β-trichlorostannylpropionic acid:

These results agree well with the data published somewhat later in reference [57] on the coupling of water-soluble iodides and bromides with $RSnCl_3$ bearing aryl, vinyl and even alkyl residues. Coupling with *p*-iodobenzoic acid can be performed either with or without a water-soluble phosphine ligand, and in most cases the influence of TPPMS is too small to be discussed:

where $R = Ph$, Me, *i*-Pr, *n*-Bu, $CH_2=CH$, *cis*-$HOOCCH=CH$.

On the other hand, the reactions with aryl bromides are very sensitive to the presence of the phosphine ligand, giving very poor results in phosphine-less mode. The range of aryl bromides that are reactive under such conditions included aryl, hetaryl and vinyl derivatives.

5.2.3   Cross-coupling with terminal acetylenes

Palladium-catalyzed cross-coupling of aryl (vinyl) halides or triflates with terminal acetylenes is a well-known and useful method of organic synthesis. Commonly, this reaction is performed in anhydrous solvents in the presence of tertiary amines and with CuI or other Cu(I) salts as co-catalysts. According to the generally adopted mechanism, this reaction presents a typical case of cross-coupling, in which acetylenes react either as acetylenides generated in situ due to high CH-acidity, or as free hydrocarbons by direct insertion of palladium across the C–H bond. The role of copper(I) is, as in many other copper-catalyzed reactions, uncertain, and the trivial explanation, that copper is involved by forming copper acetylenide, may not be true. Alternatively, this process may be regarded as a kind of Heck-type arylation, in which an arylpalladium intermediate adds to the triple bond by a carbo-palladation pathway. The possibility that both mechanisms can be operating in particular cases may explain why data on cross-coupling with acetylenes is rather controversial, especially concerning the influence of various factors (the role of copper salts, basicity, the nature of phosphine ligands, etc.). The main problem in optimizing the conditions for cross-coupling reactions with acetylenes is the relatively high reactivity of the triple bond toward palladium activation, which results in excessive formation of by-products (usually formed via nonselective oligomerization pathways) without the involvement of organic halide. As is shown below, aqueous techniques often help to eliminate this obstacle and achieve selective cross-coupling, giving high yields of the target products.

The application of aqueous mono- and biphasic techniques, both in the presence of hydrophilic phosphine ligands, and in phosphine-less mode, was investigated in detail, and turned out to be quite successful.

The reaction of both water-soluble and water-insoluble aryl iodides with hydrophilic propargyl alcohol shows the best results if run in the presence of both TPPMS and Ph$_3$P ligands, with the former being more important to achieve good yields of cross-coupling product, and the latter behaving like a typical promoter additive [58]:

where Z = p-COOH, m-COOH, p-NO$_2$, p-MeCO.

For water-insoluble acetylenes the reaction can be readily and conveniently carried out in a heterogeneous system containing water, K_2CO_3 and $10\,mol\%$ Bu_3N in the presence of the hydrophobic catalyst $PdCl_2(PPh_3)_2$ [59]:

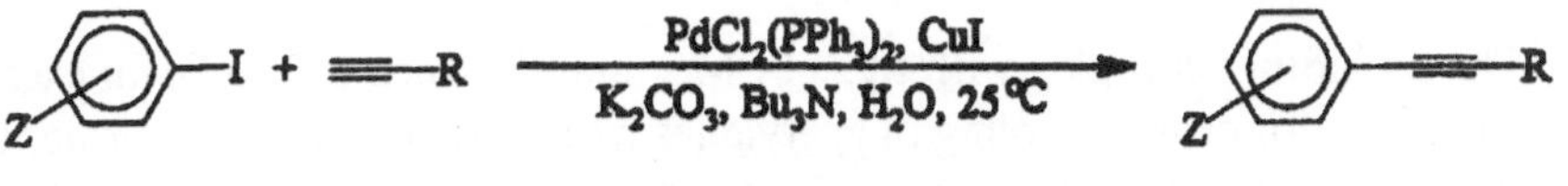

$$Z-\bigcirc\!\!\!\!\!\!\!-I + {=\!\!=}-R \xrightarrow[\text{K}_2\text{CO}_3,\ \text{Bu}_3\text{N},\ \text{H}_2\text{O},\ 25\,°\text{C}]{\text{PdCl}_2(\text{PPh}_3)_2,\ \text{CuI}} Z-\bigcirc\!\!\!\!\!\!\!-{=\!\!=}-R$$

where $Z = H$, NO_2, CN, $MeCO$; $R = Ph$, $n\text{-}C_5H_{11}$, $p\text{-}C_9H_{19}C_6H_4$.

The reactions, which are carried out under very mild conditions at room temperature, run equally easily with aryl iodides bearing both electron-donor and electron-withdrawing substituents with almost no noticeable difference in reaction time, which is in sharp contrast with similar reactions in organic solvents, which are very sensitive to substituent effects. The whole process evidently occurs in the organic phase formed by the reagents and tributylamine, with the latter playing the role of phase-transfer agent, thus allowing the use of K_2CO_3 as the stoichiometric base (Scheme 5.3).

Alternatively, the cross-coupling reaction between water-insoluble acetylene and aryl iodide can be performed in aqueous DMF with the same catalyst and base. Increase of the water content in the solvent beyond the limits of solubility of the organic reagents immediately results in a sharp increase in reaction time and decrease in yield. It should be noted that in the method employing a catalytic amount of Bu_3N, where the reaction mixture is always heterogeneous, no negative influence of heterogeneity could be observed. Thus, we can conclude that the role of phase transfer of base is decisive, as is indeed often the case in other base-driven catalytic processes involving palladium (Heck arylation, hydroxycarbonylation, etc.).

Acetylene itself can be used in cross-coupling using the method described. The use of water, however, enables one to add calcium carbide to the reaction mixture to generate C_2H_2 *in situ*, thus allowing the hazards of handling gaseous acetylene to be avoided. This approach opens the route to a one-pot procedure for the assembly of disubstituted unsymmetrical acetylenes by

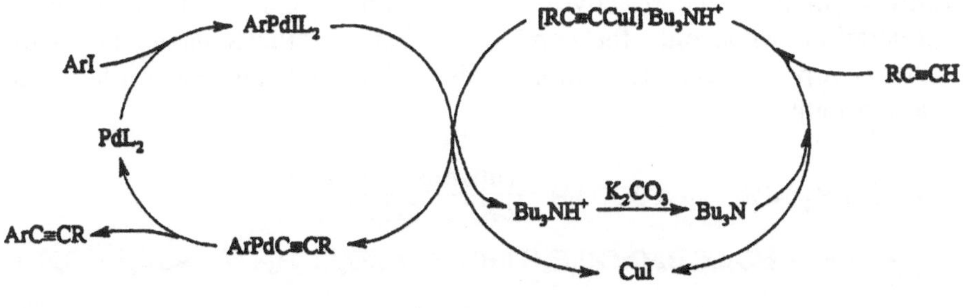

Scheme 5.3

stepwise attachment of aryl groups via cross-couplings with different halides:

$$ArI \xrightarrow[\text{K}_2\text{CO}_3, \text{ 10\% Bu}_3\text{N}]{\text{CaC}_2, \text{H}_2\text{O}, \text{PdCl}_2(\text{PPh}_3)_2} ArC\equiv CH \xrightarrow[one\text{-}pot]{+ ArI} ArC\equiv CAr$$

The palladium complex with the TPPMS ligand turned out to be an excellent catalyst for cross-coupling of acetylenes with iodo derivatives in aqueous acetonitrile solvent under very mild conditions (at room temperature or gentle heating) without copper salt promoter. The method is suitable for a wide range of highly sensitive compounds, including natural molecules and their derivatives. Thus, the method was applied for the final step (the bond formed by cross-coupling is marked by the dashed line) in the synthesis of the following chain-terminating nucleotide reagent carrying a fluorescent dye unit:

Neither the triphosphate residue nor the dye fragment interfered with the reaction, in sharp contrast with the commonly used processes in organic solvents, in which such fragile groups have no chance to survive [7].

Cross-coupling of water-soluble acetylene and aryl iodide occurs in neat water in the presence of water-soluble phosphine ligands, including TPPTS and phosphines bearing cationic guanidinium groups. This reaction was complicated by the formation of an acetylene homocoupling product. Cationic phosphines showed rather poor results with respect to both turnover number and selectivity. The addition of CuI led to a dramatic acceleration, while only the combined effect of TPPTS ligand and CuI together suppressed the formation of diyne, and made the reaction fast and selective [60]:

$$p\text{-HO}_2\text{CC}_6\text{H}_4\text{C}\equiv\text{CH} + p\text{-IC}_6\text{H}_4\text{CO}_2\text{H} \xrightarrow[\text{K}_2\text{CO}_3, \text{ H}_2\text{O}, 50°\text{C}]{\text{Pd(OAc)}_2, \text{ L}, \text{ CuI}}$$

$$\longrightarrow p\text{-HO}_2\text{CC}_6\text{H}_4\text{-C}\equiv\text{C-C}_6\text{H}_4\text{CO}_2\text{H-}p + p\text{-HO}_2\text{CC}_6\text{H}_4\text{-C}\equiv\text{C-C}\equiv\text{C-C}_6\text{H}_4\text{CO}_2\text{H-}p$$

$$\text{L} = \text{TPPTS}, [\text{Ph}_2\text{PCH}_2\text{CH}_2\text{CH}_2\text{NHC(NH}_2)_2]^+, [\text{PhP(CH}_2\text{CH}_2\text{CH}_2\text{NHC(NH}_2)_2)_2]^{2+}$$

Diaryliodonium salts can serve as a source of aryl groups in a very mild cross-coupling reaction with terminal acetylenes. When the reaction is carried out in DMF catalyzed by $PdCl_2(PPh_3)_2$ and CuI, both the aryl groups of the iodonium salts are transferred to the product. In water, or rather in the heterogeneous system described above ($PdCl_2(PPh_3)_2$, CuI, K_2CO_3, 10 mol% Bu_3N), only one aryl group is quantitatively transferred in a very fast reaction at room temperature with the ratio of reagents taken as 1:1 [61]:

$$(m\text{-}O_2NC_6H_4)_2I^+ + PhC \equiv CH \xrightarrow[\substack{10 \text{ mol}\% Bu_3N, H_2O \\ 10 \text{ min}, 100\%}]{PdCl_2(PPh_3)_2, CuI, K_2CO_3}$$

$$m\text{-}O_2NC_6H_4C \equiv CPh + m\text{-}O_2NC_6H_4I$$

Upon addition of the second equivalent of acetylene and more prolonged stirring, the liberated aryl iodide would also react, thus allowing one to transfer both aryl groups of the iodonium salt. The reaction also occurs in the absence of CuI, but only on heating to 60°C [62]. Other hypervalent iodine compounds, such as the complexes of iodosylbenzene with acids (TsOH, HBF_4, etc.), can also be used for cross-coupling with organotin compounds [63].

Recent development of preparative methods for iodonium salts with alkenyl or alkynyl groups made such compounds available as valuable reagents for direct introduction of unsaturated groups by nucleophilic substitution and cross-coupling, which may be formally regarded as catalyzed nucleophilic substitution. Such salts were shown to be reactive in phosphineless cross-coupling in homogeneous aqueous systems, thus opening a facile route to enynes and diynes [64]:

$$[PhC \equiv CIPh]^+ BF_4^- + RC \equiv CH \xrightarrow[MeCN\text{-}H_2O, \text{ r.t.}]{Pd(OAc)_2, NaHCO_3} RC \equiv C\text{-}C \equiv CP$$

Cross-coupling of iodoacetylenes with terminal acetylenes can be performed in the presence of 5 mol% $Pd(OAc)_2$ and TPPTS ligand in aqueous MeCN to afford diynes in moderate to high yields. The reactions are performed under very mild conditions, allowing one to obtain functionalized diynes, which could not endure elevated temperatures and strong bases:

$$RC \equiv CI + HC \equiv CR' \xrightarrow[MeCN\text{-}H_2O, \text{ r.t.}]{Pd(OAc)_2, TPPTS, Et_3N} RC \equiv C\text{-}C \equiv C$$

where $R = n\text{-}Bu$, $R' = C(Me)_2OH$; $R = Me_3Si$, $R' = CH(OH)CH_2Bu\text{-}n$; $R = Me_3Si$, $R' = C(Me)_2OH$; $R = C(Et)_2NH_2$, $R' = CH(OH)CH_2Bu\text{-}n$; and $R = n\text{-}Bu$, $R' = Ph$. The formation of homocoupling product was observed only for reactions with phenylacetylene.

The same technique was used for cross-coupling of terminal acetylenes with *o*-iodophenols and anilines, giving the cyclization products, 2-substituted furans and indoles, in high yields [65, 7]:

5.2.4   *Allylic substitution*

Palladium-catalyzed allylic substitution may be regarded as a special case of cross-coupling with π-allylpalladium complexes. First developed as a stoichiometric technique, this reaction was later realized in a catalytic mode, and became a valuable tool of organic synthesis, as it allows for a broad variation of both allylic substrates and nucleophiles.

Allyl acetate reacts with various N- and C-nucleophiles in aqueous MeCN (10% H_2O, v/v) in the presence of $Pd(OAc)_2$ and TPPTS, giving substitution products in moderate to high yields [66]. As this reaction is carried out in homogeneous solution containing only a small amount of water, the role of the highly hydrophilic TPPTS ligand is not clear. However, the use of water-insoluble nitrile solvents, such as benzonitrile or valeronitrile, in place of MeCN makes the system biphasic, and the application of TPPTS in this case allows an easy separation of catalyst, which may be crucial as typically very high loads of catalyst (up to 4 mol% of palladium acetate) and up to 20 mol% of the expensive TPPTS ligand are introduced [67, 68].

The reaction was extensively studied for cinnamyl esters (acetate and carbonate) as allylic substrates, with some examples given for allyl acetate and its homologs, and a wide range of nucleophiles including N-nucleophiles (primary and secondary amines, hydroxylamine and its derivatives, and sodium azide), C-nucleophiles (malonates, ethyl acetoacetate, acetylacetone, sodium tetraphenylborate) and an S-nucleophile (sodium *p*-toluenesulfinate) (Scheme 5.4).

It is quite remarkable that in most cases CH-acids are added without either primary transformation into enolate forms or even the addition of a base capable of *in situ* deprotonation. The reaction with CH-acids without prior deprotonation was described for non-aqueous media for the palladium-catalyzed allylation with allyltin derivatives [69], though the mechanism proposed is quite specific and requires the presence of an organotin compound. By adjusting this mechanism to the reaction in water, the process shown in Scheme 5.5 may be considered.

Pd(0) generated by the action of TPPTS may react in oxidative addition across the C–H bond of CH-acids, and the resulting hydridopalladium intermediate reacts with allylating agent with the formation of either σ- or π-allyl-palladium intermediate. It is quite noteworthy that the use of aqueous solvent may help the formation of allylpalladium intermediate in an S_N1-like process.

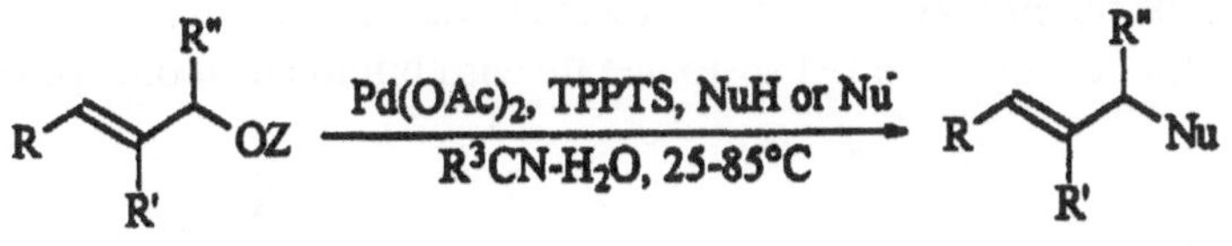

$$R = Ph \quad\quad R' = R'' = H \quad\quad Z = Ac, COOEt$$

R = Ph   R' = R" = H   Z = Ac, COOEt

R = Pr   R' = R" = H   Z = COOEt

R = OH   R' = R" = H   Z = Ac

R, R" = H   R' = Me   Z = Ac

R, R' = H   R" = Me   Z = Ac

Scheme 5.4

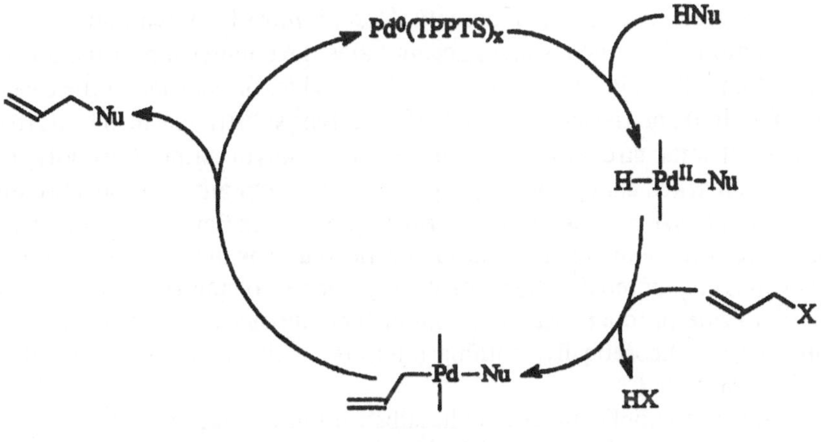

Scheme 5.5

The reaction with sterically unhindered primary amines and hydroxyl-amine often leads to bisallylation occurring with remarkable ease. Again note that hydroxylamine is used as the hydrochloride salt, and the liberation of hydrochloric acid must not have interfered with double allylation in near-quantitative yield:

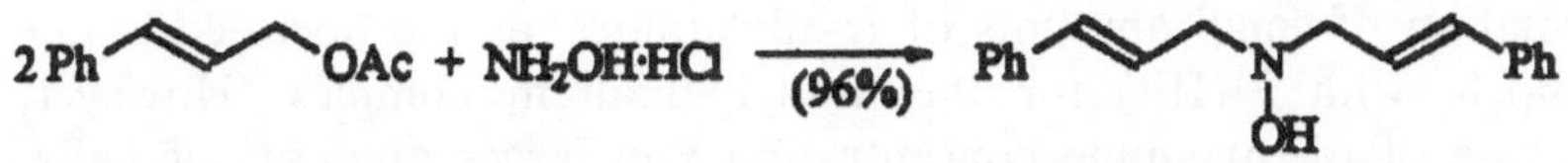

Monoallylation of hydroxylamine can only be achieved with a diprotected derivative BocNH–OBoc.

It is to be noted that allylic rearrangement occurs only in rare cases, one of which is the reaction of ethyl acetoacetate with butadiene monoepoxide:

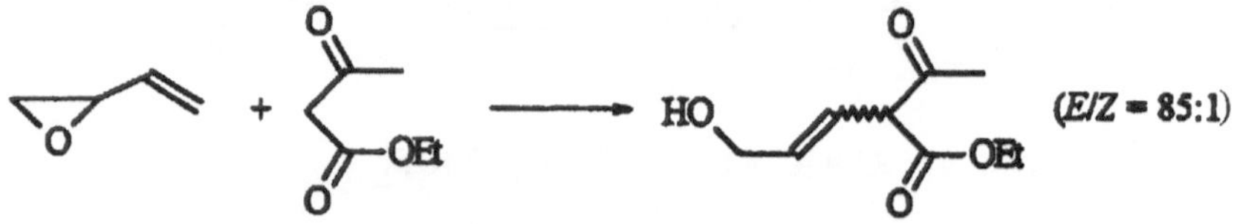

$(E/Z = 85:1)$

The reaction with 2-butenediol-1,4-diester is an example of double substitution leading to cyclization:

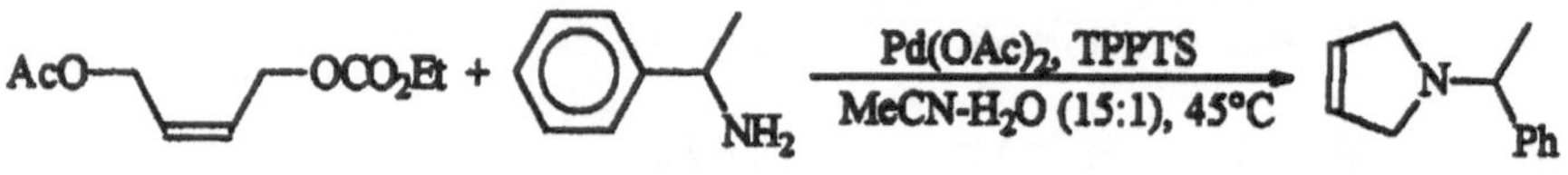

An experiment with repeated use of the aqueous layer as a catalyst system for allylations showed that no noticeable deterioration of catalyst occurred in the first and second recycling, while further reuse of the catalyst solution showed a gradual decrease of activity. Here, it must be noted that in allylation reactions the aqueous layer accumulates the counter-ion of the nucleophile (if any, like Na^+ in the case of sodium azide, or chloride in the case of $NH_2OH \cdot HCl$) and the leaving group of the allylic substrate, which leads to an increase of ionic strength and acidity of the catalytic brine. Certainly, this leads to a marked change of the properties of the aqueous solution, and must result in catalyst poisoning. Thus, workup and isolation of palladium and ligand from the aqueous layer after several reuses is unavoidable. No such problem is encountered in the hydroformylation of olefins or hydrogenation, where no side-products accumulating in the aqueous layer are formed, and thus, at least theoretically, nothing interferes with the infinite reuse of the catalyst solution.

An ingenious application of allylic substitution was suggested for the selective removal of allyloxycarbonyl (Alloc) protective group from hydroxyls and amines, and the allyl protective group from carboxyls. The reaction with an excess of simple nucleophile, such as diethylamine, catalyzed by Pd(OAc)$_2$/TPPTS either in homogeneous solution of aqueous MeCN or in the biphasic system of water–organic solvent (higher nitriles, ether, CH$_2$Cl$_2$, etc.), can be run under very mild conditions at room temperature, which allows one to spare the other protective groups removed by hydrolytic cleavage as in the examples shown in Scheme 5.6.

Deprotection of N-allyloxycarbonylamines can sometimes lead to the formation of small amounts of N-allylamines, as the liberated amine can compete with NHEt$_2$ for the π-allylpalladium complex. However, the increase of diethylamine concentration and larger amounts of palladium catalyst help to eliminate this problem [70]. Another useful method of deprotection applied to O-allyl derivatives of carbohydrates and natural polyols

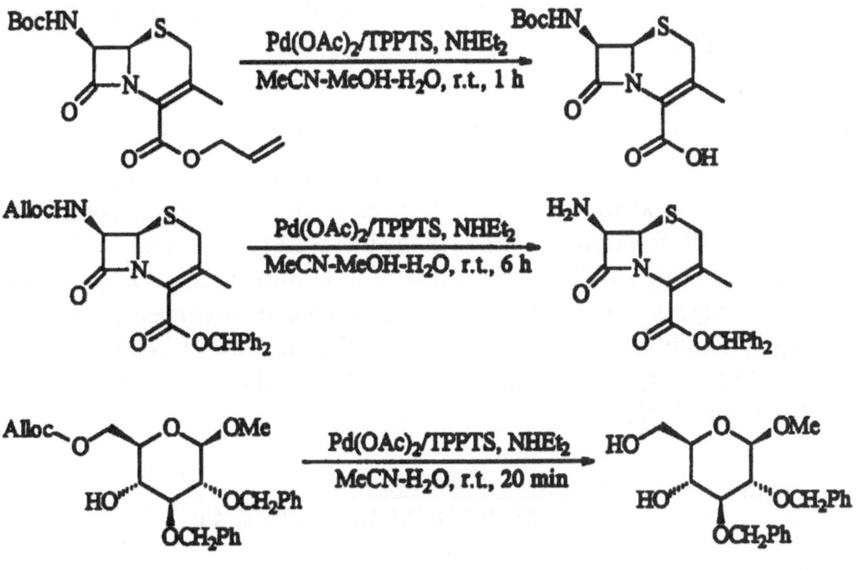

Scheme 5.6

under aqueous conditions employs Wacker-type oxidation assisted hydrolysis of the unsaturated group using the system $PdCl_2$–$CuCl$–O_2 in aqueous DMF [71].

The allylation of uracils and thiouracils was shown to be highly dependent on the solvent and the catalytic system. While in DMSO or dioxane the reaction in the presence of $Pd(OAc)_2$–Ph_3P gave a mixture of substitution products at N(1) and N(3), the reactions in aqueous MeCN catalyzed by the TPPTS complex of palladium gave selectively 1-N-allyluracils and S-allylthiouracils, respectively [72]:

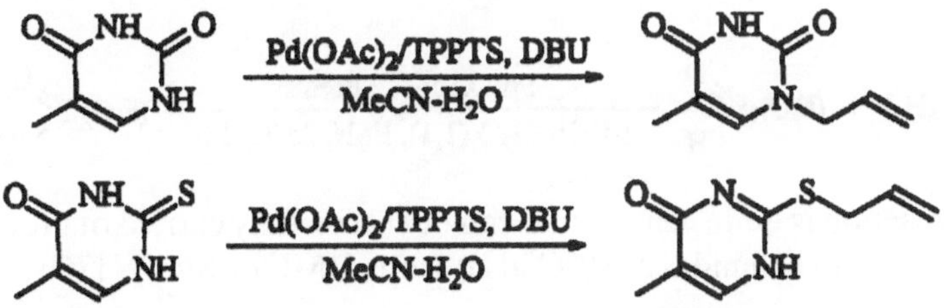

The difference was attributed to the operation of kinetic control in the aqueous system versus thermodynamic control in DMSO or dioxane, though, probably, there may be another explanation, as the reactions in organic solvents were run without the addition of base, while the reactions in the aqueous system were run with DBU, and thus, in the latter case, the nucleophiles were actually the anions of uracils. For ambident nucleophiles, the influence of the counter-ion and other factors associated with the ionic state of the species in solution are key factors determining the selectivity of substitution [73].

5.2.5   Carbon–heteroatom bond formation cross-coupling reactions

(a) Formation of C–P bonds.   Palladium-catalyzed cross-coupling is a convenient method for the preparation of organophosphorus compounds [74]. Many such reactions described in the literature were carried out in anhydrous organic solvents. The application of PTC turned out to be quite effective for the arylation of diethylphosphite, which may be carried out either in the liquid–liquid system benzene–water in the presence of NaOH, Bu₄NCl and TPPMS, or in phosphine-less mode in the solid–liquid system K₂CO₃–Bu₄NCl [75]:

The latter method excludes water and thus falls outside of the scope of this review. It is very interesting that in the former case two large ions, Bu_4N^+ and $Ph_2P(C_6H_4SO_3)^-$, should readily form a lipophilic ion pair that helps to transfer the otherwise hydrophilic phosphine ligand, as well as palladium complexes formed with the TPPMS ligand, to the organic phase, where the reaction should actually take place. The role of quaternary ammonium salts in aqueous palladium catalysis was noted earlier, though it obviously evades rational explanation if we suppose that the reaction should have taken place in an aqueous layer.

In homogeneous systems the arylation of diethylphosphite can be carried out under even milder conditions. Thus, the reaction with *p*-iodotoluene gives a quantitative yield of the respective diethylphosphonate at room temperature [7]:

The reaction is quite general and can be successfully carried out for various alyl iodides and bromides at 80°C in aqueous DMF or MeCN [76]:

where Z = H, Me, MeO, COOH; X = Br, I; Solv = DMF, MeCN.

Water must indeed play an essential role as a solvent, as the decrease of water content from 50% to 10% (v/v) with all other parameters kept invariant leads to a sharp decrease of the yield (e.g. for reaction with PhI, from quantitative to a meager 30%). Halobenzoic acids, which react in basic solutions as water-soluble carboxylates, are quite reactive even in the absence of the TPPMS ligand in aqueous MeCN. The reactions in the presence of TPPMS can be carried out even in neat water, either in a biphasic system (e.g. for halobenzenes) or in a homogeneous solution (e.g.for p-iodobenzoic acid). It is worthy of note that, in aqueous MeCN, the reaction for PhI can be carried out with Ph_3P as ligand, giving practically the same results as the much more expensive TPPMS. However, this is only an exclusion, as Ph_3P proved to be inefficient in other reactions, even in MeCN without water.

An interesting example of palladium-catalyzed phosphorylation was described for the substitution of the triflate group in the tyrosine moieties of peptides, giving modified molecules likely to be used as anticancer drugs. Both the transformation of the OH group into triflate and the palladium-catalyzed phosphorylation can be done in aqueous ethanol using bases such as K_2CO_3 or $KHCO_3$ [77]:

(b) Formation of C–H bonds. The reductive cleavage of carbon–heteroatom bonds, and in the first place carbon–halogen bonds, is an important method in organic chemistry. It is applied in preparative synthesis for e.g. the deprotection of certain sites mostly in aromatic molecules, and is used to control the regioselectivity of electrophilic substitution reactions. Reductive cleavage may also be applied to remove phenol groups through their primary transformation into such better leaving groups as triflate, other sulfonates, phosphates, etc. Reductive dehalogenation becomes of prime interest as a method for removal of chlorine atoms from polychlorinated molecules, which are among the most environmentally hazardous pollutants [78].

Certainly, the development of dechlorination techniques in aqueous media is particularly appealing, as it would open the route of economic detoxification of polluted water without the isolation of pollutants.

Water-soluble aryl bromides and iodides are readily and quantitatively hydrogenolysed in aqueous alkaline solutions by $NaBH_4$ and palladium chloride as catalyst in a phosphine-less mode. However, the application of this reaction to chlorides gave poor yields not exceeding 30%. The increase of temperature required to facilitate oxidative addition of Pd(0) across the C–Cl bond resulted in a rapid decomposition of $NaBH_4$.

The use of water allows use of such a cheap reducing agent as NaH_2PO_2, routinely used for the replacement of an amino group by hydrogen through diazotization. There are some contradictory data in the earlier literature on the use of this reductant for dechlorination of aromatic chlorides hetero-geneously catalyzed by Pd/C [79, 80]. It turned out that under homogeneous conditions NaH_2PO_2 is a very efficient reductant in the process catalyzed by $PdCl_2$ at 50–70°C in alkaline solutions:

where $Z = o\text{-},p\text{-COOH}; p\text{-OH}; p\text{-NH}_2$.

The reaction can be applied both for readily water-soluble chloropheno-lates and chlorobenzoates, and for chloroanilines, which possess only a limited solubility in aqueous alkali, though obviously sufficient for achieving high yields of dechlorinated product.

As in some other palladium-catalyzed reactions, *ortho*-substituted sub-strates showed a low reactivity, possibly because of the formation of a chelated organopalladium intermediate, in which a key coordination site in a *cis* position to the aryl group is blocked by a strongly bonded ligand. In the case of *o*-chlorobenzoic acid, the addition of iodide ions allows the influence of the *ortho* substituent to be suppressed, and an 86% yield of benzoic acid to be obtained, possibly because in this case chelation involves a five-membered palladacycle with carboxylate group retaining a certain degree of freedom. In the case of *o*-chlorophenolate, *o*-chloroaniline and similar substrates, chela-tion involves a rigid flat four-membered palladacycle, which, for geometric reasons, has no capabilities for dechelation and letting in another ligand [81].

The same reducing agent can be used for a convenient method of removal of the phenolic hydroxy group [82]. Phenols are first transformed into monoaryl sulfates by reaction with the complex of sulfur trioxide with N,N-dimethylaniline, with subsequent reduction by NaH_2PO_2 catalyzed by $PdCl_2$:

$$\text{ArOH} \xrightarrow[\text{2. KOH, H}_2\text{O}]{\text{1. PhNMe}_2\cdot\text{SO}_3} \text{ArOSO}_3\text{K} \xrightarrow[\text{H}_2\text{O, NaOH, 50-70°C}]{\text{NaH}_2\text{PO}_2, \text{PdCl}_2} \text{Ar}$$

where $Ar = p\text{-PhC}_6\text{H}_4, p\text{-PhOC}_6\text{H}_4, p\text{-MeOC}_6\text{H}_4, p\text{-MeC}_6\text{H}_4, 1\text{-C}_{10}\text{H}_7, 2\text{-C}_{10}\text{H}_7$, etc.

It should be noted that, in comparison with other known methods of reductive removal of the hydroxy group by first transforming it into a better leaving group (triflate, uretane, dialkylphosphate, etc.), this method requires a very simple and cheap sulfate group formed and removed in a one-pot method without the isolation of the intermediate. The sulfate group can be otherwise removed by means of Raney alloy reduction, which is a definitely inferior method as it is noncatalytic, requires a large amount of Ni–Al alloy, and is less tolerant to other functions.

The reduction of allylic halides or acetates, as well as benzyl halides, can be achieved by the palladium-catalyzed reaction with formate ions [83] in the biphasic systems toluene–water or heptane–water in the presence of hydrophilic phosphine ligands, such as either standard TPPMS or the more rare sodium 3-(diphenylphosphino)benzoate and phosphine bearing hydrophilic oligoethyleneglycol tails:

R = Alk, Ph; X = Cl, OAc;
L = TPPMS, m-Ph$_2$PC$_6$H$_4$COONa, P

The reaction is accompanied by extensive allylic double-bond shift, leading to the formation of both normal and rearranged reduction products, with the ratio being very sensitive to all the factors involved (the nature of the organic solvent, ligand, substrate, etc.) without any clear trends. Such behavior is quite typical for aqueous catalytic systems, in which the observable net result (yield, selectivity, catalytic efficiency) is the sum of a dozen elementary effects, often having opposite responses. The steps of the catalytic process may take place in either of the phases, as well as in the interfacial region. In this reaction the heptane–water system gave higher yields and larger turnovers than toluene–water, which can be interpreted as if the process is taking place mainly in the aqueous layer, and it is the distribution equilibrium that controls the rate of reaction, because the reverse extraction of reagents (e.g. substrates, or π-allylpalladium complex) into water is less favorable from toluene than from heptane. Indeed, in the experiment with the reduction of unsubstituted allyl acetate in an apparatus allowing separate detection of the reduction product forming in each of the phases, propene evolved from the aqueous phase when the reaction was carried out in the presence of hydrophilic ligand. However, this factor, though it might contribute, is definitely not the most important one. Indeed, it was observed that the ethoxylated phosphine ligand was highly efficient in this process, though the hydrophilicity of this compound might be overestimated. Indeed, at the reaction temperature (100°C) ethoxylated tails are strongly dehydrated to lose most of their interaction with water. Generally, at such temperatures only those PEG derivatives with tails including 40 and more ethoxy groups can retain a reasonable degree of hydrophilicity, while shorter compounds behave as typical hydrophobic oils possessing poor solubility in water and forming separate phases miscible with other nonpolar liquids. Thus, it may be argued with certainty that the most efficient ligand used in this work is hydrophobic, though potentially possessing an ability to bind inorganic cations, and thus behaving like a phase-transfer agent. Thus, the effect of this ligand may be accounted for by the ability to transport sodium formate to the organic layer by a liquid–liquid or solid–liquid phase-transfer mechanism. Separately added PEG derivatives not bonded with a phosphine ligand should have a similar influence, which indeed has been observed [84]. However, the idea to combine ligand and phase-transfer agent into one molecule is quite attractive, and should be employed in other processes [85].

Palladium-catalyzed reduction of carbon–halogen bonds can also be achieved by the formate anion in a biphasic system in the presence of cyclodextrins, which form a hydrophilic host–guest complex with hydrophobic substrates fitting into their cavities, and thus transfer them into the aqueous phase [86]:

(c) Formation of C–N bonds. There are only a few examples of palladium-catalyzed formation of C–N bonds, possibly because nitrogen-containing compounds may serve as ligands, thus altering or altogether blocking the productive pathways in catalytic cycles driven by palladium complexes. Water may indeed help, as it substantially lowers the basicity and nucleophilicity of nitrogen compounds, and may be expected to deblock the catalyst. Indeed, it was shown that a palladium-catalyzed variant of the Ullmann reaction in aqueous media can be accomplished. The arylation of diphenylamine with water-insoluble aryl iodides can be achieved in aqueous microemulsions in the presence of both palladium catalyst and copper(I) iodide [87].

The arylation of nitrogen-containing heterocycles, such as benzotriazole, with aryl iodides can be done only in anhydrous media under rather harsh conditions (DMSO, DBU, 120°C, Pd$_2$(DBA)$_3$·CHCl$_3$) No reaction occurred in the presence of the water-soluble TPPMS ligand. However, the arylation can be accomplished with diaryliodonium salts in neat water under phosphine-less conditions and in the absence of copper salt promoter [88]. Both products of substitution are formed, as the anion of benzotriazole is an ambident nucleophile:

Aqueous media also facilitate the Heck reaction with iodo derivatives of various nitrogen-containing heterocycles [52].

(d) Formation of C–S bonds. A very interesting reaction leading to the formation of a C–S bond via a palladium-catalyzed reaction, formally analogous to hydroformylation, was recently described. Olefins can be transformed into sulfinic acids or their esters by reaction with SO_2 and hydrogen in the presence of cationic complexes of Pd(II) such as [Pd(MeCN)$_2$-(DPPE)](BF$_4$)$_2$ in aqueous or alcoholic solutions [89].

5.3 Heck reaction

The addition of an organopalladium intermediate to the double bond with subsequent cleavage of palladium hydride and restoration of unsaturation is the basis of another important process involving organopalladium compounds, the Heck reaction. In a catalytic version, the organopalladium intermediate is formed by oxidative addition to carbon–halogen or other carbon–heteroatom bonds, and the hydridopalladium complex is believed to be recycled by base-promoted reductive elimination (Scheme 5.7).

However, the role of base might be more complicated than just to launch a new turn of the catalytic cycle. Bases used in the Heck reaction are either tertiary amines or anions like hydroxide, acetate, carbonate, etc., which may well serve as ancillary ligands for palladium and change the properties of the organopalladium intermediates. An accelerating effect was recently observed for acetate ions, which are involved in ligand exchange and enhance the reactivity of palladium species [90]. In water and aqueous media, there may be yet another effect, as such ligands may change the hydrophilicity of palladium intermediates by, for example, forming negatively charged complexes and help to transfer the species involved in the catalytic cycle across phase boundaries.

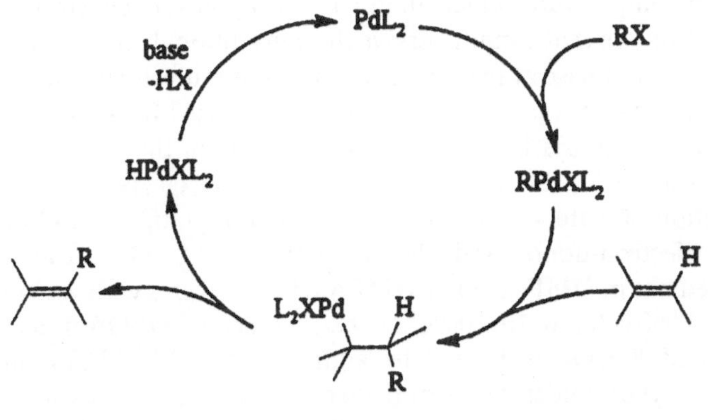

Scheme 5.7

Thus, it may be concluded that the Heck reaction, being a base-driven process not involving any species that cannot tolerate the presence of water, is an appealing case to be tried in aqueous modes. However, historically, water has been allowed to enter Heck chemistry only relatively recently. The traditional technique for carrying out the Heck reaction is to use anhydrous polar solvents (DMF and MeCN are the most frequently used) and tertiary amines as bases.

The role of water in the Heck reaction, as well as in other reactions catalyzed by Pd(0) in the presence of phosphine ligands, can be many-fold. Recent studies confirmed that it may play an important role in the transformation of catalyst precursor into Pd(0) species driving the catalytic cycle. The generation of zero-valent palladium species capable of oxidative addition by oxidation of phosphine ligands by the Pd(II) catalyst precursor can be strongly affected by the water content in the reaction mixture, as is shown below for the reaction with BINAP ligand (BINAP-O is BINAP monoxide) [91]:

$$Pd(OAc)_2 + 3BINAP + H_2O + 2Et_3N \rightarrow Pd(BINAP)_2 + BINAP\text{-}O + 2Et_3NHOAc$$

Taking into account that the palladium catalyst is typically introduced into reaction mixtures in ca. 1 mol% amounts, trace amounts of water in a purportedly anhydrous solvent may play a key role in triggering the catalytic cycle. In this respect it is suspicious that Heck arylations are usually carried out in solvents like DMF or acetonitrile, which are famous for their hygroscopicity and reluctance to be freed of residual water.

The decisive evidence that water indeed helps to reduce Pd(II) to Pd(0) by phosphine was obtained in experiments with ^{17}O-enriched water. The reaction of PdCl$_2$ with TPPTS in water gave a formally cationic complex $[PdCl(TPPTS)_3]^+$, and the respective amount of TPPTS oxide containing ^{17}O that could have come only from the water molecule.

It was shown that the Heck reaction can be accomplished under PTC conditions [92] with inorganic carbonates as bases under very mild conditions even at room temperature, which allows it to be applied to substrates such as methyl vinyl ketone that cannot survive the conventional conditions of Heck arylation (action of base at high temperature). Later it was shown that water and aqueous organic solvents can be successfully used for carrying out the Heck reaction catalyzed by simple palladium salts in the presence of inorganic bases such as K_2CO_3, Na_2CO_3, $NaHCO_3$, KOH, etc. [93].

The reaction of water-soluble acrylic acid with aryl iodides and bromides with both electron-donor and electron-withdrawing substituents can be easily carried out in DMF–H$_2$O or HMPA–H$_2$O mixtures with water content from 10 to 80% (v/v) at 70–100°C in the presence of Pd(OAc)$_2$ as catalyst precursor and K_2CO_3 as base. The addition of $(o\text{-MeC}_6H_4)_3P$ ligand is required for aryl bromides. Some of these reactions can even be carried out in neat water:

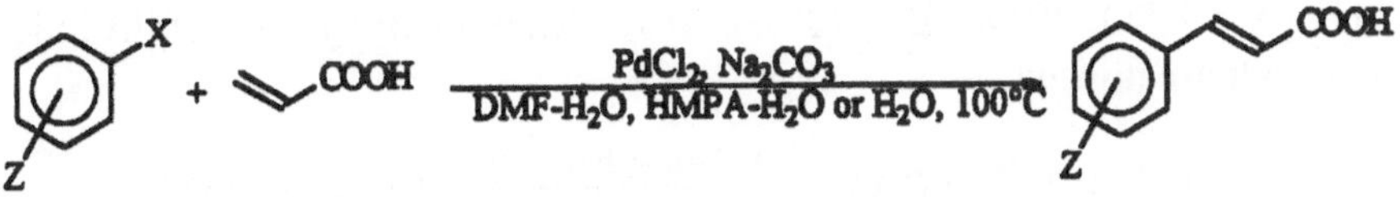

where $X = I$, Br; $Z = H$, p-Cl, p-MeO, p-Me, p-Ac, p-NO$_2$, p-CHO, p-OH, m-COOH, etc.

Acrylonitrile can also be successfully used. The reaction with water-insoluble aryl iodides is carried out in HMPA–H$_2$O mixture (9:1, v/v), giving high yields of the corresponding cinnamonitriles. Water-soluble aryl iodides like iodobenzoate and iodophenolate ions readily react in neat water. Palladium acetate (1 mol%) is usually used as the catalyst precursor. Unlike most other Heck reactions, which give almost exclusively (E) isomers of the products, the reactions with acrylonitrile yield a mixture of (E) and (Z) isomers with 3:1 to 4:1 ratio, close to that observed under conventional anhydrous conditions [94].

The reaction may be accelerated and performed under milder conditions upon the addition of acetate ion, as in:

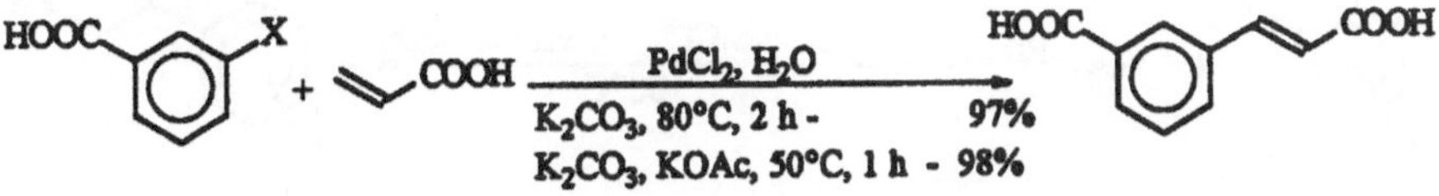

The presence of water in the solvent has a definite positive influence on reaction rate and efficiency of the catalytic system. Thus, the reaction of bromobenzene with acrylic acid catalyzed by Pd(OAc)$_2$ in the presence of tri(o-tolyl)phosphine and K$_2$CO$_3$ gave only 12% of cinnamic acid in DMF, while on the addition of 10% (v/v) water the yield increased to nearly quantitative. Further addition of water had no noticeable influence on the yield.

The reaction with water-soluble iodoarenes, such as o-iodobenzoic acid, showed record-making turnover numbers. As little as 0.0005 mol% Pd(OAc)$_2$ was sufficient to achieve full conversion, which is equivalent to 200 000 catalytic cycles. Therefore, it is not surprising that even palladium black can be successfully used as the catalyst precursor, as even traces of soluble palladium can drive the catalytic cycle.

Heck reaction with acrylic acid can be carried out at room temperature if diaryliodonium salts are taken as the arylating agents in water [95]. At room temperature, only one aryl group of the iodonium salt is transferred to the product:

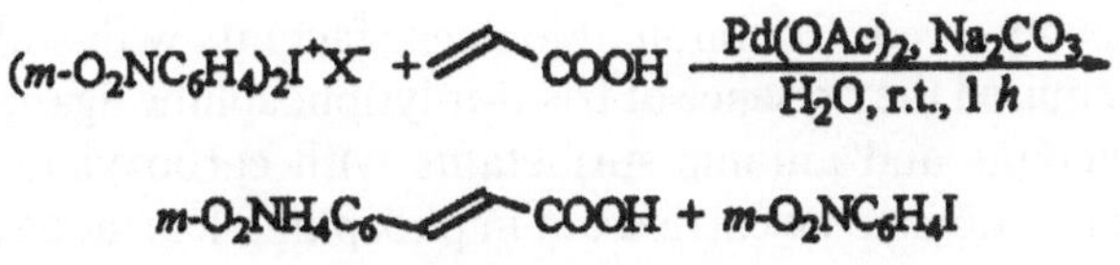

where $X = HSO_4^-$ (97%), BF_4^- (80%). At 100°C, both aryl groups of the iodonium salt are utilized:

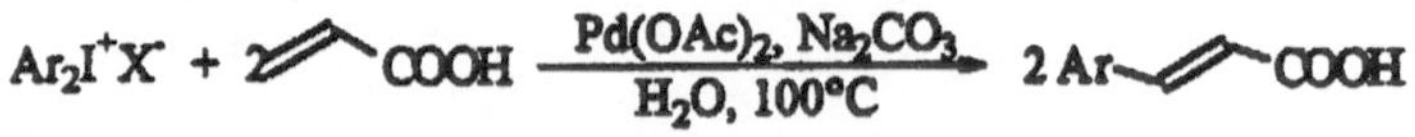

$$Ar_2I^+X^- + \text{2} \diagup\diagdown COOH \xrightarrow[\text{H}_2\text{O, 100°C}]{\text{Pd(OAc)}_2,\ \text{Na}_2\text{CO}_3} 2\,Ar\diagdown\diagup\diagdown COOH$$

where $X = HSO_4^-$, BF_4^-.

Arylation of a hydrophobic olefin such as styrene requires a significant change of procedure. In this case the use of the system $Pd(OAc)_2$–inorganic base, which is highly efficient for arylation of hydrophilic olefins, proved to be much less useful. Only when the reaction was carried out in a biphasic system (with an aqueous solution of K_2CO_3 forming one phase, and styrene plus aryl halide forming the other phase, in the presence of either 10 mol% of Bu_3N or 5 mol% Bu_4NBr) was the arylation of styrene with a broad range of aromatic and heteroaromatic bromides and iodides performed successfully. In most cases (except only the reactive aryl iodides) the addition of phosphine ligands (PPh_3 or $(o\text{-}MeC_6H_4)_3P$) was necessary to achieve high yields of stilbenes:

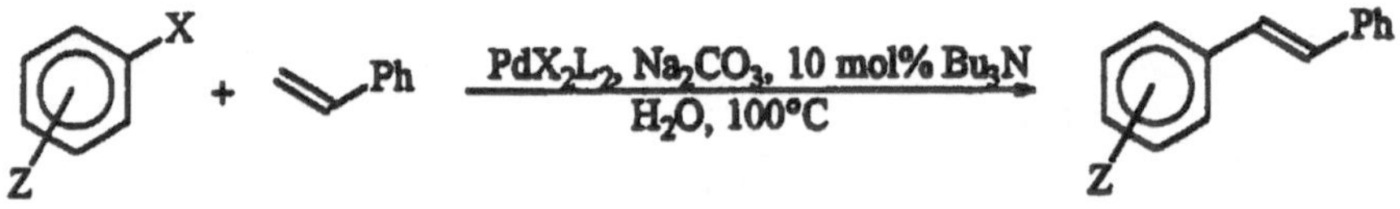

The reaction with some of the less-reactive aryl bromides bearing electron-donor substituents, such as o-bromoaniline or 5-bromo-2-aminopyridine, is better carried out in aqueous DMF with high content of water, in the presence of an inorganic base such as K_2CO_3 and palladium catalyst with (o-$MeC_6H_4)_3P$ ligands:

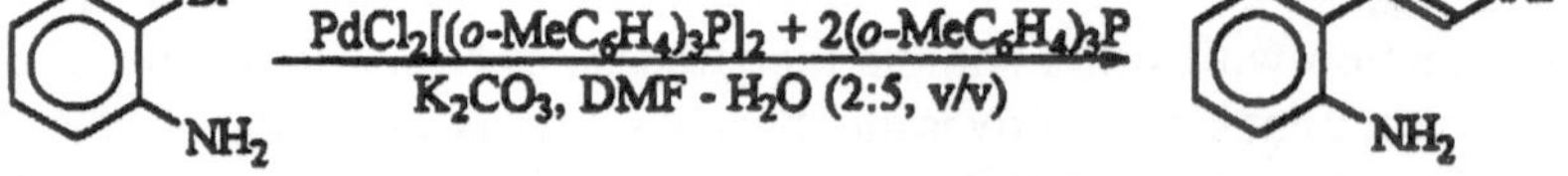

An alternative and quite general method to carry out the Heck reaction with water-insoluble reagents is the application of solubilization in microemulsions. Microemulsions formed by both cationic and anionic surfactants were highly efficient as media for arylation of styrenes, 4-vinylpyridine, acrylic acid esters and other olefins with a wide range of aryl iodides, including solid high-melting compounds like p-diiodobenzene, which are almost absolutely insoluble not only in water but in most aqueous solvents. A strong dependence of the product yield on the nature of the main surfactant was observed. Common anionic surfactants with sulfo or sulfate headgroups required the presence of tris(o-tolyl)phosphine ligand, while both cationic surfactants and anionic surfactants with carboxylate headgroups permitted the reactions to be carried out in phosphine-less mode [96].

The arylation of allyl- and vinylphosphonates can be performed not only using a traditional technique in organic solvents or without solvents in the presence of tertiary amine, but also in water–organic media with sodium hydroxide or carbonate as base. This approach allows one not only to lower the reaction temperature from 100°C to 70–80°C and to increase the yield of products, but also to shorten markedly the reaction time, as high yields are typically achieved in 1 h as compared with 12–16 h needed for the conventional procedure. The reactions in the absence of phosphines and tertiary amines lead to the selective formation of (E)-phosphonates [97]:

$$(EtO)_2P\overset{\displaystyle O}{\underset{R}{\|}}\diagdown\diagup + ArI \xrightarrow[\text{DMF - H}_2\text{O (9:1, v/v), 80°C, 1 h}]{\text{Pd(OAc)}_2,\ \text{Na}_2\text{CO}_3\text{ or NaOH}} (EtO)_2P\overset{\displaystyle O}{\underset{R}{\|}}\diagup\diagdown Ar$$

where R = H, Me, Ph.

The arylation of allylphosphonates in the presence of Et$_3$N at 100°C is complicated by the intervention of allylic rearrangement. On the other hand, the reaction in aqueous DMF in the presence of NaOH at 75–80°C leads exclusively to γ-arylallylphosphonates. Thus, the procedure employing aqueous solvent is both markedly milder and more selective:

$$(EtO)_2P\overset{\displaystyle O}{\|}\diagdown\diagup\diagdown + ArI \xrightarrow[\text{DMF - H}_2\text{O (9:1, v/v), 80°C}]{\text{Pd(OAc)}_2,\ \text{Na}_2\text{CO}_3\text{ or NaOH}} (EtO)_2P\overset{\displaystyle O}{\|}\diagdown\diagup\diagdown\diagup Ar$$

The use of quaternary ammonium salts allowed the Heck arylation of methylacrylate to be performed in a biphasic system in which organic reagents formed one phase, and water with potassium carbonate formed the other phase. The reactions were run in the presence of tetrabutylammonium salts, and thus may be regarded as an example of a phase-transfer reaction. It was shown that not only the phase-transfer agent, but also water, was essential for the reaction to give reasonable yields. Even the small amounts of water contained in the hydrated sample of tetrabutylammonium chloride were successful in accelerating the process and increasing the yield. Under such aqueous and phosphine-less conditions the reactions can be carried out at lower temperatures, and almost quantitative yields of methyl cinnamate are obtained at 50°C after 2 h, and even at room temperature after 24 h. However, the procedure described cannot allow correct comparison with other experimental techniques of Heck arylation, as too few examples for only methyl acrylate, which is known as one of the most reactive olefins in the Heck reaction, have been given [98].

A marked positive effect of water was observed in Heck reactions of 2,3-dihydrofuran with iodo and bromo derivatives of pyrimidines and other nitrogen-containing heterocycles [52]. In several cases reactions that could not be achieved under conventional conditions (DMF, Et$_3$N + NaOAc as base) took place in aqueous ethanol (1:1, v/v) with $NaHCO_3$–Et_3N mixture in the presence of Bu_4NCl (Scheme 5.8). Another notable feature of the reactions in aqueous solvent was a partial or full suppression of double-bond migration, which is the most common complication in Heck arylation of olefins having the ability to isomerize through the addition–elimination mechanism. This procedure, applied to protected glycal, can be used as an approach to C-nucleosides (Scheme 5.8).

The Heck reaction in aqueous media can be applied to reactions with aryl iodide bound to a polymeric resin support through an ester or amide link [99]. Organic reactions with immobilized reagents are now extensively studied as a tool for automated preparation of a large number of potentially bioactive molecules (the so-called combinatorial synthesis [100]). Heck arylation turned out to be potentially suitable for this purpose as, when run in aqueous media, the reaction is quite selective, gives high yields of target products under very mild conditions, and is applicable to a large number of functionalized olefins, thus generating a considerable variety of supported molecules suitable for further modification:

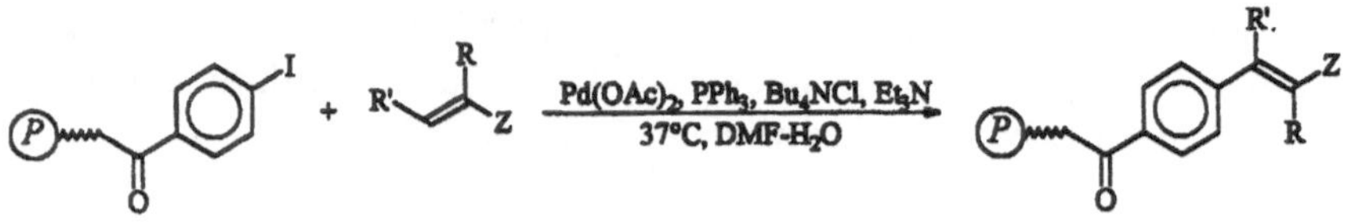

where Z = COOEt, CN, $CONH_2$, $CONR_2$, CHO, etc.

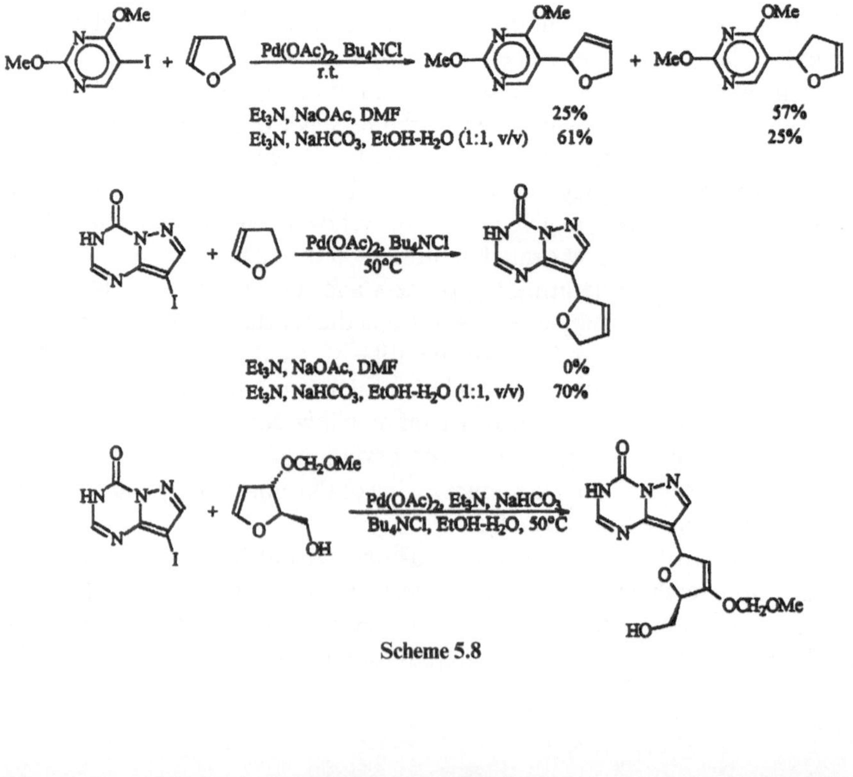

Scheme 5.8

Heck reactions were also conducted in homogeneous aqueous systems in the presence of the water-soluble phosphines TPPMS [7] and TPPTS [70], though the examples published are quite scarce and permit only the general conclusion that this technique is also suitable. Very recently it was shown that the Heck reaction in homogeneous aqueous solvent can be applied to intra-molecular cyclizations, as in the example below:

PdCl$_2$, TPPTS, (iPr)$_2$NEt
MeCN-H$_2$O, 70°C

Moreover, the use of an aqueous catalytic system leads to a dramatic change of the regioselectivity of ring closure. While the standard method of Heck-type cyclization (anhydrous solvent, Ag$_2$CO$_3$ as the base) is well known to give the exo-cyclization product, the aqueous method gave predominantly the endo-product [101]:

Pd(OAc)$_2$, Ph$_3$P, Ag$_2$CO$_3$, (iPr)$_2$NEt
MeCN, 90°C

PdCl$_2$, TPPTS, (iPr)$_2$NEt
MeCN-H$_2$O, 70°C

At high temperatures liquid water loses its famous structure and behaves more like an organic solvent with lower polarity. Thus, water heated to high temperatures either below or above the critical point can serve as a solvent for reactions with even highly hydrophobic substrates. This approach was tested for Heck reactions [102, 103]. The reactions can be run in both super-heated water (260°C) and supercritical water (400°C), though with the latter solvent all kinds of side-reactions (hydrogenation, hydrogenolysis, etc.) occurred. Olefins can be generated *in situ* from the corresponding bromo derivatives by elimination of HBr, but not from chlorides, alcohols, or esters, with the sole exception of β-phenylethanol, which behaved as a source of styrene. All kinds of palladium derivatives, either with or without phosphine ligands, catalyzed the reactions. Aryl iodides, bromides, chlorides and triflates, as well as vinyl halides, were all reactive, and the range of olefins that took part in the reaction was also quite broad and included styrene, methacrylic acid and esters, allyl halides, etc., though under such harsh conditions the initial products of arylation underwent follow-up partial destruction, e.g. by ester hydrolysis and decarboxylation (Scheme 5.9).

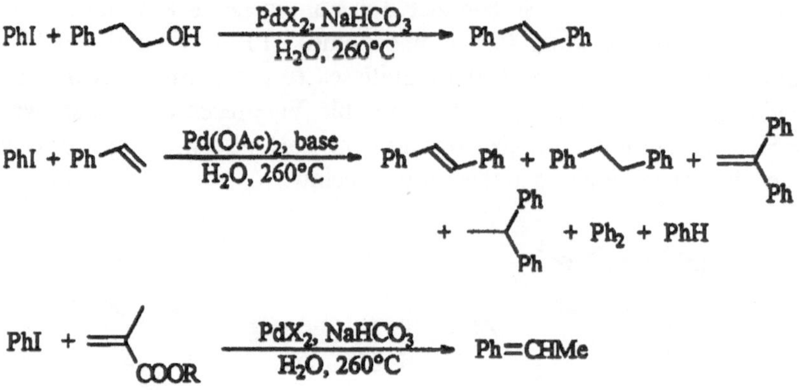

$$PhI + Ph\diagup\diagdown\diagup OH \xrightarrow[\text{H}_2\text{O, 260°C}]{\text{PdX}_2, \text{NaHCO}_3} Ph\diagdown\diagup Ph$$

$$PhI + Ph\diagdown\diagup\kern-4pt\diagdown \xrightarrow[\text{H}_2\text{O, 260°C}]{\text{Pd(OAc)}_2, \text{base}} Ph\diagdown\diagup Ph + Ph\diagup\diagdown\diagup Ph + \underset{Ph}{\overset{Ph}{=}\kern-6pt\diagup}$$

$$+ \underset{Ph}{\overset{}{\diagup}\kern-8pt\diagdown} + Ph_2 + PhH$$

$$PhI + \underset{COOR}{\overset{}{=}\kern-8pt\diagup} \xrightarrow[\text{H}_2\text{O, 260°C}]{\text{PdX}_2, \text{NaHCO}_3} Ph{=}CHMe$$

Scheme 5.9

5.4 Carbonylation

Catalytic carbonylation is one of the most important basic types of process, widely used for the preparation of oxygen-containing organic compounds, such as carboxylic acids and their derivatives, aldehydes, ketones, even alcohols, which may be formed under reductive conditions.

There are two basic types of carbonylation process that may have relevance in the context of this review. The first is closely related to cross-coupling reactions, being formally the cross-coupling of an organic halide (or triflate, or other electrophilic substrate with suitable leaving group) with the 'fictitious' anion $CONu^-$ which is actually represented by the pair CO + nucleophile:

$$RX + CO + Nu^- \rightarrow RCONu + X^-$$

The other involves the activation of double and triple bonds, being formally the addition of formic acid or a formic acid derivative broken at the C–H bond:

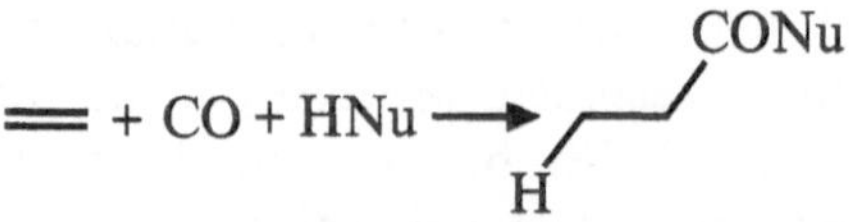

$$= + CO + HNu \longrightarrow \overset{\displaystyle CONu}{\underset{\displaystyle H}{\diagup\kern-6pt\diagup}}$$

The first type of process can be catalyzed by derivatives of palladium, nickel and cobalt, according to basically the same scheme of the catalytic cycle, involving oxidative addition of the reduced form of the metal catalyst across the R–X bond, insertion of CO to form a σ-acyl complex, which is then cleaved by the nucleophile to form the final product and regenerate the catalyst. Alternatively, the nucleophile may attack the carbonyl ligand with subsequent reductive elimination. This mechanism is also reasonably supported by close analogs, and first of all the elementary steps involved in the water-gas shift reaction.

Both types of mechanism have a considerable base of proven examples, but in particular cases the two pathways are usually indistinguishable, and possibly competitive. All types of intermediates involved are tolerant to water, and, like other types of catalytic processes in which nucleophiles or bases are taking part, carbonylation should benefit from aqueous techniques.

The most natural choice for exploiting the application of water is carbonylation leading to the formation of carboxylic acids, as in this case the nucleophile must be either water itself or its conjugate base hydroxide ion. As in all other reactions requiring hydroxide base, the application of phase-transfer catalysis is a common approach. Indeed, PTC is a well-established technique for carbonylation of organic halides [104]. Carbonylation of organic halides using various modifications of the phase-transfer technique was extensively studied and reviewed by Alper [105].

Carbonylation of aryl halides can be achieved under very mild conditions in aqueous DMF solvent in the presence of inorganic bases such as alkali-metal hydroxides, carbonates, acetates, etc. Various palladium salts and complexes ($Pd(OAc)_2$, K_2PdCl_4, $Pd(NH_3)_4Cl_2$, $PdCl_2(PPh_3)_2$, $PdCl_2(DPPF)$, etc.) can be used as catalyst precursors. The best results are achieved with simple palladium salts without phosphine ligands, and K_2CO_3 as the base, which usually allows the reaction to be carried out at room temperature and high yields of the corresponding benzoic acids to be obtained. Phosphine ligands definitely retard the reaction, which is especially well seen in the carbonylation of less-reactive aryl iodides with donor substituents. Thus, p-iodoaniline altogether refused to react in the presence of $PdCl_2(PPh_3)_2$, while the reaction catalyzed by 1 mol% of $Pd(OAc)_2$ gave 68% yield of p-aminobenzoic acid after 4 h [106]:

$$ArI \quad \xrightarrow[\text{DMF-H}_2\text{O (2:1), 25-50°C}]{\text{Pd(OAc)}_2,\ \text{CO, K}_2\text{CO}_3} \quad ArCOOH$$

$$Ar = p\text{-}ZC_6H_4\ (Z = NO_2,\ Cl,\ CN,\ Me,\ NH_2,\ \text{etc.});\ 2\text{-}C_{10}H_7,\ 2\text{-thienyl}.$$

The influences of electron-donor and electron-withdrawing substituents on the steps involved in the catalytic cycle of carbonylation are entirely different. Electron-withdrawing substituents facilitate oxidative addition across the carbon–halogen bond and possibly ligand exchange in the arylpalladium intermediate, whereas the electron-donor groups facilitate insertion of CO and reductive elimination. Experimental results clearly show that the net rate of the catalytic carbonylation process (catalytic efficiency expressed by turnover number per unit time) is limited by either the oxidative addition or ligand exchange steps. As Pd(II) derivatives are commonly introduced into reaction mixtures as catalyst precursors, the catalytic cycle (Scheme 5.10)

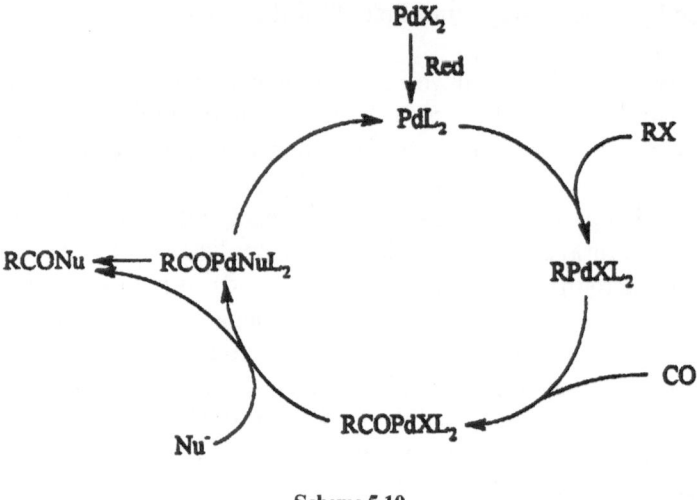

Scheme 5.10

must be seeded by any process leading to primary formation of catalytically active Pd(0) species. It is now reasonably well established that, in the presence of phosphines, the latter serve both as primary palladium reductant and as supporting ligands, which keep Pd(0) species from aggregation to clusters, nucleation and growth of inactive palladium metal particles (palladium black). In phosphine-less systems, it remains a matter of controversy what makes Pd(II) be reduced, as it may be any stray process with e.g. CO. But it is more important why in some cases phosphine-less palladium catalyst drives the catalytic cycle to completion without deactivation of unsupported Pd(0) and the precipitation of Pd black. Without phosphine ligands, Pd(0) species must be highly reactive and react with aryl halides faster than the rate of aggregation. Besides, in aqueous systems in the presence of inorganic base, Pd(0) must bear weakly bonded electron-rich ligands like OAc^-, C_3^{2-} and OH^-, which should promote oxidative addition, though at the same time may retard the cleavage of acylpalladium complex by nucleophile.

With water-soluble aryl iodides the carbonylation can be run in neat water in the presence of soluble palladium salts or complexes such as $Pd(OAc)_2$, K_2PdCl_4 or $Pd(NH_3)_4Cl_2$ and K_2CO_3 as the base under very mild conditions (25–50°C, 1 atm CO), though all aryl iodides studied must have reacted in the form of the respective salts (phenolates or carboxylates) and thus were bearing electron-donor groups that are known to retard oxidative addition [106]:

where $Z = m,p\text{-}COOH$, $o,m,p\text{-}OH$.

Ortho-iodobenzoic acid is the only exception, as it gave no carbonylation product under the same conditions as used for the reaction with the *para* isomer. It may be suggested that a carboxylate group in the *ortho* position chelates the palladium atom, and thus occupies the place of the carbonyl ligand suitable for migration (or, which leads to the same outcome, reductive elimination in the alternative mechanism). However, in the presence of excess iodide ions, which might have dechelated palladium, carbonylation took place and gave a good yield of phthalic acid:

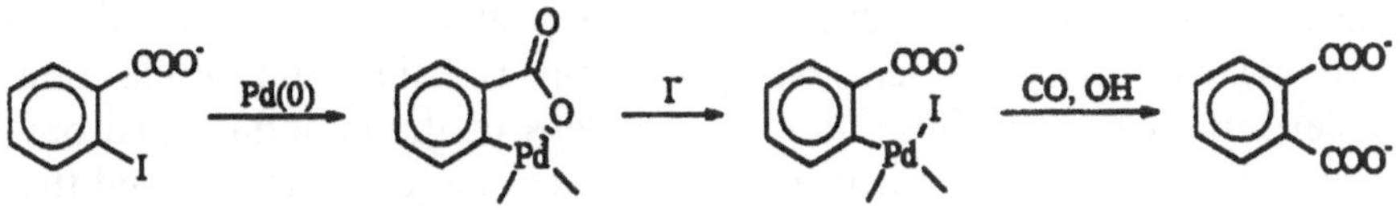

The carbonylation of *p*-bromophenol can be achieved in aqueous solution in the presence of NaOH. With K_2PdCl_4 as catalyst precursor the reaction takes 6 h at 0°C to give *p*-hydroxybenzoic acid in 82% yield. In the presence of $Pd(NH_3)_4Cl_2$ the conversion is even higher, but phenol is formed as a side-product owing to reductive debromination, possibly due to a water-gas shift-like reaction, occurring because ammine ligands may block some of the coordination places favorable for carbonyl migration:

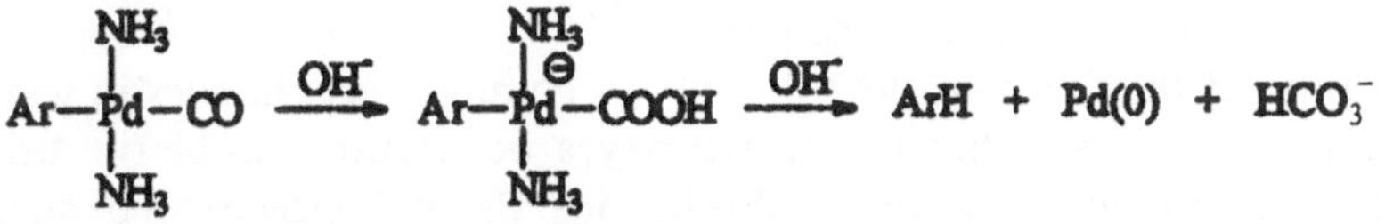

The use of tributylamine as a second phase and simultaneously the base, as in the previously described methods of Heck arylation and cross-coupling, can also be of use in carbonylation of *o*-bromoacetamides in a two-phase system Bu_3N–water at 100–130°C and 2–3 atm CO, affording anthranylic acids [107]:

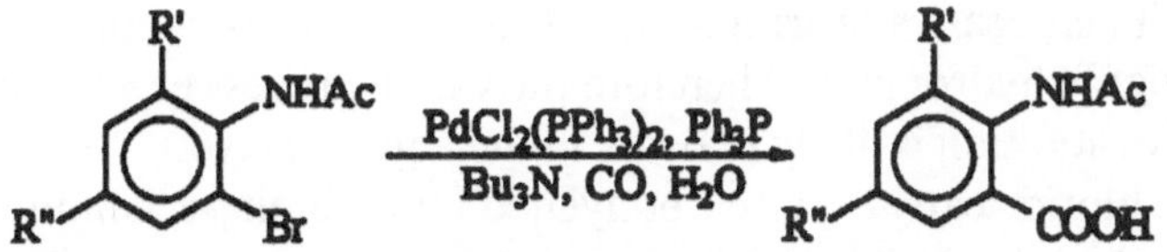

In an ingenious indirect approach toward the carbonylation of water-insoluble aryl iodides in neat water, the iodine atom is first oxidized to the iodyl group by as cheap an oxidant as commercial bleach solution. Because of either slightly enhanced solubility in water, or higher reactivity of iodyl derivatives toward oxidative addition, these compounds are readily carbonylated under very mild conditions with Na_2PdCl_4 as catalyst precursor and Na_2CO_3 as the base in neat water at 40–50°C [108]:

$$ArI \xrightarrow{\text{NaClO}} ArIO_2 \xrightarrow[\text{H}_2\text{O, 40-50°C}]{\text{Na}_2\text{PdCl}_4,\ \text{NaOH, CO (1 atm)}} ArCOOH$$

Interestingly enough, iodosylarenes were not carbonylated, but instead underwent reduction to iodoarenes, though the derivatives of I(III) can be reactive in other palladium-catalyzed reactions, e.g. cross-coupling (*vide supra*).

Certainly, such a method is not applicable to any aryl iodides that bear readily oxidizable groups, though for more rugged molecules it can be of considerable preparative value, as iodyl derivatives can be formed and carbonylated using the same setup without isolation and purification. Besides, this example shows that all the steps of the important hydroxy-carbonylation catalytic cycle can be performed in aqueous media, and that the key problem in making carbonylation run in water is to enable oxidative addition with water-insoluble aryl iodides.

Water-insoluble aryl iodides can be hydroxycarbonylated directly using solubilized media, such as canonical microemulsions or Shinoda's swollen micelles. Microemulsions formed by cationic and anionic surfactants can be used for both liquid and solid aryl iodides, giving high yields of benzoic acids in the presence of palladium salts in phosphine-less mode and inorganic bases. Though the microemulsions always contain aliphatic alcohols used to adjust the hydrophile–lipophile balance of the surfactant system, the formation of esters was never observed [109].

In spite of the fact that the solubility of common aryl halides in water is extremely low, it was shown that hydroxycarbonylation can be run using a biphasic technique with water-soluble ligands. Bromobenzene can be successfully converted into benzoic acid in the system containing TPPTS ligand [110]. It was shown that, unlike in the case of coordination with Ph_3P, Pd(0) forms a tricoordinated complex $Pd(TPPTS)_3$, which shows no measurable ability to bind the fourth TPPTS ligand, possibly because TPPTS is more bulky while being less basic than Ph_3P. Two key intermediates of the catalytic cycle – $PhPdBr(TPPTS)_2$ and $PhCOPdBr(TPPTS)_2$ – were obtained and characterized. Each bears a large number of ionic groups per molecule and thus must be highly hydrophilic. Therefore the whole process except the first step of oxidative addition is likely to occur in the aqueous layer.

Benzyl chloride can also be carbonylated under biphasic conditions, in the system in which the bulk organic phase is formed by an inert solvent, such as heptane. It is quite interesting that both hydrophobic triphenylphosphine and hydrophilic TPPMS can serve as ligands for the process. Even more intriguing are data on the influence of amphiphiles. It was shown that the addition of surfactants such as $n\text{-}C_7H_{15}SO_3Na$ or $n\text{-}C_7H_{15}COONa$ leads to a marked acceleration of the carbonylation of benzyl chloride using a water-soluble catalyst, $PdCl_2(TPPMS)_2$, while no influence was observed if the catalyst was formed by a hydrophobic ligand [111]. This effect cannot be explained by a traditional approach to micellar effects, as here both key reagents and surfactant are anionic, and thus no binding of the ligand and catalyst in micelles can happen. Besides, the surfactants used are very short-tailed, and thus are too hydrophilic and possess very large values of critical micelle concentration. Micelles of such surfactants are quite small and too rigid to solubilize considerable amounts of hydrophobic reagents. It may be argued that here the sole role of amphiphiles is to increase the rate of mass transfer of benzyl chloride to the aqueous phase by decreasing the interfacial tension and increasing the area of the interface between the immiscible phases, as soon as we believe that the reactions catalyzed by hydrophilic phosphine complexes must occur in the aqueous phase.

Highly reactive alcohols can serve as substrates for hydroxycarbonylation, which in this case ought to be run in acidic solutions to facilitate the oxidative addition to the C–O bond. Thus, 5-hydroxymethylfurfural (HMF) was carbonylated to (5-formylfuryl-2)acetic acid using a water-soluble palladium complex with TPPTS ligand as the catalyst at 70°C and 5 bar CO pressure [112]:

A very interesting example of reversed biphasic carbonylation of simple liquid chloroarenes has been described [113]. In this method, chloroarene itself is used as the organic phase in a system containing aqueous alkali and palladium salt in the presence of tricyclohexylphosphine without any phase-transfer or solubilizing agents. Though aryl chlorides are notoriously reluctant to take part in palladium-catalyzed reactions involving oxidative addition, and usually require special techniques to reveal a noticeable reactivity [114], this approach was quite successful. As the final product-forming step of the catalytic cycle requires the presence of OH⁻ nucleophile in free or bound form, and it is well known that it is virtually impossible to transfer the hydroxide ion to a nonpolar organic phase, it remains to conclude that palladium complexes themselves must have the ability to perform as phase-transfer agents and to travel across phase boundaries. There is no direct experimental confirmation of this ability, but if the opposite were true, catalytic processes in heterogeneous systems without explicit addition of phase-transfer agents would not be feasible.

Besides carboxylic acids, carbonylation can give their derivatives or ketones if other nucleophiles were used to cleave the acylpalladium complex. Thus, esters and amides are formed with alcohols and amines, while ketones can be obtained in the presence of such carbanion synthons as organometallic compounds. Certainly, these processes leave a small margin for the intervention of water in any form, as in the presence of water the competition between the different nucleophiles would lower the selectivity, as, for example, in the co-carbonylation of organoboron compounds and aryl iodides, which gave mixtures of acids and ketones [115]:

$$PhB(OH)_2 + PhI \xrightarrow[\text{DMF-H}_2\text{O (1:2,v/v)}]{\text{PdL}_4,\ CO} Ph_2CO + PhCOOH$$

The higher nucleophilicity of amines allows them to win the competition, as only amides are usually formed in carbonylation in the presence of amides and hydroxide ions or small amounts of water. However, for less-reactive arylamines, water puts up a stronger competition. Thus, in the carbonylation of *p*-nitroiodobenzene in the presence of *p*-iodoaniline, which gives only *p*-$O_2NC_6H_4CONHC_6H_4I$-*p* in DMF, the addition of only 5 vol% H_2O results not only in sharp acceleration of reaction, but also in the formation of considerable amounts of nitrobenzoic acid. No amide was formed in the carbonylation with such a weak nucleophile as diphenylamine [116].

Carbonylation of unsaturated compounds (olefins and acetylenes) involves water at several steps of the catalytic cycle (Scheme 5.11). Besides a similar role of nucleophilic cleavage of acyl complex, water is believed to play a major role in initiating the catalytic cycle by forming metal hydride due to the water-gas shift reaction.

As in other processes involving the addition of transition-metal complexes to double bonds (carbometallation, hydroformylation, etc.), regioselectivity may vary both because of addition to the double bond, and because of follow-up elimination–addition reactions leading to the migration of unsaturation along carbon chains. As is evident from the catalytic cycle, selectivity

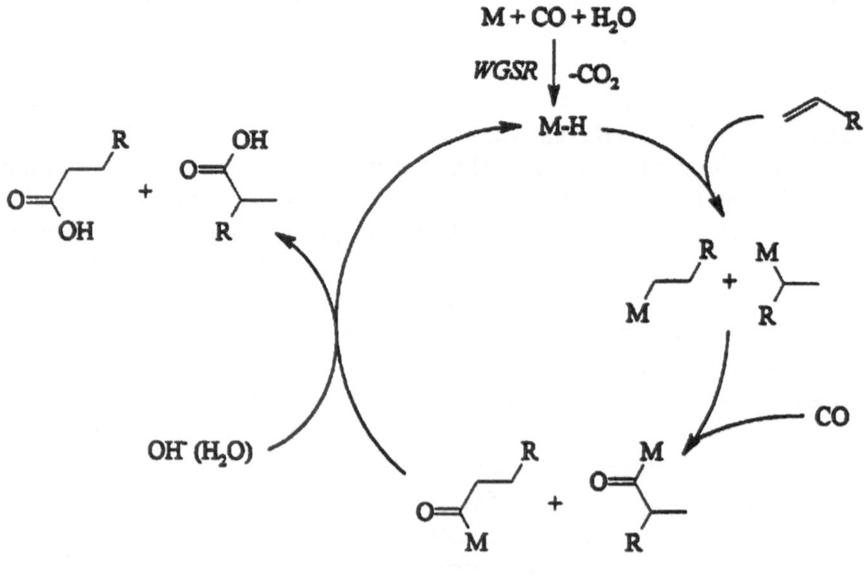

Scheme 5.11

problems in such systems may arise not only from different regioselectivity, but also from other well-known reactions that can take part between the species involved – hydrogenation, hydration, telomerization, etc., to mention only the most important. Selectivity issues are usually so serious that the application of this, potentially highly appealing, process, is severely undermined, except in the simplest cases, like the $Ni(CO)_4$-catalyzed hydroxycarbonylation of acetylene (Reppe synthesis), which still has some industrial importance. Recent successes in the design of ligands for carbonylation of unsaturated compounds may eventually disclose a large synthetic potential of this reaction [117]. Formic acid derivatives, such as methyl formate, in the presence of transition-metal complexes can trigger a WSGR-type chemistry, e.g. in reactions with olefins [118].

Interesting cyclizations sometimes occur under the conditions of hydroxycarbonylation. For example, the carbonylation of internal acetylenes in the presence of rhodium carbonyls and water gave 3,4-substituted 2,5-dihydro-2-furanones in high yields with excellent selectivity. Using D_2O it was shown that it was indeed water that gave up hydrogen atoms for reduction of one of the carbonyl groups to a methylene group [119]:

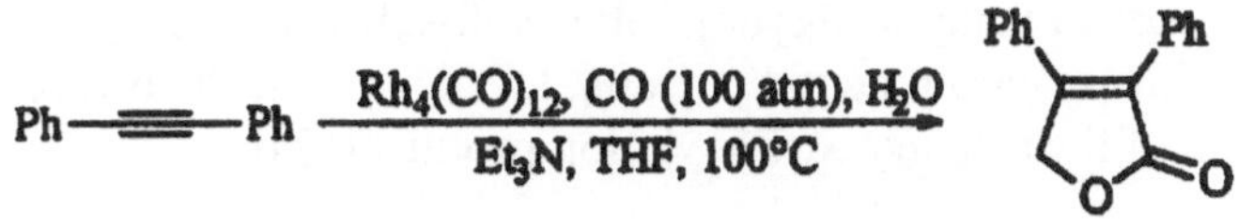

The reaction can be extended to terminal alkynes, and even to parent acetylene, giving 2(5H)-furanone in good yield [120]. Additional cyclization can happen with 2-alkynylbenzaldehydes, giving interesting tricyclic lactone products in good yields [121].

A small change of reaction conditions, or a variation of substrate structure, often results in the formation of entirely different products. For example, the carbonylation of internal acetylenes with the trimethylsilyl group at one terminus in the presence of Wilkinson catalyst gave indenones, thus involving the activation of the aromatic C–H bond, regardless of whether electron-donor or electron-withdrawing substituents were in the aromatic ring. On the other hand, the reaction with acetylenes substituted at the *meta* position of the benzene ring was highly regioselective to yield only the less sterically hindered 5-substituted indenones [122]:

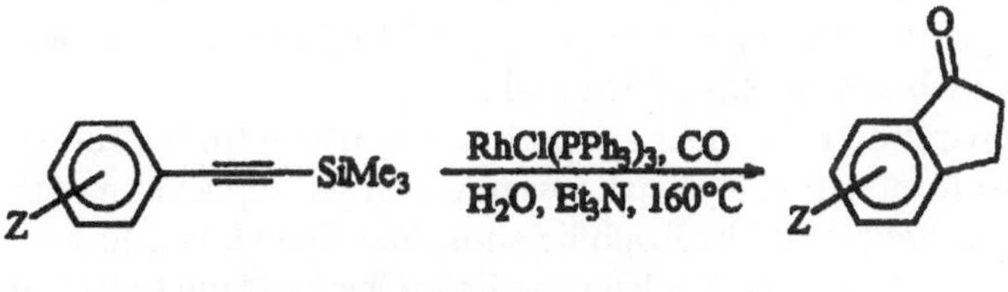

Z = H, Me, OMe, Ac, CN, CF$_3$, Cl, COOEt.

Similar products are obtained from internal acetylenes in the presence of cobalt catalyst [119]:

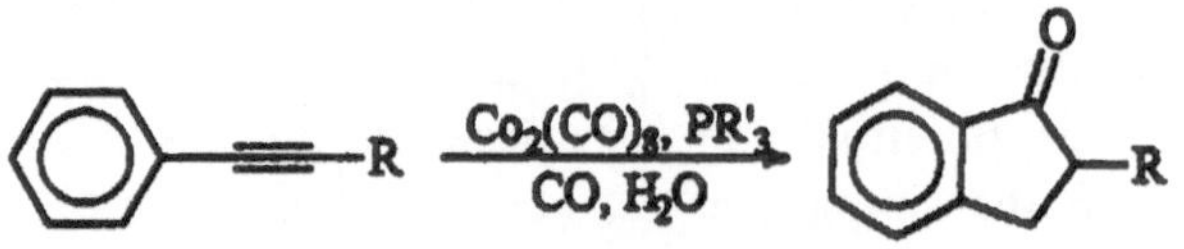

Carbonylation of acetylenes using palladium catalyst can show another case of dramatic alteration of selectivity depending on the nature of the catalyst and other factors. Thus, the carbonylation of naphthylacetylenes can be run under very mild conditions, even under atmospheric pressure of CO, to give α-naphthylacrylic acids, thus opening a convenient approach to the non-steroidal anti-inflammatory drug naproxen [123]:

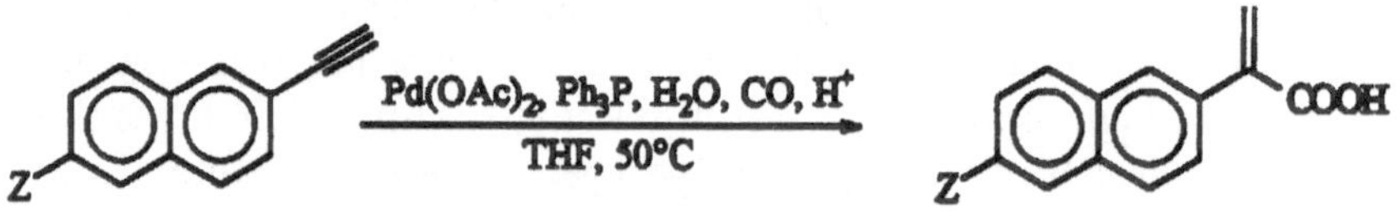

Meanwhile, the application of cationic palladium complex to the carbonylation of readily available alkynols allowed β-substituted acrylic acids to be obtained. The reaction actually leads to dienoic acids owing to facile dehydration of the primary carbonylation product [124]:

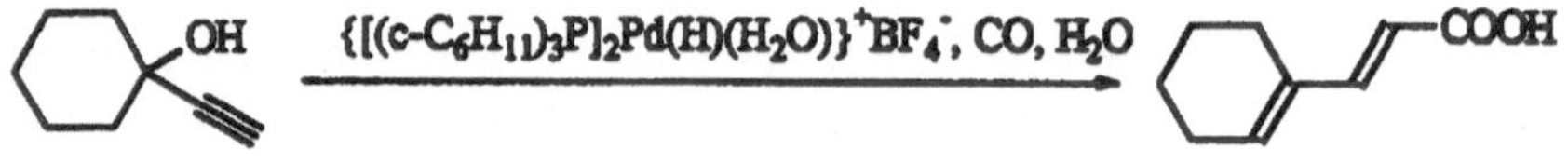

5.5 Hydroformylation

Hydroformylation, unlike the processes discussed above, is a clean process requiring only the substrate, two gaseous reagents, the catalyst and nothing else, and producing a single product (or a mixture of isomeric products) without co-formation of any other organic or inorganic species. In this respect hydroformylation and hydrogenation are similar, and it is not surprising that similar approaches and technologies have been developed for these catalytic processes. It seems natural that water can play a sole role in hydroformylation, serving as a technological tool used for reaction mixture workup and separation of products and catalysts.

Phase-separation techniques are particularly useful for hydroformylation. Olefin hydroformylation is the process in which the application of the phase-separation technique and hydrophilic phosphine ligands was first realized on an industrial scale. Rhone-Poulenc and RuhrChemie claim to have manufactured an average 300 000 tons of butyric aldehyde per year by biphasic rhodium-catalyzed hydroformylation of propene since the mid-1970s [125]. How can it be that the history of aqueous transition-metal catalysis reviewed in this book appears to begin only after the industrial implementation of a catalytic process utilizing one of the basic ideas of this branch of chemistry? The paradox is an illusion, as we can see and discuss only results belonging to public-domain research. It is perfectly evident that the state of corporate research devoted to aqueous catalysis must be different.

Indeed, sulfonated phosphine ligands appear to have been first developed and studied for industrial application in hydroformylation. An early example of hydroformylation in a two-phase system (1-hexene–water) with the TPPMS analogue of the Wilkinson complex appeared in the public-domain literature in the 1970s [126].

Hydroformylation can be performed with cobalt and rhodium catalysts. The latter are extremely expensive, but their use is justified because they usually show much higher efficiency and selectivity, with respect to both regioselectivity (the ratio of normal to branched aldehydes in hydroformylation of terminal olefins) and the absence of side-reactions, e.g. the reduction of aldehydes to alcohols. The selectivity of catalyst is particularly important for aqueous methods to avoid the concurrent processes triggered by the water-gas shift reaction. Catalyst recycling is a vital task only for rhodium-catalyzed processes, and that is clearly reflected by research efforts.

The hydroformylation of propene in a biphasic system using rhodium complex with TPPTS ligand can be thought of as a perfect implementation of the ideal phase-separation technique. All new hydrophilic phosphine ligands are usually first tried in hydroformylation, with two primary goals: (i) to improve selectivity with respect to the ratio of normal to branched products; and (ii) to enhance productivity of the biphasic system. The latter goal depends on an intrinsic limitation of the biphasic system, in that the reaction takes place in the aqueous layer and the rate (turnovers per unit time) is limited by the sparse solubility of olefins in water and by mass transfer of olefin across the very small interfacial boundary between the organic and aqueous layers.

The productivity of the ideal biphasic system in a batch reactor is appropriate only for olefins having reasonable solubility in water, such as propene and several other lower alkenes, as well as some functionalized olefins like acrylic acid esters, etc. However strange it may seem, the solubility of the lower alkenes in water is quite good. (The measure of hydrophile–lipophile balance of a given compound, i.e. the logarithm of octanol–water partition coefficient ($\log p$), is equal to 1.77 for propene [127], which means that the concentration of propene in aqueous solutions can be much higher than 0.01 M even under normal conditions.) Thus, we may safely conclude that the hydroformylation of propene is carried out in a homogeneous solution of propene in water, and the second organic phase functions as a mere feedstock of olefin to keep the concentration of reagents in aqueous solution roughly constant during the process. Besides, the organic layer may be needed as an extractant for the product, though again, the solubility of lower aldehydes in water is so high that these compounds can be classified as hydrophilic.

This circumstance is quite typical for the so-called biphasic catalysis. The cleaner the system, i.e. the better the separation of hydrophilic and hydrophobic components, the lower is the productivity of the system. Therefore, all improvements to the biphasic technique pursue the main goal of enhancing the degree of contact between immiscible phases, and thus move away from the ideal model. Thus, the ideas in this area deal mostly with the technical aspects of finding an economic compromise between catalyst recovery and productivity.

The industry needs cheap and efficient methods for the hydroformylation of long chain olefins, the solubilities of which in water are negligible. (The log p values for C_8–C_{10} olefins are lower than 4.5, which means that the concentration of these hydrocarbons in water is lower than the detection limit of common analytical methods, and the values of partition coefficients are not measured directly, but estimated by, for example, HPLC retention times.) As may be expected, such olefins are either totally unreactive, or react with inappropriately low rates under pure biphasic conditions.

To improve the productivity, several basic approaches are utilized. The most straightforward solution is to add a cosolvent like lower alcohols (MeOH, EtOH) miscible with water, thus improving the solubility of olefins in the aqueous phase. A certain additional effect may be obtained by varying the nature of the catalyst precursors, though, under the conditions used for hydroformylation, rhodium complexes are rapidly transformed into [HRh(CO)L$_3$] thus masking the differences [128, 129].

In order to obtain a considerable effect, the use of catalysts that are more soluble in aqueous alcohols than in neat water is required. One such ligand described recently is sulphos, forming zwitterionic complexes with rhodium, having net zero charge:

With this ligand the reactions can be carried out in the biphasic systems MeOH–hydrocarbon or MeOH–H$_2$O–hydrocarbon, which can afford reasonable productivity at least with such alkenes as hexene. It is quite interesting that, in the system without water, the reaction gave alcohols as the main products, whereas in the aqueous system the only oxygen-containing products were aldehydes. The yields are given below each of the main products of the reaction, the values in parentheses corresponding to the data obtained in the system containing water [14, 130]. Note that, as a commercial hexene fraction (containing mostly *trans*-2-hexene, as well as other isomers) was taken for hydroformylation in this reaction, no reliable conclusions about the regioselectivity of this reaction and the comparative activity of the catalyst can be made. The formation of large amounts of alkanes by hydrogenation can possibly be due to the reaction of internal olefins, which are much less reactive in hydroformylation than the terminal ones:

$$C_6H_{12} \xrightarrow[\text{80°C, MeOH-(H}_2\text{O)-isooctane}]{\text{Rh(SULPHOS)(CO)}_2,\ CO,\ H_2}$$

5 (37)   9 (17)

43 (traces)   14 (no data)

The extreme of such an approach is to exclude water altogether, but still use the idea of two immiscible liquid media, separately holding the catalyst and the substrate. It is well known that some fluorinated organic compounds, owing to similar solvency properties, may behave as good substitutes for water in some areas. Thus, the system composed of a hydrocarbon, a fluorinated liquid and a rhodium complex with long-chain branched polyfluoroalkylphosphine ligands ('fluorous ponytails'), rendering to the catalyst the affinity to the fluorinated phase, may function like a water-based biphasic system. However, this system has a crucial advantage, as, at the elevated temperatures at which the hydroformylation is carried out, both phases become miscible and the reaction runs in a homogeneous solution. Unfortunately, insufficient data have been published to date to evaluate the perspectives of this elegant idea [131].

The rate of hydroformylation in a biphasic system can be increased by the addition of a hydrophobic co-ligand. However simple such an approach may be, the results are spectacular and hard to explain. The addition of triphenylphosphine to a biphasic system for the hydroformylation of l-octene catalyzed by [HRh(CO)(TPPTS)$_3$] resulted in an increase of catalytic activity (turnover frequency) by 1–2 orders of magnitude. Meanwhile, the use of specially prepared mixed complexes [HRh(CO)(TPPTS)$_{3-x}$(Ph$_3$P)$_x$] gave only a marginal increase of activity. Free Ph$_3$P ligand is likely to promote the reaction at the interfacial boundary, operating, to some extent, like a phase-transfer quaternary ammonium catalyst in reactions with hydroxide ions, which cannot be transferred into the bulk of the nonpolar organic phase, but can be bound by lipophilic counter-ions at the interface [132].

Another general approach utilizes the systems with a variable hydrophile–lipophile balance (HLB). A catalyst suitable for this technique must have the ability to change its HLB depending on the variation of external factors. The reaction is carried out in organic solvent with catalyst, the HLB of which is adjusted to make it hydrophobic. After the reaction, the HLB is changed toward the hydrophilic end, and the catalyst is extracted into water, and thus separated from the products. Recycling is achieved by re-adjusting the HLB back to hydrophobicity.

In order to behave in such a manner, the catalysts (or the ligands) must possess amphiphilic properties. Ionic amphiphiles can change their HLB by changing their ionic state: hydrophobic neutral amines become hydrophilic surfactants on quaternization or protonation. Such examples were already discussed in section 5.1.5 devoted to water-soluble phosphines.

More interesting is the case of non-ionic surfactants bearing polyoxyethylene chains. Such ligands can be obtained by polymerization of ethylene oxide induced by nucleophilic anions, exactly as done in the preparation of industrial surfactants (polyethoxylated nonylphenols, etc.):

At room temperature such surfactants are mostly hydrophilic owing to the hydration of the polyethoxy chains, though the exact value of HLB strongly depends on the number of ethoxy units in the chain. At higher temperatures, owing to partial dehydration, the chains gradually lose their hydrophilicity, and HLB becomes more hydrophobic. At some temperature hydrophilicity and hydrophobicity become balanced, and after this point the hydrophobicity prevails, and the surfactant becomes soluble in nonpolar organic liquids (for brevity called 'oils') and insoluble in water. In reality the process is more complex, as we are dealing not with the surfactant itself, but rather with its solution in water or oil, in which the surfactant is aggregated to form micelles or more complex aggregates. Thus, at lower temperatures we have normal micelles in water solution, while at high temperature we have inverted micelles in oil. Phase inversion thus occurs when HLB changes from hydrophilic to lipophilic. Visually, the transparent solution of surfactant in water at some temperature (called the 'cloud point') becomes opaque, and then an immiscible phase of oil appears and separates. It is well known that most such surfactants with number of ethoxy units lower than 30–50 have a phase inversion temperature below the boiling point of water, which is quite convenient. The phenomenon of phase inversion is well known in surfactant science, and has a very broad range of uses, including the concentration of impurities from water, etc.

In hydroformylation using a phase-separation technique, this idea was realized quite recently. Tris(p-hydroxyphenyl)phosphine was used for modification with three polyoxyethylene chains of moderate lengths, possibly in order to increase the net hydrophilicity without the need to grow a single long chain. This approach deserves a warning, as such multi-head surfactants are much less studied, but are known to form more complex aggregates than simple micelles, the formation of which is impossible in such cases for purely geometric reasons. More complex aggregates would not have a distinct phase inversion behavior, and would have unpredictable solubilization properties. The rhodium complexes formed by such a ligand were efficient hydroformylation catalysts for higher olefins in the organic phase at temperatures above 100°C. On cooling, the catalysts were claimed to redissolve in water, and could be recovered [133].

This approach may have a number of problems, which are difficult to overcome. Polyethoxylated surfactants are famous for their very low critical micelle concentrations, and thus have outstanding interfacial properties, being extremely efficient solubilizers, detergents and emulsifiers. In a realization of the biphasic technique, these properties may ruin the whole scheme, as the emulsions formed may be very hard to destroy, and solubilization may decrease the recovery of catalyst.

The surfactants formed by block copolymers of ethylene and propylene oxides, built of the following units, can be even more versatile:

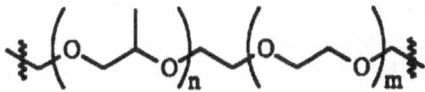

In this case, there is an additional parameter allowing one to control the HLB in a more precise and predictable way, as the number of propylene oxide units allows fine-tuning of the lipophilicity of the surfactant tailored to the phosphine ligand [134].

Hydroformylation of olefins, especially long-chain ones, in aqueous systems with hydrophilic catalysts has another specific feature, not relevant for reactions in anhydrous media. As is evident from the generally adopted scheme of the catalytic process (Scheme 5.12), the intermediates are typical amphiphilic compounds, being built of a lipophilic part from the olefin and a hydrophilic part from the ligand. The products of reaction, long-chain aldehydes, also possess a certain degree of amphiphilicity in aqueous media. Thus, it is evident that interfacial effects associated both with the influence of added surfactants and with the amphiphilic properties of the species involved in the reaction must have a strong influence on hydroformylation in aqueous media.

The picture in Scheme 5.12 can be read in a metaphorical sense: hydroformylation catalyzed by hydrophilic catalysts promises some hair-raising findings. Indeed, it was shown that the solid complex [HRh(CO)(TPPTS)$_3$] is an efficient catalyst for hydroformylation of 1-octene in hexane solution, giving quantitative conversions. As the complex is claimed to be absolutely

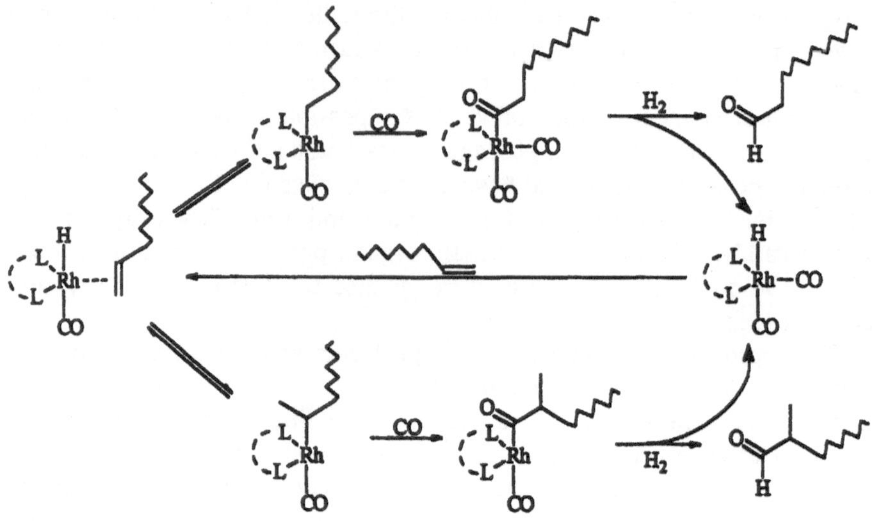

Scheme 5.12

insoluble in nonpolar solvents, it may be concluded that the reaction proceeds at the liquid–solid interface [135]. If this is true, the catalyst should have a tremendous catalytic activity to run the reaction with only tiny amount loaded into the reactor, as in this case only molecules located at crystal edges would be accessible. However, it is commonly accepted that the activity of catalysts with sulfonated phosphine ligands under comparable conditions in homogeneous solution is somewhat less than the activity of the basic triphenylphosphine complex. Therefore, we may either suppose that the catalyst packed in crystal planes is more reactive due to some cooperative effects (too beautiful to be true!), or we are left with a trivial hypothesis that, at the temperatures used to carry out the hydroformylation reactions, the catalyst is solubilized in nonpolar media when the highly hydrophilic rhodium complex binds to an olefin molecule and becomes an amphiphilic intermediate. So we actually have the situation discussed above for numerous reactions catalyzed by palladium complexes, when the intermediates formed along the pathways of the catalytic cycle have the ability to cross phase boundaries and behave like phase-transfer agents. Certainly, this hypothesis may be valid for reactions in biphasic systems.

Anyway, it is well known that micellar effects are common in hydroformylation conducted in all varieties of aqueous systems. The addition of surfactants often helps the reactions in biphasic media, possibly due to solubilization of hydrophobic reagents in micelles, which thus function like cooperative phase-transfer agents. For example, the hydroformylation of unsaturated fats or methyl esters of unsaturated long-chain natural fatty acids (oleic, linolenic acids) can be successfully done using biphasic techniques with the TPPTS complex of rhodium. As can be expected, cationic surfactants are more efficient due to their interaction with anionic catalyst or intermediates of the catalytic cycle [136].

The solubilization and transfer of hydrophobic olefins to the aqueous phase can also be achieved using host–guest complexes with cyclodextrin molecules, though this approach is too expensive to be practicable for large-scale reactions. The application of cyclodextrins can be justified in cases when it can enhance the selectivity of reactions by a mechanism of molecular recognition. In this case, only those substrates which can fit into the cavities of cyclodextrin would be selectively solubilized and converted in a biphasic process [137, 138].

In order to avoid the limitations of low solubility and inefficient mass transfer across a small interfacial boundary, a group of methods based on immobilization was developed. The simplest system used for the biphasic process consists of two liquid phases – water and organic liquid. One of the liquid phases can be supported on a solid carrier with a proper affinity to give rise to 'supported liquid-phase catalysis' techniques, of which supported aqueous-phase catalysis (SAP or SAPC) is the most extensively elaborated method, in spite of quite a short history. The basis of this method is as contradictory as aqueous organometallic chemistry in general. Indeed, in this method the aqueous phase is absorbed on the surface of supports such as silica, taken either in the form of silica gel or as specially treated forms known as controlled porosity glasses (CPG). The use of silica as a support for numerous reagents is a powerful and versatile method in organic chemistry, but in most cases the aim of such methods is just the opposite – to exclude water and provide a simple and reproducible way to use those reagents which are highly sensitive to water, and are deactivated by traces of moisture. The silica used for such purposes is carefully dehydrated. In the SAP approach silica is used just as a support for a thin film of water kept on the surface due to strong interaction with silanol groups. However, this is not just a layer of tightly bonded molecules, but a real layer of liquid water, which can behave similarly (but not exactly) as bulk water; for example, it dissolves ionic species, such as transition-metal complexes with sulfonated ligands. In this way the surface area of contact between the phases would increase by orders of magnitude, and thus the rate of mass transfer would also be dramatically enhanced. The catalyst is thus immobilized in a two-dimensional way, but not bound to a single point at the surface. Thus prepared, the SAP system shows a dramatic increase of activity, and the ability to process highly hydrophobic olefins, even the altogether less-reactive molecules with internal double bonds, such as oleic alcohol or dicyclopentadiene [135]. Experiments on ^{31}P spin-lattice relaxation times, which are indicative of the degree of mobility of a given species in a given medium, showed that $[HRh(CO)(TPPTS)_3]$ complex in the water layer impregnated over a silica surface is only slightly less mobile than the same species in bulk water phase.

Moreover, the mobility was shown to depend on the water content in the supported phase, and the catalyst with more easily diffusing catalyst was the most active in olefin hydroformylation [139].

Though SAP systems usually exhibit a better activity than biphasic systems, both are in most cases inferior to simple homogeneous reactions. The hydroformylation of methyl acrylate is an example of the entirely different behavior. Hydroformylation of acrylic acid derivatives can be of considerable industrial importance [140], as it gives mainly formylpropionic acids, which are viable precursors of methacrylate monomers and can be used for preparation of several important pharmaceuticals:

Biphasic systems with rhodium–TPPTS catalysts were shown to have activities about two times higher than the activity of the homogeneous system with organic solvent. The application of SAP catalysis gave more than an order-of-magnitude increase of catalyst activity, though the optimum performance was strongly dependent on the degree of hydration of the support [141].

The optimal thickness of water film on the surface of the carrier depends on whether the olefin being hydroformylated is more or less water-soluble. Olefins such as methyl acrylate, which are evenly distributed between water and organic phase, show better results for higher degrees of hydration, as the reaction occurs in the whole volume of the supported aqueous phase. Hydrophobic olefins show better reactivity with moderate or low degrees of hydration, probably because in this case the reaction can take place only at the interface, and more of the catalyst is exposed if the aqueous layer is thin [142]. An SAP catalyst may be further modified by the addition of inert electrolytes, which, as in the processes in liquid–liquid biphasic systems, exerted a profound effect on the activity of the catalyst and the selectivity of the reaction [143]. The addition of amphiphilic quaternary ammonium salts, either physically adsorbed on the silica surface, such as Aliquat 336, or bonded to the surface as $Me_3N(CH_2)_3Si(OMe)_3{}^+Cl^-$, enabled the preparation of supported rhodium catalysts, similar to SAP systems. The catalysts proved to be leach-resistant and capable of delivering a reasonable level of activity upon recycling [144].

The recycling and reuse of SAP catalysts are theoretically very simple, and require only filtration. However, data on the leaching of rhodium catalysts from the support are quite controversial, and while usually no catalyst activity could be found in the separated organic phase, the activity of the supported catalyst is subject to degradation. A major drawback of the SAP method is the loss of aqueous phase from the surface of the support during the reaction, carried out at rather high temperatures. The catalyst in the SAP technique is too weakly bound to the support.

Further steps to improve the method are actually steps leading back to the well-acclaimed methods of supported catalysis, in which not the solvent but the catalyst itself is bound to a carrier. The use of catalysts with water-soluble ligands, however, permits one to bridge the gap between SAP and canonical supported catalysis, not in one step, but in a series of smaller steps.

For example, the surface of silica can be covered not by water, but by a viscous liquid such as hydrated polyethyleneglycol (PEG), tightly bound to the support by numerous hydrogen bonds. The film of PEG dissolves the catalyst [HRh(CO)(TPPTS)$_3$], and this 'supported homogeneous film catalyst' (SHFC) works well in hydroformylation of hydrophobic olefins, having a longer lifetime than SAP catalysts [145], though certainly more detailed studies are needed to fully assess the perspectives of this approach. Hydrated PEG molecules are well known to suffer from dehydration at elevated temperatures, which should lead to a change of properties of the supported film, degradation of the catalyst and increased leaching of rhodium.

A step further (and at the same time a step aside from aqueous chemistry) is to use a modified silica support by attaching a layer of hydrophilic polar polymer, either by depositing a crosslinked poly(hydroxyethyl methacrylate) network or by transforming the silanol groups by attaching propylethylenediamine groups. The modified silica can bind phosphines bearing groups capable of hydrogen bonding, e.g. PEG endcapped at both ends with Ph$_2$P groups, to give immobilized catalysts that can catalyze the hydroformylation of hydrophobic olefins in organic solvents, and that of hydrophilic olefins (such as salts of 10-undecenoic acid and N-vinylimidazole) in water [146, 147].

5.6 Hydrogenation

The hydrogenation of C=C and C=O bonds (as well as other kinds of unsaturated groups) can be achieved by a huge variety of both catalytic (mostly heterogeneous) and noncatalytic methods. Therefore, the application of costly homogeneous catalysts is not justified in most cases for hydrogenation of simple substrates with isolated unsaturated fragments. However, in more complex cases the development of efficient techniques based on soluble catalysts is perfectly motivated. The most important is undoubtedly asymmetric hydrogenation, which is now regarded as one of the key methods for the synthesis of chiral molecules from achiral starting reagents. Asymmetric hydrogenation is done with the help of chiral catalysts, the chirality of which in the overwhelming majority of cases is introduced by chiral phosphine ligands. Thus, the cost of catalysts used in asymmetric hydrogenation is the sum of the cost of the precious metal (rhodium is by far the most actively used metal) and the cost of the chiral ligand (which can be much more expensive). The need for catalyst regeneration and recycling is thus obvious, though it may not be so pressing as in hydroformylation processes, which is evident just by comparison of research efforts devoted to the development of such methods for hydrogenation and hydroformylation. The latter is applied for the production of basic chemicals consumed in huge amounts in highly competitive markets. Asymmetric hydrogenation, on the other hand, is applied for the production of unique fine chemicals (mostly pharmaceuticals) in much smaller quantities, where the price is not so strictly limited. In future the situation may change, as the need for next-generation drugs, plant protection agents, food additives, etc., is rapidly changing the chemical industry.

Good homogeneous catalysts may be worth their cost also for selective hydrogenation of multifunctional substrates, such as α,β-unsaturated carbonyl compounds, or natural molecules.

Hydrogenation of C=O bonds usually requires much more rigorous conditions (high pressures of H_2, higher temperatures, higher loads of catalysts, etc.) and is therefore less selective than the hydrogenation of C=O bonds. The development of new ruthenium catalysts may change the now prevailing attitude to the homogeneous catalytic hydrogenation of C=O bonds as an inferior method of organic synthesis.

Catalytic hydrogenation by H_2, like hydroformylation, is almost ideally suited for the application of phase-separation techniques, as it involves only two reagents (the substrate and hydrogen), requires no additional ionic or non-ionic components and, most importantly, produces only the single target product without any by-products that may accumulate in the reaction system and poison the catalyst.

The hydrogenation of olefins catalyzed by a water-soluble Wilkinson-type rhodium catalyst with sulfonated phosphine ligands was described as early as 1978 [126]. Early results on hydrogenation using hydrophilic catalysts can be found in reference [148]. RhCl(TPPTS)$_3$ is indeed an excellent catalyst for very mild hydrogenation of olefins. Hydrophilic olefins, such as unsaturated carboxylic acids, can be hydrogenated in homogeneous solutions in aqueous ethanol or neat water [149]. Water-insoluble olefins are hydrogenated in a biphasic system consisting of an aqueous solution of the catalyst and the olefin itself, without any other components, thus making the system ideally suited for easy separation and repeated reuse of the active aqueous layer [150]. However, as in many other cases of phase-separation technique the number of reuses is limited, because the catalyst is not left intact during the reaction. It was shown that soluble rhodium catalyst is transformed into a hydrosol of rhodium metal, protected from further aggregation and sedimentation by phosphine ligands bonded to the surface of metal colloid particles [151, 152]. The formation of metal colloids during catalytic reactions is an intriguing and largely unexplored area, which may bring interesting results especially in aqueous phosphine-less catalysis. However, with regard to the ideal biphasic technique, the formation of colloid would have rather a deleterious effect on the behavior of the catalytic system, as the system is thus switched into heterogeneous mode with lower selectivity and poorer reproducibility. Besides, hydrosols of this type are usually quite unstable, thus limiting the number of reuses of the catalytic system.

The formation of a hydride complex of rhodium RhH(TPPMS)$_3$ was observed during the the the hydrogenation of olefins by a TPPMS analog of the Wilkinson catalyst in aqueous systems [153, 154].

A water-soluble ruthenium complex, HRu(CO)Cl(TPPMs)$_3$·2H$_2$O, was used for hydrogenation of olefins in the biphasic system decalin–water [155]. Ruthenium catalyst immobilized on a hydrophilic microporous resin turned out to be more selective for partial hydrogenation of benzene to cyclohexene than the same catalyst immobilized on charcoal.

The hydrophilic catalyst was apparently more uniform, thus providing more catalytically active sites [156]. The use of water-soluble catalysts permits one to perform hydrogenation in homogeneous aqueous solution, which can be very convenient for natural compounds such as carbohydrates. The reduction of CHO groups in aldoses was performed using the TPPTS complex of ruthenium, using either molecular hydrogen or transfer hydrogenation by formic acid derivatives [157].

This complex showed good stability during the reaction, which permits efficient recycling. Hydrogenation in biphasic media is not limited to complexes with water-soluble phosphines. Ruthenium forms a lot of water-soluble cationic complexes, e.g. with nitrogen-containing ligands such as bipyridines or phenanthroline, which are highly efficient catalysts in the hydrogenation of both C=C and C=O bonds. For example, the complex cis-[Ru(6,6'-Cl$_2$bpy)$_2$(H$_2$O)$_2$](CF$_3$SO$_3$)$_2$ where 6,6'-Cl$_2$bpy is the 6,6'-dichloro-2,2'-bipyridine ligand, is a good and easily recycled catalyst for the hydrogenation of aldehydes, ketones and olefins in biphasic systems. Water is not only a solvent, but actually a reagent in this process, as it is involved in the equilibrium between η^2-dihydrogen and hydrido complexes of ruthenium, and thus can catalyze an isotope exchange between water and molecular hydrogen. The hydrogenation of acetophenone by H$_2$ in the presence of D$_2$O produced a mixture of deuterated and nondeuterated alcohols [158]. Several mononuclear and binuclear ruthenium complexes with bipyridine or phenanthroline ligands were used as catalysts in the hydrogenation of both C=C and C=O bonds in a range of polar solvents including water. The stability of the complexes was high enough to be comparable with the results obtained with sulfonated ligands [159]. Some recent results on hydrogenation using water-soluble rhodium complexes with amphiphilic ligands bearing quaternized amino groups can be found in reference [160].

Hydrogenation can be performed not only by molecular hydrogen, but also by a multitude of reagents capable of forming hydride complexes in situ. This technique is usually called transfer hydrogenation. The hydride complexes of many late transition metals are formed by water-gas shift reaction from water and CO or its precursors, and thus WGSR conditions can be used for hydrogenation. However, the use of such systems for double-bond hydrogenation is limited because hydrogenation is not the only, and usually not the main, pathway possible in this system. Formic acid and its derivatives (salts, esters) are well-known precursors of CO, as many transition metals catalyze their decarbonylation, and thus generate the pathways involving metal hydrides. Owing to the very low concentration of CO in such systems, hydrogenation processes usually prevail, and can be made selective under mild conditions. It is essential that most catalysts suitable for hydrogenation with molecular hydrogen are usually capable of catalyzing transfer hydrogenation.

Rhodium catalysts with sulfonated chiral phosphine ligands were tried for asymmetric hydrogenation in aqueous systems. Tetrasulfonated cyclobutanediop was used for asymmetric transfer hydrogenation of acrylic acid derivatives by formates with low to moderate enantiomeric excess [161]. The same ligand as well as sulfonated (S,S)-BDPP, (S,S)-chiraphos and (R)-prophos were used for rhodium-catalyzed hydrogenation by H_2 in biphasic systems. The C=C double bonds in various derivatives of acrylic acid, including α-acetamidocinnamic acids, can be hydrogenated under 1–15 atm of H_2 giving moderate to high enantiomeric excesses. The degree of sulfonation of ligands apparently had no influence on the efficiency of asymmetric induction. The hydrogenation of C=O and C=N double bonds can be performed at higher H_2 pressures (up to 70 atm), while the enantiomeric excesses obtained were usually poor.

The sulfonated BINAP ligand was used for hydrogenation of the double bonds in acrylic acids using ruthenium complexes both in homogeneous systems (aqueous and non-aqueous) and using an SAP technique. The hydrogenation of naproxen precursor in organic solvents gave an enantioselectivity matching that obtained with the parent BINAP ligand:

However, both homogeneous aqueous systems and the catalyst immobilized on hydrated silica were inferior in activity and enantioselectivity giving ee not exceeding 70%, though the use of the SAP technique permitted a facile recycling of catalyst without the leaching of ruthenium into the organic phase [162].

The same catalyst showed higher ee values approaching 90% in the hydrogenation of acetamidocinnamic and methylenesuccinic acids in aqueous solutions, which is still somewhat less than those obtained with standard BINAP [163]. A new approach to the immobilization of the catalyst utilized a technique similar to the supported aqueous phase method. Sulfonate surfactant residues were tailored to the surface of silica, thus providing a layer capable of solubilizing both chiral rhodium catalyst and the substrate. The approach therefore attempts to combine the micellar effects with the ability to recycle the immobilized catalyst [164].

Aqueous systems have allowed the development of new selective methods for the hydrogenation of α,β-unsaturated carbonyl compounds. Selective reduction of the carbonyl group can be done in the presence of water-soluble ruthenium or iridium complexes, while the double bond can be hydrogenated in the presence of rhodium complexes. The reactions are done under very mild conditions, which is essential because complexes of the same metals are known to catalyze decarbonylation. Thus, the complex $RuCl_2(TPPMs)_2$ catalyzes the reduction of the carbonyl group by transfer hydrogenation using formate salts [165]. The complexes of TPPTS can catalyze the reduction of both C=O and C=C bonds by molecular hydrogen in biphasic systems [166]:

$$\text{(structure)} \xleftarrow{\substack{H_2,\ [RhCl(COD)]_2,\ TPPTS \\ PhMe-H_2O,\ r.t.}} \text{(structure)}$$

$$\xrightarrow{\substack{H_2,\ RuCl_3\ or\ [IrCl(COD)]_2,\ TPPTS \\ PhMe-H_2O,\ r.t.}} \text{(structure)}\ OH$$

The ligand 1,3,5-triaza-7-phosphaadamantane (PTA) can be similarly used for transfer hydrogenation of both C=C and C=O groups by formate salts:

$$\text{(structure)} \xleftarrow{\substack{RhCl(PTAH)(PTA)_2,\ HCOONa \\ H_2O}} \text{(structure)}$$

$$\xrightarrow{\substack{RuCl_2(PTA)_4,\ HCOONa \\ H_2O}} \text{(structure)}\ OH$$

where PTAH is the monoprotonated PTA ligand. The reduction of α,β-unsaturated carbonyl compounds is also quite selective. Unlike TPPTS, which is a much weaker ligand that usually must be taken in excess to keep the metal properly ligated, PTA forms stable complexes, and the excess of ligand inhibits the reaction. Besides, PTA is much more stable to oxidation, and is not a surfactant (as e.g. TPPMS), thus making the recovery simpler by avoiding emulsification [25].

Hydrogenation using the rhodium complex may involve the formation of colloid particles, as it was shown that the reaction is inhibited by metallic mercury [167]. Selective hydrogenation of the carbonyl group in α,β-unsaturated carbonyl compounds can be done with immobilized ruthenium or iridium complexes by using either the supported aqueous phase technique with complexes such as $RuCl_2(TPPTS)_3$ and $RuH_2(TPPTS)_4$ absorbed on hydrated silica, or an iridium complex chemically bonded to the silica surface, with excellent chemoselectivity and extent of recovery [168].

5.7   Other reactions

Besides the reactions already discussed, the application of water can be found in a number of other reactions, which can only be briefly mentioned because of the limited format of this review.

Water-gas shift chemistry is applied not only for carbonylation of unsaturated molecules, but also for several other important reaction types. A review on basic WGSR chemistry can be found in reference [169]. The metal hydride complexes formed in WGSR can be involved in reduction, e.g. of the nitro group. In the absence of water, reductive carbonylation of the nitro group gives the most valuable isocyanates. Under aqueous conditions, anilines become the principal products of WGSR reduction achieved by CO in the presence of water and usually ruthenium or rhodium complexes. The process is chemospecific, leaving intact all other types of reducible groups [170–174]. The reaction can be catalyzed by polymer-immobilized rhodium complexes [175]. Hydrophilic methyl formate can be used as an alternative reducing agent for the reduction of the nitro group [176].

Reduction under WGSR conditions can be accompanied by cyclizations, as in the following example of the formation of indoles and quinolines from readily available o-nitrochalcones [177]:

Water is so extensively used in catalytic oxidation reactions that usually this fact is regarded as a natural feature and remains unnoticed. Wacker oxidation of olefins by palladium complexes involves water as a nucleophilic reagent, and thus the whole Wacker-type chemistry, which has developed into a powerful and versatile method of organic synthesis, is derived from aqueous catalysis [178]. The role of the nature of the co-oxidant and the mechanism of deactivation of the palladium catalyst due to aggregation and growth of inactive metal particles were recently investigated, and such study may have relevance for other processes catalyzed by phosphine-less palladium catalysts [179].

Catalytic epoxidation of olefins in systems containing metal catalyst and stoichiometric oxidant is often carried out in the presence of water. Water is often present in such systems just because some of the most popular stoichiometric oxidants like *t*-butylhydroperoxide, and especially sodium hypochlorite, are available in aqueous solutions. It may be rigorously stated that no nonhydrated form of sodium hypochlorite is known, and that all studies on oxidations by this reagent deal with aqueous systems by default. In most published examples no discussion of the possible role or influence of water in such systems is given, though it is often evident from the description of experimental procedures that these reactions run in truly biphasic systems, with oxidant residing in the aqueous layer, and substrate and transition-metal catalyst residing in the organic phase. In some cases, however, an explicit discussion of water-soluble catalysts for olefin epoxidation can be found, e.g. for processes catalyzed by ruthenium complexes [180–183].

Other types of oxidative processes include the activation of saturated CH bonds. Such processes are especially interesting, from the practical point of view of processing the hydrocarbons from oil refining, and because such processes can occur in living organisms, and thus are definitely catalyzed by transition-metal complexes in aqueous environments. Among recent work on the development of model systems for such a purpose is the following. Water-soluble Ru(III)–EDTA complexes are usable for activation of $C(sp^3)$–H bonds, e.g. in the oxidation of adamantane to a mixture of 1- and 2-adamantanols and adamantanone [184].

Catalytic hydration of unsaturated bonds is to a certain extent a complementary process to Wacker-type oxidation. Hydration catalyzed by transition-metal complexes has been rather little studied [185]. One process involving hydration and having a good industrial perspective is the telomerization of unsaturated compounds. For example, the aqueous telomerization of butadiene can afford octadienols. The process may be performed in a micellar system in the presence of surfactants with remarkable selectivity [186]:

Conjugated dienes can be selectively hydrated to ketones in the presence of cationic ruthenium complexes with bipyridyl ligands. The role of ruthenium is to catalyze the isomerization of allylic alcohols formed by the addition of water to diene. This method allows one to convert butadiene to methyl ethyl ketone in high yield [187]. Hydration of triple bonds is one of the oldest catalytic processes of organic chemistry. Though this reaction has no industrial value, it can serve as a tool of fine organic synthesis. The hydration can be catalyzed by rhodium salts under phase-transfer conditions [188]. The more exotic process of the hydrolysis of phenylacetylene to toluene and carbon monoxide catalyzed by ruthenium complex should also be mentioned [189]:

$$PhC\equiv CH + H_2O \xrightarrow[\text{THF, }60°C]{\text{(PNP)RuCl}_2\text{(Ph}_3\text{P)}} PhCH_3 + CO$$

$$PNP = n\text{-PrN(CH}_2\text{CH}_2\text{PPh}_2)_2$$

Aqueous microemulsions have proved to be excellent media for polymerization of unsaturated compounds, giving valuable monodisperse polymers. Though polymerization in such systems is usually initiated by radical means, the so-called ring-opening methathesis polymerization (ROMP) relies on transition-metal complexes. First, it was shown that water-soluble aqua-complexes of ruthenium are good catalysts for ROMP of hydrophilic caged olefins, such as exo,exo-5,6-bis(methoxycarbonyl)-7-oxabicyclo[2.2.1]hept-2-ene [190] (however see reference [191] on the deficiencies of this method). The method was later extended to water-insoluble monomers, and in the first place to norbornene, which are polymerized in emulsion and microemulsion media. The water-soluble bis(allyl)ruthenium complexes $[Ru(\eta^3:\eta^3\text{-}C_{10}H_{16})(OH_2)(O_2CCH_3)]BF_4$ and $Ru(\eta^3:\eta^3\text{-}C_{10}H_{16})(OH_2)(O_3SCF_3)_2$ were used for ROMP of emulsified norbornene to give an unusual polymer with *cis* double-bond bridges and very high relative molecular mass exceeding 10^6 [192]:

Norbornene can also be polymerized in aqueous emulsion by palladium chloride to give fine microlatex polymers unavailable by polymerization in anhydrous solvents [193].

Several examples of poly- and oligomerization of acetylenes are also quite typical. Water-soluble rhodium complexes with hydrophilic phosphine ligands catalyze the polymerization of arylacetylenes into stereoregular *cis*-oriented poly(arylacetylenes), which can further be selectively depolymerized on heating to provide a convenient method for the preparation of triarylbenzenes [194]. Different types of product formed by regioselective di- and trimerization can be obtained by the reaction of such catalysts with other kinds of terminal acetylenes [195].

Supported aqueous phase catalysis can be used not only for hydrogenation and hydroformylation, but also for more specific transformations, as in the addition of diphenylacetylene to azobenzene catalyzed by TPPTS complex of rhodium immobilized on hydrated SiO_2 [196]:

or the isomerization of olefins [197].

Water-soluble rhodium complexes catalyze the reduction of CO_2 to formates even in such media as aqueous solutions of amines used to capture carbon dioxide from flue gases [198].

Finally, two relevant recent reviews can be mentioned, one dealing with catalysis by the colloids and clusters often involved in aqueous processes [199], and the other dealing with the common methodology of water-soluble catalysts [200].

References

Owing to limitations on the length of this review, the list of references is not comprehensive. We have tried in each case to provide at least the most recent relevant reference, so that other important work on the subject discussed can be obtained by backtracking.

[1] Fendler, J.H.; Fendler, E.J. (1975) *Catalysis in micellar and macromolecular system*; Academic Press, New York.

[2] Fendler, J.H. (1982) *Membrane mimetic chemistry: characterizations and applications of micelles, microemulsions, monolayers, bilayers, vesicles, host–guest systems, and polyions*; Wiley, NewYork.

[3] Sinou, D. (1986) Phosphines hydrosolubles. Syntheses et applications en catalyse, *Bull. Soc. Chim. Fr.*, 480–6.

[4] Ding, H.; Hanson, B.E. (1994) Reaction activity and selectivity as a function of solution ionic-strength in oct-1-ene hydroformylation with sulfonated phosphines, *Chem. Commun.*, 2747–8.

[5] Ding, H.; Hanson, B.E. (1995) Spectator cations and catalysis with sulfonated phosphines: the role of cations in determining reaction selectivity in the aqueous-phase hydroformylation of olefins, *J. Mol. Catal. A – Chem.*, **99**(3), 131–7.

[6] Ding, H.; Hanson, B.E.; Glass, T.E. (1995) The effect of salt on selectivity in water-soluble hydroformylation catalysts, *Inorg. Chim. Acta*, **229**, 329–33.

[7] Casalnuovo, A.L.; Calabrese, J.C. (1990) Palladium catalysed alkylations in aqueous media, *J. Am. Chem. Soc.*, **112**, 4324–30.

[8] Bartik, T.; Bunn, B.B.; Bartik, B.; Hanson, B.E. (1994) Synthesis, reactions, and catalytic chemistry of the water-soluble chelating phosphine 1,2-bis[bis(*m*-sodiosulfonatophenyl)-phosphino]ethane (dppets): complexes with nickel, palladium, platinum, and rhodium, *Inorg. Chem.*, **33**, 164–9.

[9] Fell, B.; Papadogianakis, G. (1994) Rhodium-catalyzed 2-phase hydroformylation of hex-1-ene with sulfonated tris(4-fluorophenyl)phosphine as water-soluble complex ligands, *J. Prakt. Chem. – Chem. Ztg.*, **336**, 591–5.

[10] Bartik, T.; Bartik, B.; Hanson, B.E.; Guo, I.; Toth, I. (1993) Water-soluble electron-donating phosphines: sulfonation of tris(omega-phenylalkyl)phosphines, *Organometallics*, **12**, 164–70.

[11] Bartik, T.; Ding, H.; Bartik, B.; Hanson, B.E. (1995) Surface-active phosphines for catalysis under 2-phase reaction conditions: $P(menthyl)[(CH_2)_8C_6H_4\text{-}p\text{-}SO_3Na]_2$ and the hydroformylation of styrene, *J. Mol. Catal. A – Chem.*, **98**, 117–22.

[12] Herrmann, W.A.; Kohlpaintner, C.W.; Bahrmann, H.; Konkol, W. (1992) A new efficient water-soluble catalyst for two-phase hydroformylation of olefins, *J. Mol. Catal. A*, **73**, 191–201.

[13] Herrmann, W.A.; Kohlpaintner, C.W.; Manetsberger, R.B.; Bahrmann, H.; Kottmann, H. (1995) Water-soluble metal-complexes and catalysts. 7. New efficient water-soluble catalysts for 2-phase olefin hydroformylation: BINAS-Na, a superlative in propene hydroformylation, *J. Mol. Catal. A – Chem.*, **97**, 65–72.

[14] Bianchini, C.; Frediani, P.; Sernau, V. (1995) Zwitterionic metal complexes of the new triphosphine $NaO_3S(C_6H_4)CH_2C(CH_2PPh_2)_3$ in liquid biphasic catalysis, *Organometallics*, **14**, 5458–9.

[15] Herd, O; Hessler, A.; Langhans, K.P.; Stelzer, O.; Sheldrick, W.S.; Weferling, N. (1994) Water-soluble phosphanes. 2. New method for synthesis of water-soluble secondary and tertiary phosphanes with sulfonated aromatic residues: crystal structure of P(para-$C_6H_4SO_3K)_3$·KCl·0.5H_2O, J. Organomet. Chem., 475, 99–111.

[16] Kiji, J.; Okano, T. (1994) 2-Phase catalysis by water-soluble phosphine complexes of palladium, J. Synth. Org. Chem. Jpn., 52, 276–84.

[17] Heesche-Wagner, K.; Mitchell, T.N. (1994) Approaches to water-soluble phosphines. 2. Free-radical addition-reactions of phenylphosphines, J. Organomet. Chem., 468, 99–106.

[18] Mitchell, T.N.; Heesche-Wagner, K. (1992) Approaches to new water-soluble phosphines, J. Organomet. Chem., 436, 43–53.

[19] Avey, A.; Schut, D.M.; Weakley, T.J.R.; Tyler, D.R. (1993) A new water-soluble phosphine for use in aqueous organometallic systems – products from the reactions of 2,3-bis(diphenylphosphino)maleic anhydride with water and oxygen, Inorg. Chem., 32, 233–6.

[20] Lelievre, S.; Mercier, F.; Mathey, F. (1996) Phosphanorbornadienephosphonates as a new-type of water-soluble phosphines for biphasic catalysis, J. Org. Chem., 61, 3531–3.

[21] Mercier, F.; Mathey, F. (1993) A new-type of water-soluble phosphine for biphasic catalysis, J. Organomet. Chem., 462, 103–6.

[22] Schull, T.L.; Fettinger, J.C.; Knight, D.A. (1995) The first examples of an aryl ring-substituted by both phosphine and phosphonate moieties: synthesis and characterization of the new highly water-soluble phosphine ligand $Na_2[Ph_2P(C_6H_4\text{-}p\text{-}PO_3)]$·1.5$H_2O$ and platinum(II) complexes, Chem. Commun., 1487–8.

[23] Ravindar, V.; Schumann, H.; Hemling, H.; Blum, J. (1995) Synthesis and structure determination of some platinum(II) complexes with hydrophilic carboxylated tertiary phosphine-ligands, Inorg. Chim. Acta, 240, 145–52.

[24] Reddy, V.S.; Katti, K.V.; Barnes, C.L. (1995) Chemistry in environmentally benign media. 1. Synthesis and characterization of 1,2-bis(bis(hydroxymethyl)phosphino)ethane (HMPE): X-ray structure of $[Pt((HOH_2C)_2PCH_2CH_2P(CH_2OH)_2)_2]Cl_2$, Inorg. Chim. Acta, 240, 367–70.

[25] Darensbourg, D.J.; Joo, F.; Kannisto, M.; Katho, A.; Reibenspies, J.H.; Daigle, D.J.(1994) Water-soluble organometallic compounds. 4. Catalytic-hydrogenation of aldehydes in an aqueous 2-phase solvent system using a 1,3,5-triaza-7-phosphaadamantane complex of ruthenium, Inorg. Chem., 33, 200–8.

[26] Bitterer, F.; Kucken, S.; Stelzer, O. (1995) Water-soluble phosphanes. 4. Tertiary alkylphosphanes with ammonium groups in the side-chains: amphiphiles with basic P-atoms, Chem. Ber., 128, 275–9.

[27] Hessler, A.; Kucken, S.; Stelzer, O.; Blotevogelbaltronat, J.; Sheldrick, W.S. (1995) Water-soluble phosphanes. 5. Complexes of amphiphilic tertiary alkylphosphanes with ammonium groups in the side-chains, J. Organomet. Chem., 501, 293–302.

[28] Brauer, D.J.; Fischer, J.; Kucken, S.: Langhans, K.P.; Stelzer, O.; Weferling, N. (1994) Water-soluble phosphanes. 3. Water-soluble primary phosphanes with ammonium groups NR_2R' in the side-chain: donor-functionalized amphiphiles, Z. Naturforsch. B, 49, 1511–24.

[29] Kovacs, I.; Baird, M.C. (1995) Neutral and cationic iron carbonyl-complexes substituted with the water-soluble phosphines $Ph_2P(CH_2)_nPMe_3^+$ (n = 2, 3, 6 and 10), J. Organomet. Chem., 502, 87–94.

[30] Buhling, A.; Kamer, P.C.J.; Van Leeuwen, P.W.N.M. (1995) Rhodium-catalyzed hydroformylation of higher alkenes using amphiphilic ligands, J. Mol. Catal. A – Chem., 98, 69–80.

[31] Baird, I.R.; Smith, M.B.; James, B.R. (1995) Nickel(II) and nickel(0) complexes containing 2- pyridylphosphine ligands, including water-soluble species, Inorg. Chim. Acta, 235, 291–7.

[32] Buhling, A.; Elgersma, J.W.; Nkrumah, S.; Kamer, P.C.J.; Van Leeuwen, P.W.N.M. (1996) Novel amphiphilic diphosphines: synthesis, rhodium complexes, use in hydroformylation and rhodium recycling, Dalton Trans., 2143–54.

[33] Malmstrom, T.; Weigl, H.; Andersson, C. (1995) Coupling of the triphenylphosphine moiety to water-soluble polymers: a new method to achieve water-soluble metal phosphine complexes, Organometallics, 14, 2593–6.

[34] Wallow, T.I.; Novak, B.M. (1991) In aqua synthesis of water-soluble poly(p-phenylene) derivatives, J. Am. Chem. Soc., 113, 7411–12.

[35] Miyaura, N.; Yamada, K.; Suginome, H.; Suzuki, A. (1985) Novel and convenient method for the stereo and regiospecific synthesis of conjugated alkadienes and alkenynes via the palladium catalyzed cross-coupling reaction of 1-alkenylboranes with bromoalkenes and bromoalkynes, *J. Am. Chem. Soc.*, **107**, 972–80.

[36] Miyaura, N.; Ishiyama, T.: Sasaki, H.; Ishikawa, M.; Sato, M.; Suzuki, A. (1989) Palladium catalyzed inter- and intramolecular cross-coupling reaction of *B*-alkyl-9-borabicyclo[3.3.1]-nonane derivatives with 1-halo-1-alkanes or haloarenes, *J. Am. Chem. Soc.*, **111**, 314–21.

[37] Hoshino, Y.; Miyaura, N.; Suzuki, A. (1988) Novel synthesis of isoflavones by the palladium catalysed cross-coupling reaction of 3-bromochromones with arylboronic acids or their esters, *Bull. Chem. Soc. Jpn.*, **61**, 3008–10.

[38] Thompson, W.J.; Gaudino, J. (1984) A general synthesis of 5-arylnicotinates, *J. Organomet. Chem.*, **49**, 5237–43.

[39] Anderson, J.C.; Namli, H. (1995) Ambient-temperature unsymmetrical bialyl synthesis using Suzuki methodology, *Synlett*, 765–6.

[40] Bumagin, N.A.; Bykov, V.V.; Beletskaya, I.P. (1989) Synthesis of biaryls from phenylboronic acid and aryl iodides in aqueous media, *Izv. Akad. Nauk SSSR, Ser. Khim.*, 2394.

[41] Campi, E.M.; Jackson, W.R.; Marcuccio, S.M.; Naeslund, C.G.M. (1994) High yields of unsymmetrical biaryls via cross-coupling of arylboronic acids with haloarenes using a modified Suzuki–Beletskaya procedure, *Chem. Commun.*, 2395.

[42] O'Keefe, D.F.; Dannock, M.C.; Marcuccio, S.M. (1992) Palladium catalyzed coupling of halobenzenes with arylboronic acids. Role of the triphenylphosphine ligand, *Tetrahedron Lett*, **33**, 6679–80.

[43] Wallow, T.I.; Novak, B.M. (1994) Highly efficient and accelerated Suzuki aryl couplings mediated by phosphine-free palladium sources, *J. Org. Chem.*, **59**, 5034–7.

[44] Bumagin, N.A.; Bykov, V.V.; Beletskaya, I.P. (1989) Reaction of sodium tetraphenylborate with aryl halides in aqueous medium, *Metalloorg. Khim.*, **2**, 1200.

[45] Bykov, V.V.; Bumagin, N.A.; Beletskaya, I.P. (1995) Palladium-catalyzed phenylation of chloroarenes by tetraphenylborate sodium in aqueous-media, *Dokl. Akad. Nauk*, **340**, 775–8.

[46] Portnoy, M.; Milstein, D. (1993) Mechanism of aryl chloride oxidative addition to chelated palladium(0) complexes, *Organometallics*, **12**, 1665–73.

[47] Bumagin, N.A.; Luzikova, E.V.; Sukhomlinova, L.I.; Tolstaya, T.P.; Beletskaya, I.P. (1995) Palladium-catalyzed cross-coupling of symmetrical diaryliodonium salts with sodium tetraphenylborate in water, *Russ. Chem. Bull.*, **44**, 385–6.

[48] Darses, S.; Jeffery, T.; Genet, J.P.; Brayer, J.L.; Demoute, J.P. (1996) Cross-coupling of arenediazonium tetrafluoroborates with arylboronic acids catalyzed by palladium, *Tetrahedron Lett.*, **37**, 3857–60.

[49] Stille, J.K. (1986) Palladium-catalyzed coupling reactions of organic electrophiles with organic tin compounds, *Angew. Chem.*, **98**, 504–19.

[50] Bumagin, N.A.; Bumagina, I.G.; Beletskaya, I.P. (1983) The reaction of RSnMe₃ with R'X catalysed by PhPdI(PPh₃)₂ in HMPA, *Dokl. Akad. Nauk SSSR*, **272**, 1384–8.

[51] Tueting, D.R.; Echavarren, A.M.; Stille, J.K. (1989) Palladium catalysed coupling of organostannanes with vinyl epoxides, *Tetrahedron*, **45**, 979–92.

[52] Zhang, H.C.; Daves, G.D. (1993) Water facilitation of palladium-mediated coupling reactions, *Organometallics*, **12**, 1499–500.

[53] Kikukawa, K.; Kono, K.; Wada, F.; Matsuda, T. (1983) Palladium catalysed carbon–carbon coupling of arenediazonium salts with organotin compounds, *J. Org. Chem.*, **48**, 1333–6.

[54] Bumagin, N.A.; Sukhomlinova, L.I.; Tolstaya, T.P.; Vanchikov, A.N., Beletskaya, I.P. (1990) Palladium-catalyzed cross-coupling of arenediazonium salts with tetramethyltin in aqueous media, *Izv. Akad. Nauk SSSR, Ser. Khim.*, 2665–6.

[55] Bumagin, N.A.; Sukhomlinova, L.I.; Igushkina, S.O.; Tolstaya, T.P.; Vanchikov, A.N.; Beletskaya, I.P. (1992) Palladium-catalyzed cross-coupling of diaryliodonium salts with organotin compounds, *Izv. Akad. Nauk SSSR, Ser. Khim.* 2683–4.

[56] Roshchin, A.I.; Bumagin, N.A.; Beletskaya, I.P. (1995) Palladium-catalyzed cross-coupling reaction of organostannoates with aryl halides in aqueous-medium, *Tetrahedron Lett.*, **36**, 125–8.

[57] Rai, R.; Aubrecht, K.B.; Collum, D.B. (1995) Palladium-catalyzed Stille couplings of aryl-

trichlorostannanes, vinyltrichlorostannanes and alkyltrichlorostannanes in aqueous-solution, *Tetrahedron Lett.*, **36**, 3111–14.

[58] Bumagin, N.A.; Bykov, V.V.; Beletskaya, I.P (1995) Palladium-catalyzed reaction of propargyl alcohol with aryliodides in water, *Zh. Org. Khim.*, **31**, 385–7.

[59] Luzikova, E.V.; Bumagin, N.A.; Beletskaya, I.P. (1993) Palladium-catalyzed reaction of phenylacetylene with aryl iodides in aqueous-medium, *Russ. Chem. Bull.*, **42**, 585–6.

[60] Dibowski H.; Schmidtchen. F.P. (1995) Synthesis and investigation of phosphine-ligands containing cationic guanidino functions in aqueous Heck reactions, *Tetrahedron*, **51**, 2325–30.

[61] Sukhomlinova, L.I.; Luzikova, E.V.; Tolstaya, T.P.; Bumagin, N.A.; Beletskaya, I.P. (1995) Palladium-catalyzed condensation of symmetrical diaryliodonium salts with phenyl-acetylene in aqueous-media, *Russ. Chem. Bull.*, 44, 769–70.

[62] Kang, S.K.; Lee, H.W.; Jang, S.B.; Kim, T.H.; Kim, J.S. (1996) Palladium-catalyzed coupling of organostannanes with hypervalent iodonium salts, *Synth Commun.*, **26**, 4311–18.

[63] Kang, S.K.; Lee, H.W.; Kim, J.S.; Choi, S.C. (1996) Palladium-catalyzed cross-coupling of organostannanes with iodanes, *Tetrahedron Lett.*, **37**, 3723–6.

[64] Kang, S.K.; Lee, H.W.; Jang, S.B.; Ho, P.S. (1996) Palladium-catalyzed cross-coupling reactions of aryl-iodonium, alkenyl-iodonium and alkynyl-iodonium salts and iodanes with terminal alkynes in aqueous-medium, *Chem. Commun.*, 835–6.

[65] Amatore, C.; Blart, E.; Genet, J.P.; Jutand, A.; Lemaireaudoire, S.; Savignac, M. (1995) New synthetic applications of water-soluble acetate Pd/TPPTS catalyst generated in-situ: evidence for a true Pd(0) species intermediate, *J. Org. Chem.*, **60**, 6829–39.

[66] Genet, J.P.; Blart, E.; Savignac, M. (1992) Palladium catalysed cross-coupling reactions in a homogeneous aqueous media, *Synlett*, 715–17.

[67] Safi, M.; Sinou, D. (1991) Palladium(0)-catalyzed substitution of allylic substrates in a two-phase aqueous-organic medium, *Tetrahedron Lett.*, **32**, 2025–8.

[68] Blart, E.; Genet, J.P.; Safi, M.; Savignac, M.; Sinou, D. (1994) Palladium(0)-catalyzed substitution of allylic substrates in an aqueous-organic medium, *Tetrahedron*, **50**, 505–14.

[69] Yamamoto, Y.; Fujiwara, N. (1995) Palladium catalysed direct allylation of pronucleo-philes with allylstannanes, *Chem. Commun.*, 2013–14.

[70] Genet, J.P.; Blart, E.; Savignac, M.; Lemeune, S.; Lemaireaudoire, S.; Paris, J.M.; Bernard, J.M. (1994) Practical palladium-mediated deprotective method of allyloxycarbonyl in aqueous media, *Tetrahedron*, **50**, 497–503.

[71] Mereyala, H.B.; Guntha, S. (1993) A novel, mild palladium-mediated deprotection of *O*-allyl and prop-1-enyl ethers, *Tetrahedron Lett.*, **34**, 6929–30.

[72] Sigismondi, S.; Sinou, D.; Perez, M.; Morenomanas, M.; Pleixats, R.; Villarroya, M. (1995) Palladium(0)-catalyzed allylation of uracils and 2-thiouracils. Drastic effect of an aqueous reaction medium on the regioselectivity, *Tetrahedron Lett.*, **35**, 7085–8.

[73] Reutov, O.A.; Beletskaya, I.P.; Kurts, A.L. (1983) *Ambident Anions*; Consultants Bureau, New York.

[74] Hirao, T.; Masunaga, T.; Ohshiro, Y.; Agawa, T. (1981) A novel synthesis of dialkyl arenephosphonates, *Synthesis*, 56–7.

[75] Novikova, Z.S.; Demik, N.N.; Agarkov, A.Y.; Beletskaya, I.P. (1995) Palladium-catalyzed arylation of dialkylphosphite, *Zh. Org. Khim.*, **31**, 142.

[76] Demik, N.N.; Kabachnik, M.M.; Novikova, Z.S.; Beletskaya, I.P. (1994) Palladium-catalyzed arylation of *O*,*O*-diethylylphosphonate, *Zh. Org. Khim.*, **30**, 876–81.

[77] Petrakis, K.S.; Nagabhushan, T.L. (1987) Palladium catalysed substitutions of triflates derived from tyrosine-containing peptides and simpler hydroxyarenes forming 4-(diethoxy-phosphinyl)phenylalanines and diethyl arylphosphonates, *J. Am. Chem. Soc.*, **109**, 2831–3.

[78] Hoke, J.B.; Gramiccioni, G.A.; Balko, E.N. (1992) Catalytic hydrodechlorination of chlorophenols, *Appl. Catal. B – Environm.*, **1**, 285–96.

[79] Sala, R.; Doria, G.; Passarotti, C. (1984) Reduction of carbon–carbon double bonds and hydrogenolysis by sodium hypophosphite, *Tetrahedron Lett.*, **25**, 4565–8.

[80] Boyer, S.K.; McKenna, J.; Karliner, J.; Nirsberger, M. (1985) A mild and efficient process for detoxifying polychlorinated biphenyls, *Tetrahedron Lett.*, **26**, 3677–80.

[81] Davydov, D.V.; Beletskaya, I.P. (1993) PdCl$_2$-catalyzed hydrogenolysis of an Ar–Cl bond by sodium phosphinate in an aqueous alkaline-medium, *Russ. Chem. Bull.*, **42**, 575–7.

[82] Davydov, D.V.; Beletskaya, I.P. (1993) PdCl₂-catalyzed hydrogenolysis of a C–O bond in monoaryl sulfates by sodium phosphinate in an aqueous alkaline-medium, *Russ. Chem. Bull.*, **42**, 573–5.

[83] Okano, T.; Moriyama, Y.; Konishi, H.; Kiji, J. (1986) Counter phase transfer catalysis by water-soluble phosphine complexes. Catalytic reduction of allyl chlorides and acetates with sodium formate in two-phase system, *Chem. Lett.*, 1463–6.

[84] Pätzold, E.; Öhme, G. (1993) Water-soluble palladium(II) phosphine complexes as catalysts in the hydrodehalogenation of allyl and benzyl halogenides under biphasic and phase-transfer conditions, *J. Prakt. Chem. – Chem. Ztg.*, **335**, 181–4.

[85] Okano, T.; Iwahara, M.; Suzuki, T.; Konishi, H.; Kiji, J. (1986) Transition metal phosphine complexes possessing a phase transfer function. Preparation and catalytic reactivity of palladium phosphine complexes containing crown-ethers, *Chem. Lett.*, 1467–70.

[86] Shimizu, S.; Sasaki, Y.; Hirai, C. (1990) Inverse phase-transfer catalysis by cyclodextrins. Palladium-catalyzed reduction of bromoanisoles with sodium formate, *Bull. Chem. Soc. Jpn.*, **63**, 176–8.

[87] Davydov, D.V.; Beletskaya, I.P. (1995) Palladium-catalyzed and copper-catalyzed synthesis of triarylamines in an aqueous-organic emulsion, *Russ. Chem. Bull.*, **44**, 1141.

[88] Beletksya, I.P., Davydov, D.V.; *et al.* In press.

[89] Herwig, J.; Keim, W. (1994) Palladium complex-catalyzed synthesis of sulfinic acids, sulfinic acid-esters, sulfonic-acids and S-alkyl alkanethiosulfonates, *Inorg. Chim. Acta*, **222**, 381–5.

[90] Amatore, C.; Jutand, A.; Meyer, G. (1995) Evidence for the ligation of palladium(0) complexes by acetate ions: consequences on the mechanism of their oxidative addition with phenyl iodide and PhPd(OAc)(PPh₃)₂ as intermediate in the Heck reaction, *Organometallics*, **14**, 5605–13.

[91] Ozawa, F.; Kubo A.; Hayashi, T. (1992) Generation of tertiary phosphine-coordinated Pd(0) species from Pd(OAc)₂ in the catalytic Heck reaction, *Chem. Lett.*, 2177–80.

[92] Jeffery, T. (1984) Palladium catalysed vinylation of organic halides under solid–liquid phase-transfer conditions, *Chem. Commun.*, 1287–9.

[93] Bumagin, N.A.; More, P.G.; Beletskaya, I.P. (1989) Synthesis of substituted cinnamic acid and cinnamonitriles via palladium catalyzed reaction of aryl halides with acrylic acid and acrylonitrile in aqueous media, *J. Organomet. Chem.*, **371**, 397–401.

[94] Bumagin, N.A.; Andryukhova, N.P.; Beletskaya, I.P. (1990) Palladium catalyzed arylation of acrylonitrile by aryl halides, *Dokl. Akad. Nauk SSSR*, **313**, 107–9.

[95] Bumagin, N.A.; Sukhomlinova, L.I.; Vanchikov, A.N.; Tolstaya, T.P.; Beletskaya, I.P. (1992) Palladium-catalyzed arylation of acrylic acid by diaryliodonium salts in water, *Bull. Russ. Acad. Sci. – Div. Chem. Sci.*, **41**, 2130.

[96] Cheprakov, A.V.; Beletskaya, I.P.; in press.

[97] Demik, N.N.; Kabachnik, M.M.; Novikova, Z.S.; Beletskaya, I.P. (1995) Palladium-catalyzed arylation of vinyl-substituted and alpha-substituted vinylphosphonates, *Zh. Org. Khim.*, **31**, 64–8.

[98] Jeffery, T. (1994) Heck-type reactions in water, *Tetrahedron Lett.*, **35**, 3051–4.

[99] Hiroshige, M.; Hauske, J.R.; Zhou, P. (1995) Formation of CC bond in solid-phase synthesis using the Heck reaction, *Tetrahedron Lett.*, **36**, 4567–70.

[100] Terrett, N.K.; Gardner, M.; Gordon, D.W.; Kobylecki, R.J.; Steele, J. (1995) Combinatorial synthesis – the design of compound libraries and their application to drug discovery, *Tetrahedron*, **51**, 8135–73.

[101] Lemaire-Audoire, S.; Savignac, M.; Dupuis, C.; Genet, J.P. (1996) Intramolecular Heck-type reactions in aqueous-medium – dramatic change in regioselectivity, *Tetrahedron Lett.*, **37**, 2003–6.

[102] Diminnie, J.; Metts, S.; Parsons, E.J. (1995) In-situ generation and Heck coupling of alkenes in superheated water, *Organometallics*, **14**, 4023–5.

[103] Reardon, P.; Metts, S.; Crittendon, C.; Daugherity, P.; Parsons, E.J. (1995) Palladium-catalyzed coupling reactions in superheated water, *Organometallics*, **14**, 3810–16.

[104] Colquhoun, D.J.; Thompson, H.M.; Twigg, M.V. (1991) *Carbonylation: Direct Synthesis of Carbonyl Compounds*, Plenum Press, New York.

[105] Alper, H. (1986) Homogeneous and phase transfer carbonylation reactions, *J. Organomet. Chem.*, **300**, 1–6.

[106] Bumagin, N.A.; Nikitin, K.V.; Beletskaya, I.P. (1990) Palladium catalysed hydroxycarbonylation of aryl halides in aqueous organic media, *Dokl. Akad. Nauk SSSR*, **312**, 1129–34.

[107] Valentine, D.; Tilley, J.W.; LeMahieu, R.A. (1981) Practical catalytic synthesis of anthranylic acids, *J. Org. Chem.*, **46**, 4614–17.

[108] Grushin, V.V.; Alper, H. (1993) Simple and efficient palladium catalysed carbonylation of iodoxyarenes in water under mild conditions, *J. Org. Chem.*, **58**, 4794–5.

[109] Cheprakov, A.V.; Ponomareva, N.V.; Beletskaya, I.P. (1995) Palladium-catalyzed carbonylation of iodoarenes in aqueous solubilized systems, *J. Organomet. Chem.*, **486**, 297–300.

[110] Monteil, F.; Kalck, P. (1994) Carbonylation of bromobenzene in a biphasic medium catalyzed by water-soluble palladium complexes derived from tris(3-sulphophenyl)phosphine, *J. Organomet. Chem.*, **482**, 45–51.

[111] Okano, T.; Hayashi, T.; Kiji, J. (1994) Synthesis of phenylacetic acid via carbonylation of benzyl-chloride in the presence of a water-soluble complex, $PdCl_2(PPh_2(m-C_6H_4SO_3Na))_2$, and surfactants under 2-phase conditions, *Bull. Chem. Soc. Jpn*, **67**, 2339–41.

[112] Papadogianakis, G.; Maat, L.; Sheldon, R.A. (1994) Catalytic conversions in water: a novel carbonylation reaction catalyzed by palladium trisulfonated triphenylphosphine complexes, *Chem. Commun.*, 2659–60.

[113] Grushin, V.V.; Alper, H. (1992) An exceptionally simple biphasic method for the metal catalysed carbonylation of chloroarenes, *Chem. Commun.*, 611.

[114] Grushin, V.V.; Alper, H. (1994) Transformations of chloroarenes, catalyzed by transition-metal complexes, *Chem. Rev.*, **94**, 1047–62.

[115] Bumagin, N.A.; Nikitin, K.V.; Beletskaya, I.P. (1990) Competitive carbonylation and cross-coupling of the aryl halide and arylboron derivatives system catalysed by palladium compounds, *Dokl. Akad. Nauk SSSR*, **320**, 619–22.

[116] Bumagin, N.A.; Nikitin, K.V.; Beletskaya, I.P. (1990) Palladium catalysed acylation of amines effected by aryl iodides and carbon monooxide, *Dokl. Akad. Nauk SSSR*, **320**, 887–90.

[117] Drent, E.; Arnoldy, P.; Budzelaar, P.H.M. (1994) Homogeneous catalysis by cationic palladium complexes – precision catalysis in the carbonylation of alkynes, *J. Organomet. Chem.*, **475**, 57–63.

[118] Jenner, G.; Bentaleb; A. (1994) Ruthenium-catalyzed ethylene–methyl formate reactions: synthesis of propanol and ketones, *J. Mol. Catal.*, **91**, 31–43.

[119] Joh, T.; Doyama, K.; Onitsuka, K.; Shiohara, T.; Takahashi, S. (1991) Rhodium catalysed carbonylation of acetylenes under water gas shift conditions. Selective synthesis of furan-2(5*H*)-ones, *Organometallics*, **10**, 2493–8.

[120] Joh, T.; Nagata, H.; Takahashi, S. (1994) Rhodium-catalyzed synthesis of 2(5*H*)-furanones from terminal alkynes and non-substituted alkynes under water-gas shift reaction conditions, *Inorg. Chim. Acta*, **220**, 45–53.

[121] Sugioka, T.; Zhang, S.-W.; Morii, N.; Joh, T.; Takahashi, S. (1996) Rhodium-catalyzed carbonylation of 2-alkynylbenzaldehyde under water-gas shift reaction conditions. Formation of novel tricyclic lactones, *Chem. Lett.*, 249–50.

[122] Takeuchi, R.; Yasue, H. (1993) Rhodium complex-catalyzed desilylative cyclocarbonylation of 1-aryl-2-(trimethylsilyl)acetylenes – a new route to 2,3-dihydro-1*H*-inden-1-ones, *J. Org. Chem.*, **58**, 5386–92.

[123] Scrivanti, A.; Matteoli, U. (1995) A convenient synthesis of 2-(6-methoxy-2-naphthyl)propenoic acid (a naproxen precursor), *Tetrahedron Lett.*, **36**, 9015–18.

[124] Huh, K.T.; Orita, A.; Alper, H. (1993) Synthesis of dienoic acids and esters by cationic palladium complex-catalyzed carbonylation of alkynols and alkynediols, *J. Org. Chem.*, **58**, 6956–7

[125] Cornils, B.; Kuntz, E.G. (1995) Introducing TPPTS and related ligands for industrial biphasic processes, *J. Organomet. Chem.*, **502**, 177–86.

[126] Borowski, A.F.; Cole-Hamilton, D.J.; Wilkinson, G. (1978) Water-soluble transition metal phosphine complexes and their use in two-phase catalytic reactions of olefins, *Nouv. J. Chim.*, **2**, 137–44.

[127] Sangster, J. (1989) Octanol–water partition coefficients of simple organic compounds, *J. Phys. Chem. Ref. Data*, **18**, 1111–229.

[128] Purwanto, P.; Delmas, H. (1995) Gas–liquid–liquid reaction-engineering: hydroformylation of 1-octene using a water-soluble rhodium complex catalyst, *Catal. Today*, **24**, 135–40.

[129] Monteil, F.; Queau, R.; Kalck, P. (1994) Behavior of water-soluble dinuclear rhodium complexes in the hydroformylation reaction of oct-1-ene, *J. Organomet. Chem.*, **480**, 177–84.

[130] Kanagasabapathy, S.; Xia, Z.G.; Papadogianakis, G.; Fell. B. (1995) Hydroformylation with water-soluble and methanol-soluble rhodium carbonyl/phenyl-sulfonatoalkyl-phosphine catalyst systems: a new concept for the hydroformylation of higher molecular olefins, *J. Prakt. Chem. – Chem. Ztg.*, **337**, 446–50.

[131] Horvath, I.T.; Rabai, J. (1994) Facile catalyst separation without water – fluorous biphase hydroformylation of olefins, *Science*, **266**, 72–5.

[132] Chaudhari, R.V.; Bhanage, B.M.; Deshpande, R.M.; Delmas, H. (1995) Enhancement of interfacial catalysis in a biphasic system using catalyst-binding ligands, *Nature*, **373**, 501–3.

[133] Jin, Z.L.; Yan, Y.Y.; Zuo, H.P.; Fell, B. (1996) 2-Phase hydroformylation of higher molecular olefins by ethoxylated tris(*p*-hydroxyphenyl)-phosphine as complex ligand for the rhodiumcarbonyl catalyst, *J. Prakt. Chem. – Chem. Ztg.*, **338**, 124–8.

[134] Karakhanov, E.A.; Kardasheva, Y.S.; Maksimov, A.L.; Predeina, V.V.; Runova, E.A.; Utukin, A.M. (1996) Macrocomplexes on the basis of functionalized polyethylene glycols and copolymers of ethylene-oxide and propylene-oxide: synthesis and catalysis, *J. Mol. Catal. A – Chem.*, **107**, 235–40.

[135] Arhancet, J.P.; Davis, M.E.; Merola, J.S.; Hanson, B.E. (1989) Supported aqueous phase catalysts, *J. Catal.*, **121**, 327–39.

[136] Fell, B.; Leckel, D.; Schobben, C. (1995) Micellar 2-phase hydroformylation of multiple unsaturated fatty substances with water-soluble rhodiumdicarbonyltert: phosphine catalyst systems, *Fett Wiss. Technol.*, **97**, 219–28.

[137] Monflier, E.; Fremy, G.; Castanet, Y.; Mortreux, A. (1995) Molecular recognition between chemically-modified β-cyclodextrin and dec-1-ene: new prospects for biphasic hydroformylation of water-insoluble olefins, *Angew. Chem. Int. Ed. Engl.*, **34**, 2269–71.

[138] Monflier, E.; Fremy, G.; Castanet, Y.; Mortreux, A. (1995) A further breakthrough in biphasic, rhodium-catalyzed hydroformylation – the use of per(2,6-di-*O*-methyl)-β-cyclodextrin as inverse phase-transfer catalyst, *Tetrahedron Lett.*, **36**, 9481–4.

[139] Bunn, B.B.; Bartik, T.; Bartik, B.; Bebout, W.R; Glass, T.E.; Hanson, B.E. (1994) P-31 NMR spin-lattice relaxation-times as an indirect probe of adsorbed water in supported aqueous-phase catalysts, *J. Mol. Catal.*, **94**, 157–61.

[140] Fremy, G.; Castanet, Y.; Grzybek, R.; Monflier, E.; Mortreux, A.; Trzeciak, A.M.; Ziolkowski, J.J. (1995) A new, highly selective, water-soluble rhodium catalyst for methyl acrylate hydroformylation, *J. Organomet. Chem.*, **505**, 11–16.

[141] Fremy, G.; Monflier, E.; Carpenter, J.F.; Castanet, Y.; Mortreux, A. (1995) Enhancement of catalytic activity for hydroformylation of methyl acrylate by using biphasic and supported aqueous-phase systems, *Angew. Chem. Int. Ed. Engl.*, **34**, 1474–6.

[142] Fremy, G.; Monflier, E.; Carpentier, J.F.; Castanet, Y.; Mortreux, A. (1996) Expanded scope of supported aqueous-phase catalysis: efficient rhodium-catalyzed hydroformylation of alpha,beta-unsaturated esters, *J. Catal.*, **162**, 339–48.

[143] Yuan, Y.Z.; Xu, J.L.; Zhang, H.B.; Tsai, K.R (1994) The beneficial effect of alkali-metal salt on supported aqueous-phase catalysts for olefin hydroformylation, *Catal. Lett.*, **29**, 387–95.

[144] Blum, J.; Rosenfeld, A.; Polak, N.; Israelson, O.; Schumann, H.; Avnir, D. (1996) Comparison between homogeneous and sol–gel-encapsulated rhodium-quaternary ammonium ion-pair catalysts, *J. Mol. Catal. A – Chem.*, **107**, 217–23 .

[145] Naughton, M.J.; Drago, R.S. (1995) Supported homogeneous film catalysts, *J. Catal.*, **155**, 383–9.

[146] Hong, L.; Ruckenstein, E. (1994) Immobilization of alkoxylated phosphine-ligands and their Rh complexes to a silica surface-coated with an organic monolayer or multilayer, *J. Mol. Catal.*, **90**, 303–21.

[147] Mdleleni, M.M.; Rinker, R.G.; Ford, P.C. (1994) Catalysis of the water-gas shift reaction by polymer-immobilized rhodium complexes, *J. Mol. Catal.*, **89**, 283–94.

[148] Joo, F.; Toth, Z. (1980) Catalysis by water-soluble phosphine complexes of transition metal ions in aqueous and two-phase media, *J. Mol. Catal.*, **8**, 369–83.

[149] Larpent, C.; Dabard, R.; Patin, H. (1987) Transition metal and water soluble ligands for the catalytic hydrogenation of aqueous solutions of unsaturated carboxylic acids, *C. R. Acad. Sci. Ser. II*, **304**, 1055–7.

[150] Larpent, C.; Dabard, R.; Patin, H. (1987) Catalytic hydrogenation of olefins in biphasic water–liquid systems, *Tetrahedron Lett.*, **28**, 2507–10.

[151] Larpent, C.; Patin, H. (1988) Catalytic hydrogenations in biphasic liquid–liquid systems. 2. Utilization of sulfonated tripod ligands for the stabilization of colloidal rhodium dispersions, *J. Mol. Catal.*, **44**, 191–5.

[152] Larpent, C.; Patin, H. (1990) Mechanistic aspects of alkene hydrogenation and deuteration catalysed by dispersion of hydroxyhydridorhodium colloids in aqueous medium, *J. Mol. Catal.*, **61**, 65–73.

[153] Joo, F.; Csiba, P.; Benyei, A. (1993) Effect of water on the mechanism of hydrogenations catalyzed by rhodium phosphine complexes, *Chem. Commun.*, 1602–4.

[154] Benyei, A.; Stafford, J.N.W.; Katho, A.; Darensbourg, D.J.; Joo, F. (1993) The effect of phosphonium salt formation on the kinetics of homogeneous hydrogenations in water utilizing a rhodium *meta*-sulfonatophenyl-diphenylphosphine complex, *J. Mol. Catal.*, **84**, 157–63.

[155] Andriollo, A.; Bolivar, A.; Lopez, F.A.; Paez, D.E. (1995) Homogeneous catalysis in water: on the synthesis and characterization of a ruthenium water-soluble complex – preliminary hydrogenation of olefins in a biphasic system, *Inorg. Chim. Acta*, **238**, 187–92.

[156] Hronec, M.; Cvengrosova, Z.; Kralik, M.; Palma, G.; Corain, B. (1996) Hydrogenation of benzene to cyclohexene over polymer-supported ruthenium catalysts, *J. Mol. Catal. A – Chem.*, **105**, 25–30.

[157] Kolaric, S.; Sunjic, V. (1996) Comparative-study of homogeneous hydrogenation of D-glucose and D-mannose catalyzed by water-soluble Ru(tri(*m*-sulfophenyl)phosphine) complex, *J. Mol. Catal. A – Chem.*, **110**, 189–93.

[158] Lau, C.P.; Cheng, L. (1993) Catalytic hydrogenation reactions by *cis*[RU(6-6'-Cl$_2$bpy)$_2$(OH$_2$)$_2$](CF$_3$SO$_3$)$_2$ in biphasic media (6,6'-Cl$_2$bpy = 6,6'-dichloro-2,2'-bipyridine), *J. Mol. Catal.*, **84**, 39–50.

[159] Frediani, P.; Bianchi, M.; Salvini, A.; Guarducci, R; Carluccio, L.C.; Piacenti, F. (1995) Ruthenium carbonyl carboxylate complexes with nitrogen-containing ligands. 3. Catalytic activity in hydrogenation, *J. Organomet. Chem.*, **498**, 187–97.

[160] Schumann, H.; Hemling, H.; Goren, N.; Blum, J. (1995) Dicarbonylbis[2-(diphenylphosphino)-*N*,*N*-dimethylethanamine]-bis-[µ-(2-methyl)-2-propanethiolato]dirhodium and dicarbonylbis[2-(diphenylphosphino)-*N*,*N*,*N*-trimethylethanaminium]bis-[µ-(2-methyl)-2-propanethiolato]dirhodium tetraphenylborate: their syntheses, characterization and application as hydrogenation catalyst, *J. Organomet. Chem.*, **485**, 209–13.

[161] Sinou, D.; Safi, M.; Claver, C.; Masdeu, A. (1991) Catalytic transfer hydrogenation of unsaturated substrates with formates in the presence of water soluble complexes of rhodium, *J. Mol. Catal.*, **68**, L9–12.

[162] Wan, K.T.; Davis, M.E. (1994) Asymmetric-synthesis of naproxen by supported aqueous-phase catalysis, *J. Catal.*, **148**, 1–8.

[163] Wan, K.T.; Davis, M.E. (1993) Ruthenium(II)-sulfonated BINAP: a novel water-soluble asymmetric hydrogenation catalyst, *Tetrahedron – Asymmetry*, **4**, 2461–8.

[164] Flach, H.N.; Grassert, I.; Oehme, G.; Capka, M. (1996) New insoluble surfactant systems as aids in catalysis: a convenient method for nonbonded immobilization of catalytically active transition-metal complexes, *Colloid Polym. Sci.*, **274**, 261–8.

[165] Joo, F.; Benyei, A. (1989) Biphasic reduction of unsaturated aldehydes to unsaturated alcohols by ruthenium complex-catalysed hydrogen transfer, *J. Organomet. Chem.*, **363**, C19–21.

[166] Grosselin, J.M.; Mercier, C.; Allmang, G.; Grass, F. (1991) Selective hydrogenation of α,β-unsaturated aldehydes in aqueous organic two-phase solvent systems using Ru or Rh complexes of sulfonated phosphines, *Organometallics*, **10**, 2126–33.

[167] Darensbourg, D.J.; Stafford, N.W.; Joo, F.; Reibenspies, J.H. (1995) Water-soluble organometallic compounds. 5. The regioselective catalytic-hydrogenation of unsaturated aldehydes to saturated aldehydes in an aqueous 2-phase solvent system using 1,3,5-triaza-

7-phosphaadamantane complexes of rhodium, *J. Organomet. Chem.*, **488**, 99–108.

[168] Fache, E.; Mercier, C.; Pagnier, N.; Despeyroux, B.; Panster, P. (1993) Selective hydrogenation of α,β-unsaturated aldehydes catalyzed by supported aqueous-phase catalysts and supported homogeneous catalysts, *J. Mol. Catal.*, **79**, 117–31.

[169] Laine, R.M.; Crawford, E.J. (1988) Homogeneous catalysis of the water-gas shift reaction, *J. Mol. Catal.*, **44**, 357–87.

[170] Nomura, K. (1995) Efficient selective reduction of aromatic nitro-compounds by ruthenium catalysis under CO/H$_2$O conditions, *J. Mol. Catal. A – Chem.*, **95**, 203–10.

[171] Kaneda, K.; Kuwahara, H.; Imanaka, T. (1994) Chemoselective reduction of nitro-groups in the presence of olefinic, ester, and halogeno functions using a reducing agent of CO and H$_2$O catalyzed by Rh carbonyl clusters, *J. Mol. Catal.*, **88**, L267–70.

[172] Nomura, K.; Ishino, M.; Hazama, M. (1993) Selective reduction of aromatic nitro-compounds affording aromatic-amines under CO/H$_2$O conditions catalyzed by phosphine-added rhodium and ruthenium carbonyl-complexes, *J. Mol. Catal.*, **78**, 273–82.

[173] Ragaini, F.; Cenini, S. (1996) Homogeneous catalysis in water without charged ligands: reduction of nitrobenzene to aniline by CO/H$_2$O catalyzed by [Rh(CO)$_4$]$^-$, *J. Mol. Catal. A – Chem.*, **105**, 145–8.

[174] Moya, S.A.; Pastene, R.; Sariego, R.; Sartori, R.; Aguirre, P.; Lebozec, H. (1996) Complexes with heterocyclic nitrogen ligands. 3. Cationic rhodium(I) derivatives and applications in catalysis, *Polyhedron*, **15**, 1823–7.

[175] Mdleleni, M.M.; Rinker, R.G.; Ford, P.C. (1994) Catalysis of the water-gas shift reaction by polymer-immobilized rhodium complexes, *J. Mol. Catal.*, **89**, 283–94.

[176] Bentaleb, A.; Jenner, G. (1993) Catalytic reduction of nitrobenzene to aniline with aqueous methyl formate, *J. Organomet. Chem.*, **456**, 263–9.

[177] Cenini, S.; Bettettini, E.; Fedele, M.; Tollari, S. (1996) Intramolecular amination catalyzed by ruthenium and palladium: synthesis of 2-acyl indoles and 2-aryl quinolines by carbonylation of 2-nitrochalcones, *J. Mol. Catal. A–Chem.*, **111**, 37–41.

[178] Tsuji, J. (1984) Synthetic applications of the palladium catalyzed oxidation of olefins to ketones, *Synthesis*, 369–84.

[179] Hronec, M.; Cvengrosova, Z.; Holotik, S. (1994) Is metallic palladium formed in Wacker oxidation of alkenes?, *J. Mol. Catal.*, **91**, 343–52.

[180] Upadhyay, M.J.; Bhattacharya, P.K.; Ganeshpure, P.A.; Satish, S. (1994) Synthesis of binuclear amide complexes of Ru(III) and study of their catalytic activity in epoxidation of alkenes, *J. Mol. Catal.*, **88**, 287–94.

[181] Bailey, A.J.; Griffith, W.P.; Savage, P.D. (1995) Studies on transition-metal nitrido and oxo complexes. 15. Oxo complexes of ruthenium with N,N'-donors as oxidation catalysts for alkenes, alkanes, alcohols, and their osmium analogs, *Dalton Trans.*, 3537–42.

[182] Khan, M.M.T.; Chatterjee, D.; Bhatt, S.D.; Rao, A.P. (1992) Kinetics and mechanism of the epoxidation of styrene and substituted styrenes with O$_2$ catalyzed by [Ru^{III}(edta)(H$_2$O)]$^-$, *J. Mol. Catal.*, **77**, 23–8.

[183] Fisher, J.M.; Fulford, A.; Bennett, P.S. (1992) Catalytic alkene epoxidation with ruthenium complexes and hydrogen-peroxide, *J. Mol. Catal.*, **77**, 229–34.

[184] Khan, M M.T.; Shukla, R.S. (1992) Ascorbate-dependent monoxygenase vs dioxygenase mechanism for Ru(III)–EDTA catalyzed oxidation of adamantane by molecular-oxygen, *J. Mol. Catal.*, **77**, 221–8.

[185] Ganguly, S.; Roundhill, D.M. (1993) Catalytic hydration of diethyl maleate to diethyl malate using divalent complexes of palladium(II) as catalysts, *Organometallics*, **12**, 4825–32.

[186] Monflier, E.; Bourdauducq, P.; Couturier, J.L.; Kervennal, J.; Mortreux, A. (1995) Highly efficient telomerization of butadiene into octadienol in a micellar system: a judicious choice of the phosphine/surfactant combination, *Appl. Catal. A – Gen.*, **131**, 167–78.

[187] Stunnenberg, F.; Niele, F.G.M.; Drent, E. (1994) Hydration of conjugated dienes to produce ketones catalyzed by ruthenium complexes, *Inorg. Chim. Acta*, **222**, 225–33.

[188] Blum, J.; Huminer, H.; Alper, H. (1992) Alkyne hydration promoted by RhCl$_3$ and quaternary ammonium-salts, *J. Mol. Catal.*, **75**, 153–60.

[189] Bianchini, C.; Casares, J.A.; Peruzzini, M.; Romerosa, A.; Zanobini, F. (1996) The mechanism of the Ru-assisted C–C bond-cleavage of terminal alkynes by water, *J. Am. Chem. Soc.*, **118**, 4585–94.

[190] Novak, B.M.; Grubbs, R H. (1988) The aqueous ROMP of 7-oxanorbornene derivatives, *J. Am. Chem. Soc.*, **110**, 7543–4.

[191] Muhlebach, A.; Bernhard, P.; Buhler, N.; Karlen, T.; Ludi, A. (1994) Ring-opening metathesis polymerization of bicyclo[2.2.1]hept-2-ene (2-norbornene) and exo,exo-5,6-bis(methoxycarbonyl)-7-oxabicyclo|2.2.1]hept-2-ene using Ru(II) and Ru(III) complexes: polymerization kinetics and ruthenium content in polymers, *J. Mol. Catal.*, **90**, 143–56.

[192] Wache, S. (1995) Organometallic complex catalysis in water. 2. Water-soluble organoruthenium(IV) catalysts for the emulsion polymerization of norbornene, *J. Organomet. Chem.*, **494**, 235–40.

[193] Eychenne, P.; Perez, E.; Rico, I.; Bon, M.: Lattes, A.; Moisand, A. (1993) First example of lattices synthesis via oligomerization of norbornene in aqueous emulsions, catalyzed by palladium-chloride, *Colloid. Polym. Sci.*, **271**, 1049–54.

[194] Amer, I.; Schumann, H.; Ravindar, V.; Baidossi, W.; Goren, N.; Blum, J. (1993) A convenient route to 1,3,5-trisubstituted benzenes via rhodium-catalyzed polymerization of arylacetylenes, *J. Mol. Catal.*, **85**, 163–71.

[195] Baidossi, W.; Goren, N.; Blum, J.; Schumann, H.; Hemling, H. (1993) Homogeneous and biphasic oligomerization of terminal alkynes by some water-soluble rhodium catalysts, *J. Mol. Catal.*, **85**, 153–62.

[196] Westernacher, S.; Kisch, H. (1996) Transition-metal complexes of diazenes. 36. Formation of indoles from azobenzene and diphenylacetylene through supported aqueous-phase catalysis by Rh(I) complexes, *Monatsh. Chem.*, **127**, 469–73.

[197] Sertchook, H.; Avnir, D.; Blum, J.; Joo, F.; Katho, A.; Schumann, H.; Weimann, R.; Wernik, S. (1996) Sol–gel entrapped lipophilic and hydrophilic ruthenium–phosphine, rhodium–phosphine, and iridium-phosphine complexes as recyclable isomerization catalysts, *J. Mol. Catal. A. – Chem.*, **108**, 153–60.

[198] Gassner, F.; Leitner, W. (1993) CO_2-activation. 3. Hydrogenation of carbon-dioxide to formic acid using water-soluble rhodium catalysts, *Chem. Commun.*, 1465–6.

[199] Lewis, L.N. (1993) Chemical catalysis by colloids and clusters, *Chem. Rev.*, **93**, 2693–730.

[200] Herrmann, W.A.; Kohlpaintner, C.W. (1993) Water-soluble ligands, metal-complexes, and catalysts: synergism of homogeneous and heterogeneous catalysis, *Angew. Chem. Int. Ed. Engl.*, **32**, 1524–44.

6 Oxidations and reductions in water

F. FRINGUELLI, O. PIERMATTI and F. PIZZO

6.1 Oxidations in water: introduction

Oxidation of organic compounds has probably been the most widely investigated process because it is of interest to both academic scientists and industrial technicians. Many oxidants and catalysts are known and a number of reaction conditions have been carefully investigated [1]. The modern chemical industry requires selective highly efficient oxidations and environmentally sound technological processes. One way to attain these objectives is to make a rational selection of the reaction medium.

The use of water as reaction medium is not new. The oxidation of arenes with $KMnO_4$ in alkaline aqueous medium is a very old reaction [2]. Also, H_2O_2 in water, in the presence or absence of catalyst, is a more frequently used oxidant owing to its peculiarities, such as low relative molecular mass, stability and solubility in many organic solvents. It is economical and environment-friendly since it forms water as a secondary product [3].

In the following sections, some recent and innovative oxidations by chemical reagents in aqueous media are illustrated and classified according to the type of bond or chemical functionality involved. Enzymic oxidations in water have been widely investigated during the last decade and significant synthetic applications are known. However, this subject will be not discussed here.

6.2 Oxidation of carbon–hydrogen bonds

The selective oxidative functionalization of C–H bonds of alkanes is a chemical problem of great practical importance. The major obstacles in the metal–oxygen-catalyzed processes are the incompatibility of catalysts with O_2 and the greater reactivity of the products (aldehydes, alcohols, alkenes, etc.) with respect to the starting alkanes.

Pt(II) salts have recently received great attention because they are oxygen-tolerant and the reaction can be carried out in water under mild conditions. The methyl group of p-toluenesulfonic acid and that of ethanol are oxidized to hydroxymethyl and chloromethyl in water at 80–100°C by the Na_2PtCl_2–Na_2PtCl_4 system in 3–6 h. Even though the conversion levels are low, it is remarkable that in water the methyl group is more reactive than the CH_2OH group [4a]. The mechanistic hypothesis invokes: (i) formation of a Pt(II) alkyl by electrophilic C–H activation; (ii) oxidation to a Pt(IV) alkyl; and (iii) nucleophilic attack by water or chloride to give alcohols and/or alkyl chlorides [4b]. A detailed study has shown that the unactivated C–H bonds of esters, alcohols, ethers and simple alkanes, such as ethane and propane, are oxidized in aqueous medium by K_2PtCl_4, whereas the C–H bonds α to an oxygen are oxidized by metallic Pt in the presence of O_2 [4c].

The oxidation of arenes to form the corresponding benzoic acids is an important industrial process executed by using dilute HNO_3 or air, under vigorous stirring, high temperature and pressure. In the laboratory, $KMnO_4$ and $K_2Cr_2O_7$ are generally used under analogous conditions. In the presence of ultrasound (US) the oxidation of arenes with aqueous $KMnO_4$ is strongly accelerated at room temperature [5]. Toluene and its analog, for instance, are oxidized in 3–4 h with 60–82% yields.

Recently it was pointed out [6] that potassium and sodium bromate oxidize arenes in good yields in a 3:2 dioxane–water solution by using cerium ammonium nitrate (CAN) as catalyst. Toluene derivatives are oxidized to ca. 1:1 mixture of benzaldehydes and benzoic acids, while ethylbenzenes are converted to acetophenones. According to the proposed mechanistic scheme, water either influences the polarity of the reaction medium or acts as the reagent, reacting with the intermediate benzylic carbonium ion to give a hydroxy derivative, which is oxidized from either Ce^{4+} or BrO_3^- ions.

6.3   Oxidation of carbon–carbon double bonds

6.3.1   Epoxidation of simple alkenes

The epoxidation of simple alkenes with peroxycarboxylic acids involves an electrophilic attack of peracid with the formation of a bicyclic or a mono-cyclic transition state (Scheme 6.1), as first proposed by Bartlett [7] or recently suggested by theoretical studies [8] respectively. The reaction takes place readily in nonpolar solvents such as dichloromethane (DCM) and benzene and one can argue that hydrogen-bonding solvents, in particular, should retard the rate by interfering with intramolecular hydrogen bonding

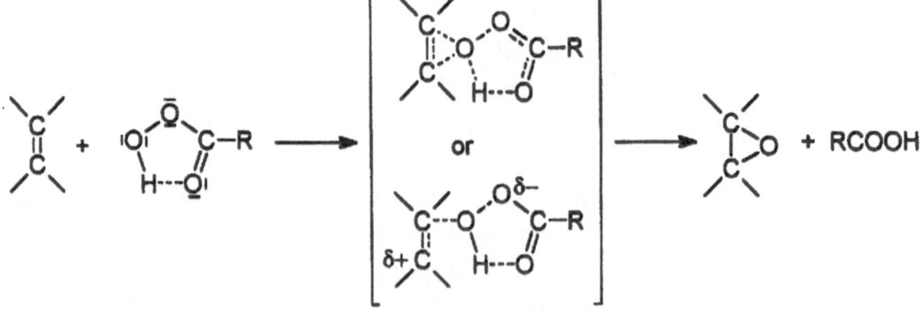

Scheme 6.1

of the peracid. In contrast, it has been observed [9] that the epoxidation of simple alkenes with *m*-chloroperoxybenzoic acid (MCPBA) in aqueous solution of NaHCO₃ (pH ca. 8.3) at room temperature is fast and pure epoxides are isolated in good or excellent yield. Some representative examples are reported in Table 6.1. The mild reaction conditions make the procedure suitable for acid-sensitive alkenes and epoxides such as styrene, methylstyrenes and aryloxiranes [10]. By using peroxybenzoic acid (PBA) the reaction time increases but highly acid-sensitive olefins give better yields.

In aqueous medium the reaction occurs in heterogeneous phase, but this does not affect the reactivity, which, on the contrary, is sometimes higher than in homogeneous organic phase. For instance, the reactivity of cyclohexene with MCPBA in water is comparable to that in DCM and is 8 and 16 times higher than in biphasic medium and in homogeneous *n*-hexane solution, respectively.

Epoxidation in water allows the perhydroxylation of the carbon–carbon double bond to be achieved by a one-pot procedure [11]. The process is completely stereospecific and 1,2-diols are generally obtained in excellent yield:

1. H₂O; MCPBA; 20° C; 0.5-8 h

2. H⁺; 20-100° C; 1-10 h

76-95%

The direct epoxidation of simple alkenes by H₂O₂ requires that the peroxide must be activated. In buffered aqueous tetrahydrofuran (THF), 50% H₂O₂ activated by stoichiometric amounts of organophosphorus anhy-

Table 6.1 Epoxidation of simple alkenes by MCPBA

Alkene	Reaction conditions		
	Temp. (°C)	Time (h)	Yield (%)
cyclopentene	20	0.5	90
cyclohexene	20	0.5	95
cycloheptene	20	0.5	90
cyclooctene	20	0.5	95
methylenecyclohexane	20	0.5	95
(+)-3-carene	20	0.5	95
1-octene	20	8	95[a]
(–)-carvone	20	1	95[b]
styrene	20	1.5	95
α-methylstyrene	0	1	63[c]
trans-β-methylstyrene	0	1	93
indene	0	1	53[d]
1,2-dihydronaphthalene	0	0.75	70

[a]Reaction carried out in the absence of NaHCO₃.
[b]Mixture (ca. 1:1) of *erythro*- and *threo*-8,9-epoxy-*p*-menth-2-en-1-one.
[c]With PBA, 80% yield.
[d]With PBA, 65% yield.

drides and halides leads to efficient epoxidation of a variety of alkenes [12]. The best results were obtained by using diphenylphosphinic anhydride:

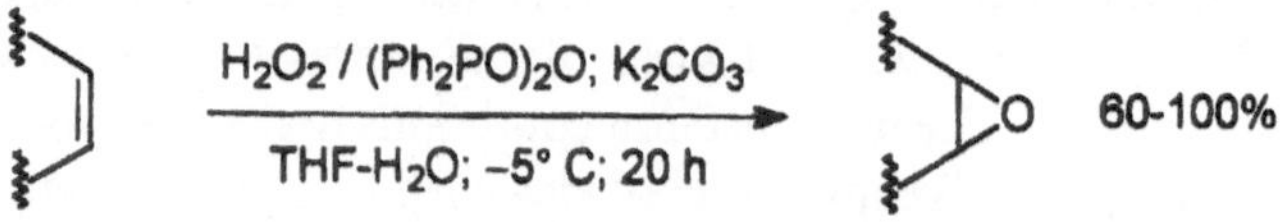

Neumann and Miller [13] catalyzed the epoxidation of alkenes by 30% aqueous H_2O_2 with an insoluble catalytic assembly consisting of a silicate xerogel covalently modified with phenyl groups and quaternary ammonium polyoxometallate ion pairs as catalytically active species. The combined presence of phenyl units and the use of the octyldimethyl-substituted ammonium salt and $[WZnMn_2^{II}(ZnW_9O_{34})_2]^{12-}$ as anionic polyoxometallate gives the best results.

Cycloalkenes and styrenes have been epoxidized by an aqueous solution of sodium bromite in the presence of $CuSO_4 \cdot 5H_2O$ at room temperature [14]. In the absence of copper ion, no epoxidation occurs. It has been hypothesized that the unstable $Cu(BrO_2)_2$ is the oxidizing species (Scheme 6.2). Manganese porphyrin complexes have been used to catalyze the epoxidation of simple alkenes in aqueous medium.

The water-soluble *meso*-tetrakis(4-*N*-methylpyridyl)porphinatomanganese(III) chloride efficiently catalyzes (2 m in at r.t., 99% conversion) the epoxidation of sodium 4-styrenesulfonate by oxone (a stable water-soluble oxidant with the approximate composition $K_2SO_4 \cdot 2KHSO_5 \cdot KHSO_4$) in water at neutral pH [15]. In the absence of catalyst, the reaction time is ca. 10 times longer.

The tetrasodium salt of 5,10,15,20-tetrakis(2,6-dichloro-3-sulfonato-phenyl)porphinatomanganese(III) chloride is water-soluble and, if bound to colloidal polymer particles, catalyzes in water the epoxidation of styrene by sodium hypochlorite but does not catalyze the oxidation of aliphatic alkenes [16].

The industrial processes of epoxidation of alkenes are carried out by using hydrogen peroxide, peracetic acid or *t*-butyl hydroperoxide (TBHP) [17]. An easy, safe and cheap method has recently been proposed [18] that uses

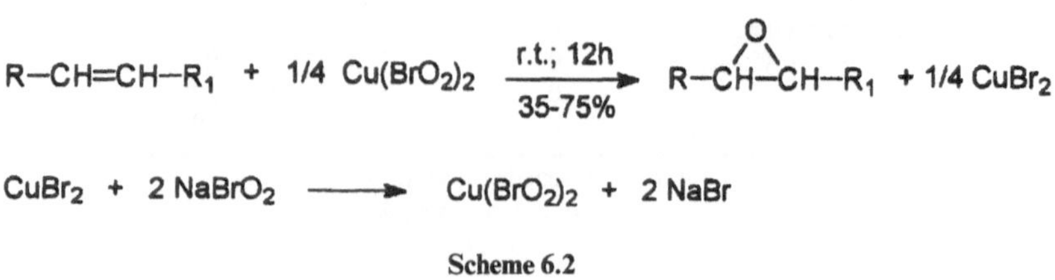

$$NaBrO_2 + 1/2\ CuSO_4 \longrightarrow 1/2\ Cu(BrO_2)_2 + 1/2\ Na_2SO_4$$

$$CuBr_2 + 2\ NaBrO_2 \longrightarrow Cu(BrO_2)_2 + 2\ NaBr$$

Scheme 6.2

nascent oxygen generated by electrolysis of water at room temperature by using Pd black as an anode. Cyclohexene, for example, is epoxidized with good yield and selectivity. This is an example of a process in which water is both the reaction medium and a reagent.

Previously we pointed out that the oxidation of alkenes in water allows one-pot multi-step reactions to be carried out involving tandem epoxidation and ring-opening reactions. Further examples are the synthesis of optically active trifluorolactic acid by aqueous nitric acid ring-opening oxidation of (–)-1,2-epoxy-3,3,3-trifluoropropane [19] in the presence of a catalytic amount of copper metal:

$$F_3C \diagdown\!\!\!\!\triangle \xrightarrow[\text{60-80°C; 4h}]{\text{60\% HNO}_3\text{; Cu; r.t.; 2 h}} F_3C \diagup\!\!\!\!\overset{OH}{\diagdown} COOH \qquad 84\%$$

and the regioselective ring opening of racemic styrene oxide by lithium azide in the presence of β-cyclodextrin (β-CD) in aqueous media to give optically active phenylazido alcohols [20]:

$$\xrightarrow[\text{H}_2\text{O; }\beta\text{-CD}]{\text{LiN}_3\text{; r.t.; 43 h}}$$

Yield: 20%		76%
ee: 56%		12%

6.3.2 Epoxidation of olefinic alcohols

The selective epoxidation of olefinic alcohols is a goal of great interest in organic synthesis. In organic solvents, oxidation with peroxycarboxylic acids is generally face-selective but poorly regioselective [21]. A remarkable regio- and stereoselectivity is obtained by using hydroperoxides in the presence of transition metals [22].

In aqueous media, epoxidation can be carried out with 30% H_2O_2 at pH 4.5 by using tungstic acid, a cheap and quite safe reagent, as catalyst [23]:

$$\xrightarrow[\text{H}_2\text{O-MeOH; pH 4.5}]{\text{H}_2\text{WO}_4\text{ / H}_2\text{O}_2\text{; 0°C-r.t.}} \qquad 22\text{-}98\%$$

n = 1, 2, 3

While acyclic allylic alcohols generally react quickly and give epoxyalcohols in good yield, the cycloallylic alcohols require longer reaction times and the yields are unsatisfactory. Homo- and bis-homoallylic alcohols are poorly reactive under these conditions and the reaction is regioselective but scarcely stereoselective.

Diphenylphosphinic anhydride has recently been suggested [12] to activate H_2O_2. The true oxidizer is the hydroperoxyphosphinic acid, which epoxidizes cycloallylic alcohols in aqueous THF with good *cis* stereoselectivity.

Interesting results have been obtained by carrying out epoxidation in water only with peroxycarboxylic acids. MCPBA gives high yields of epoxide only with allylic alcohols but the reaction is neither regio- nor stereoselective [24]. High regio- and diastereoselectivity are obtained with monoperoxyphthalic acid (MPPA) at room temperature by controlling the pH of the reaction [24, 25]. Geraniol, for example, is epoxidized regioselectively by MPPA at the more-reactive 6,7-position at pH 8.3 and at the less-reactive 2,3-position at pH 12.5 (Figure 6.1). Epoxidation of nerol imitates that of geraniol, while 2E,6E-farnesol is epoxidized with high regioselectivity at the 2,3-position only in the presence of surfactant because of the insolubility of this alcohol. Linalool is unreactive at both the 1,2- and 6,7-positions in strong alkaline medium, while at pH 8.3 the reaction occurs quickly and regioselectively at the 6,7- position and no cycloadducts are detected.

This last result is significant because the epoxidation of linalool and analogous homo- and bis-homoallylic alcohols affords tetrahydrofuran and tetrahydropyran derivatives in organic solvent:

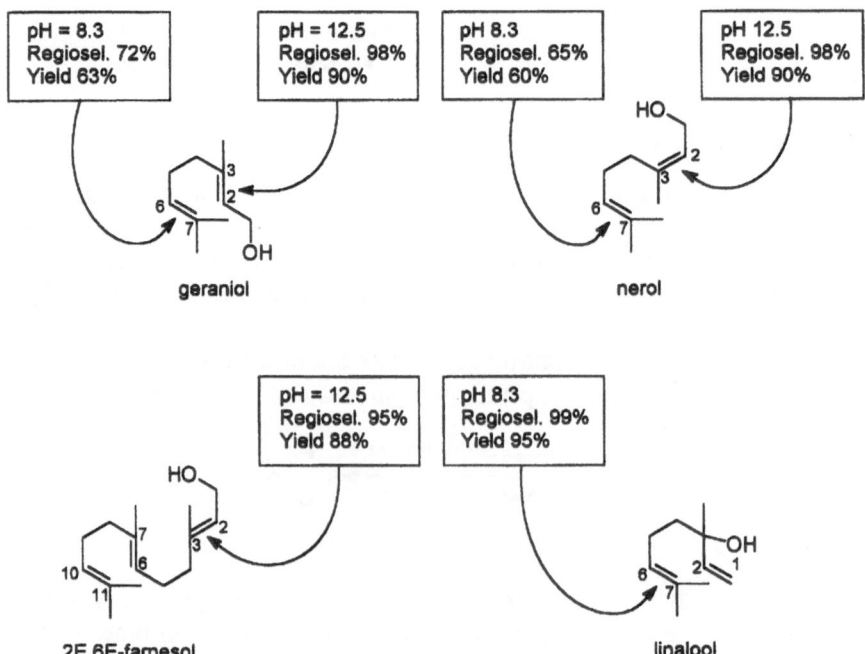

Figure 6.1 Regioselective epoxidation of polyolefinic alcohols with monoperoxyophthalic acid in water at 25°C.

In water the intramolecular nucleophilic attack occurs with greater difficulty and the epoxide is generally the main reaction product (Table 6.2) [26].

The epoxidation of allylic and homoallylic alcohols in water is also highly diastereoselective. Some examples are the epoxidations of cycloalkenols reported in Table 6.3. All of the reactions occur with high yields and, with respect to DCM, the strong alkaline medium increases the amount of *cis* adduct except in the epoxidation of 2-cycloheptenol, which gives a prevalence of *trans* compound [27].

Table 6.2 Epoxidation of homo- and bis-homoallylic alcohols with MCPBA

	Alcohol[a]		H$_2$O (pH 7–8.3)		CH$_2$Cl$_2$	
n	R$_1$ R$_2$	R$_3$	Epoxide (%)	Cycloadduct (%)	Epoxide (%)	Cycloadduct (%)
1	H H	Me	100	0	90	10
2	H Me	Me	98[b]	2	10[b]	90
2	H H	H	73	27	40	60
3	H H	H	100	0	45	55
2	Me CH=CH$_2$	Me	98[b]	2	69[b]	31

[a]For n, R$_1$, R$_2$ and R$_3$, see the relevant structural equation in the text.
[b]A 1:1 diastereoisomeric mixture.

Table 6.3 Epoxidation of cycloallylic and homo-cycloallylic alcohols with MPPA at 0–25°C (yields 80–90%)

	cis/trans		
Alcohol	CH$_2$Cl$_2$	NaHCO$_3$	NaOH
$n = 1$	5.2/1[a]	9/1	100/1
$n = 2$	5.6/1	7/1	99/1
$n = 3$	1.6/1[a]	1/1	1/1.6
$n = 1$	24/1	7/1	100/1
$n = 2$	5.6/1	1/1	100/1
$n = 3$	3/1	1.2/1	99/1

[a]With MCPBA.

The high regio- and stereo-controlled epoxidation of allylic alcohols in strong alkaline medium can be explained [24] on the basis of stabilizing secondary interactions between the hydrogen of the alcoholic group and the oxygens of the peroxycarboxylate functionality. The resulting cyclic transition state:

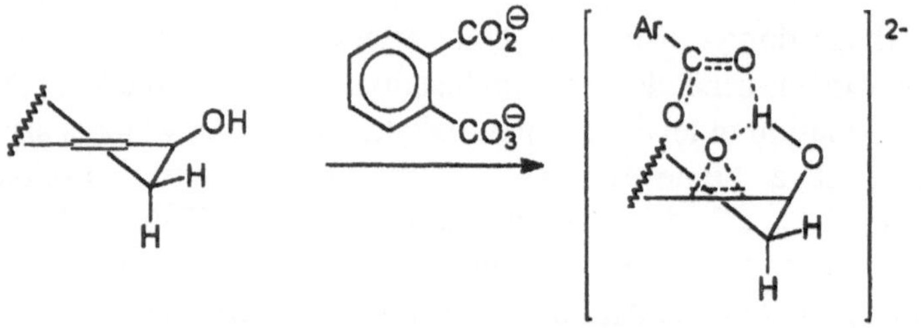

resembles that hypothesized in the epoxidation of simple alkenes by peroxy-acids, where the hydrogen atom of the peroxycarboxylic group is the driving force of the reaction. When the hydrogen atom of the hydroxy group of the alcohol is removed (OH→OMe), no epoxidation is observed in alkaline medium.

6.3.3   Epoxidation of α,β-unsaturated carbonyl compounds

A traditional way to epoxidize α,β-unsaturated carbonyl compounds in aqueous meda is the Michael reaction with alkaline hydroperoxides [1]. Baeyer–Villiger oxidation is a concomitant reaction, making double-bond epoxidation a delicate procedure.

Sodium perborate (SPB; $NaBO_3 \cdot nH_2O$, $n = 1$ to 4) is a mild and inexpensive oxidant [28] that, in aqueous media at pH 8–10, gives α,β-epoxyketones in good yields [29] (Scheme 6.3).

α,β-Unsaturated carboxylic acids are epoxidized with difficulty by peroxy-

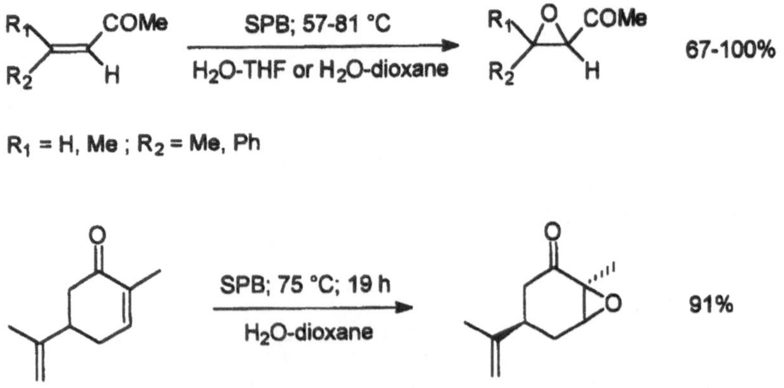

R_1 = H, Me ; R_2 = Me, Ph

Scheme 6.3

carboxylic acids because of the electron-withdrawing effect of the carboxylic group. In water, the oxidation can be performed by using Payne's reaction [30] with the modified procedure of Sharpless [31]:

R_1, R_2, R_3 = H, Me, Pr

A convenient alternative is to use the aqueous oxone–acetone system, buffering the reaction with $NaHCO_3$ [32]. Excellent yields of α,β-epoxyacids are obtained at 24–27°C in 2–3 h. By using oxone in water only at neutral pH, fumaric acid is epoxidized quantitatively but requires a long reaction time [15].

6.4   Other oxidations

6.4.1   Oxidation of carbonyl compounds

The oxidation of aldehydes to the corresponding carboxylic acids has been widely investigated and numerous procedures are known in aqueous and organic media [33]. Aromatic aldehydes have recently been oxidized at 0–4°C in aqueous performic acid produced by the addition of H_2O_2 to formic acid [34]. Generally, the carboxylic acid precipitates out of the reaction mixture and can be isolated simply by filtration. When heteroaromatic aldehydes such as formylpyridines, formylquinolines and formylazaindoles are oxidized, the formation of N-oxides is avoided. The use of cosolvents (tetrahydrofuran (THF), N,N-dimethylformamide (DMF)) gives less satisfactory results.

The formyl group can be chemoselectively oxidized, in the presence of other oxidizable functionalities, in aqueous media containing an equivalent amount of surfactant. For example, 4-(methylthio)benzaldehyde is quantitatively oxidized [35] to 4-(methylthio)benzoic acid with TBHP in a basic aqueous medium in the presence of $(CTA)_2SO_4$ (cetyltrimethylammonium sulfate).

Benzaldehydes having an electron-releasing group such as OH and OMe in *ortho* and *para* position are converted (Dakin reaction) to phenols via aryl formates when treated with H_2O_2 in sodium hydroxide solution [1]. The reaction was recently [36] carried out in high yields by using sodium percarbonate (SPC; $Na_2CO_3 \cdot 1.5H_2O_2$) in H_2O–THF under ultrasonic irradiation. Some results are reported in Table 6.4.

Baeyer–Villiger oxidation of ketones has been satisfactorily carried out in aqueous heterogeneous medium with MCPBA at room temperature [37] (Scheme 6.4). The reaction is fast and occurs with high yield. By using this

Table 6.4 Dakin reaction with sodium percarbonate under sonication irradiation

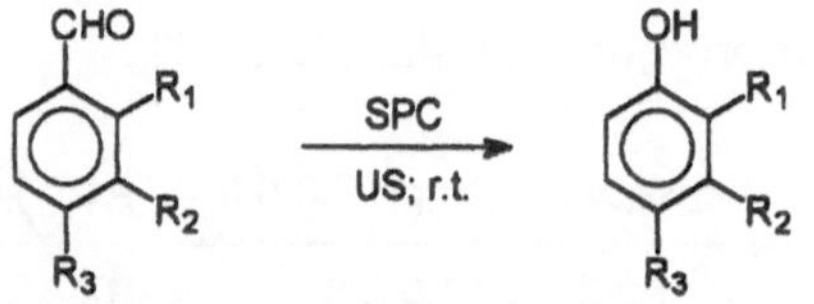

R_1	R_2	R_3	Time (h)	Yield (%)
OH	H	H	5	91
H	H	OH	8	86
OH	H	OMe	2	83
OH	OMe	H	1	95
H	OMe	OH	4	93

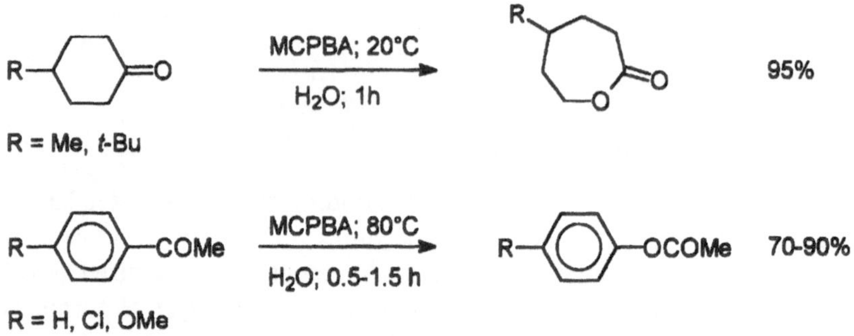

Scheme 6.4

procedure [37], peculiar ketones such as the very reactive anthrone, which usually gives anthraquinone, and pinanones, which are unreactive or give the expected lactones in organic solvents with difficulty [38] , have been oxidized:

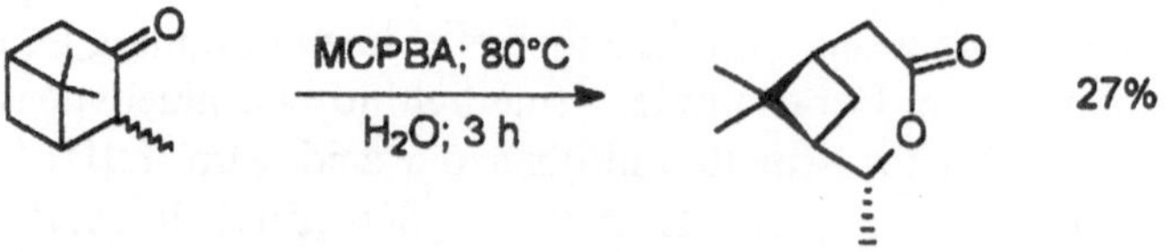

Selective oxidative cleavage of α-diketones with peroxides is also a Baeyer–Villiger oxidation which produces acids via anhydrides. SPC in H_2O–acetone cleaves aromatic and aliphatic α-diketones to give the corresponding carboxylic acids in high yields [39]:

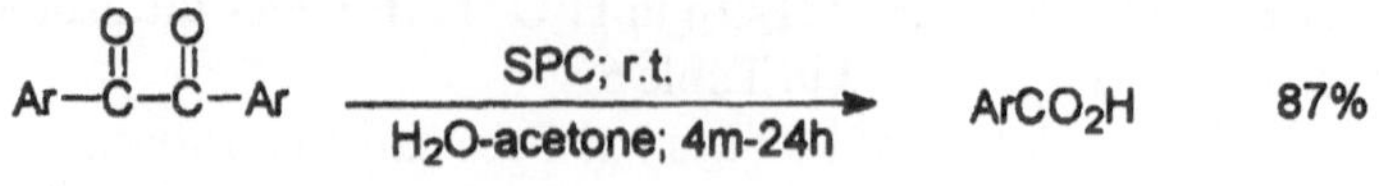

Ar = X-C_6H_4 (X = H, *p*-Me, *p*-Br, *p*-OMe)

6.4.2   Oxidation of nitriles and amines

The transformation of nitriles into amides is a well-known and well-studied reaction that can be carried out under a variety of conditions and generally requires heating, a long reaction time and metal catalysts [40]. A simple, mild and efficient method uses urea–hydrogen peroxide adduct (UHP; $H_2NCONH_2 \cdot H_2O_2$) in the presence of a catalytic amount of K_2CO_3 in water–acetone at room temperature [40, 41]. Some results are reported in Table 6.5. The procedure is efficient for both aromatic and aliphatic nitriles, and other hydrolyzable functionalities are preserved.

Nitriles are not affected by SPB in acetic acid, but undergo oxidative hydration to form amides in aqueous media such as H_2O–MeOH [42]:

R = H, Me, Cl, OH, OMe, SMe, NH$_2$, NO$_2$, CN

H_2O–acetone [43] and H_2O–dioxan [44]. When the reaction is performed on o-amidobenzonitriles, the oxidation is followed by cyclization to give quinozalin-4-(3H)-ones, which are suitable systems to build pharmaceutical compounds [45]:

R_1 = Me, Ph, NMe$_2$, 2-Et-c-C$_6$H$_{10}$; R_2 = H, I, 3,4,5-trimethoxystyryl

Table 6.5 Conversion of nitriles into amides with urea–hydrogen peroxide in an aqueous medium

$$R-CN \xrightarrow[\text{H}_2\text{O-acetone; r.t.}]{\text{UHP; K}_2\text{CO}_3} R-CONH_2$$

R	Time (h)	Yield (%)
Ph	1	85
3-Py	0.5	95
Bn	0.75	90
Me	0.5	77
CH$_2$CONH$_2$	1.25	90
CH$_2$CO$_2$Et	1.5	95
CH$_2$Cl	1	87

A study on the chemoselective oxidation of functional groups in water [35] has pointed out that the cyano group of 4-(methylthio)benzonitrile is quantitatively and selectively oxidized to amide by TBHP in strong alkaline aqueous medium in the presence of $(CTA)_2SO_4$ (Scheme 6.5). TBHP is unable to oxidize the cyano group at pH 7 even at 100°C and, in the absence of $(CTA)_2SO_4$, only the methylsulfenyl group is oxidized to methylsulfinyl. Under basic conditions TBHP turns both groups quantitatively into amide and sulfonyl groups, respectively (Scheme 6.5).

Anilines containing a carboxylic or alcoholic functionality are oxidized to nitro compounds by oxone in 20–50% aqueous acetone at 18°C in 73–84% yields [46]. When the reactions are performed in the absence of acetone, lower yields are obtained, suggesting the generation of dimethyldioxirane, which competes with oxone.

The synthesis of aminopyridine N-oxides usually occurs under acidic conditions in organic media and generally requires protection of the amino group by acylation reaction and then a final deprotection [47]. By using oxone in water under neutral or basic conditions at room temperature, the reaction can be carried out directly on aminopyridine and the N-oxide is isolated in good yield [48]. The chemoselectivity depends on the position of the amino group in the pyridine ring.

6.4.3 Oxidation of sulfides

Many chemical reagents are reported to selectively oxidize sulfides, but none seems to be suitable for synthesis on a large scale.

A protocol that sometimes is amenable to scale-up uses as oxidant oxone in aqueous acetone, buffered to pH 7.8–8.0 with sodium bicarbonate [49]. The procedure is mild, cheap and environment-friendly and the oxidation produces sulfoxides or sulfones depending on the equivalents of oxone, temperature and reaction time. When the oxidation is carried out in water only buffered with phosphate to pH 6–7, the reaction is very fast and high conversions of sulfoxides and sulfones are obtained [15].

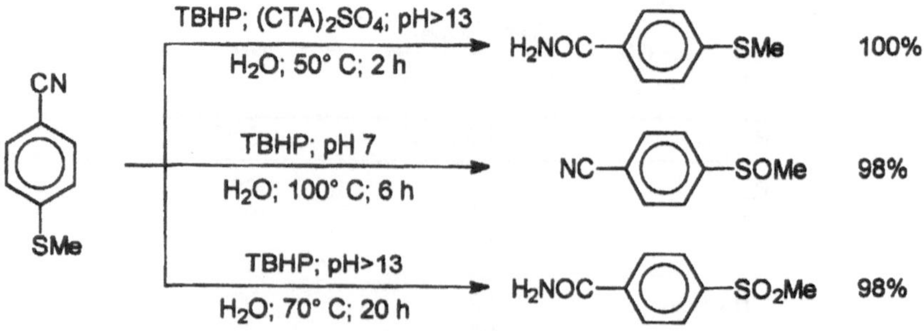

Scheme 6.5

SPB is another cheap and large-scale industrial chemical that in aqueous methanolic sodium hydroxide smoothly oxidizes sulfides into sulfones with excellent yield [50].

Oxidation of sulfides with commercial 70% aqueous TBHP in water only occurs in the heterogeneous phase at 20–70°C and affords sulfoxides selectively with very good yields [51]. Some examples are reported in Table 6.6. The reaction works better in water than in dichloroethane (DCE) and the oxidation rate is markedly increased when the reaction is executed in aqueous H_2SO_4. The higher reactivity of sulfides in aqueous medium can be explained considering that water facilitates O–O bond fission, which promotes hydrogen transfer in the transition state of the rate-determining step:

In acid medium, protonated TBHP is the oxidizing agent.

Selective oxidation of the methylsulfenyl group in the presence of other oxidizable functionalities has been achieved in aqueous media [35]. TBHP at 70°C in water at pH 7 selectively oxidizes the SMe group of 4-(methylthio)benzaldehyde to a sulfinyl group, while under strong basic conditions either the formyl or the methylsulfenyl group are oxidized to their highest oxidation state (Scheme 6.6). By using MCPBA under basic conditions, oxidation of the formyl group is prevented and 4-(methylsulfonyl)benzaldehyde is obtained in excellent yield.

Asymmetric oxidations of sulfides in water have been obtained by carrying out the reaction in the presence of cyclodextrin (CD). The solvent effect, the oxidizing agent and the nature of the sulfide have been widely investigated [52] and reviewed [53]. Sulfides are oxidized selectively to sulfoxides. The reaction yields are generally good but the enantiomeric excesses are rarely acceptable if compared with those obtained with other procedures. The best

Table 6.6 Oxidation of sulfides to sulfoxides by TBHP at 20°C

Sulfide	$CH_2Cl–CH_2Cl$		H_2O		$H_2O–H_2SO_4$	
	t (min)	Yield (%)	t (min)	Yield (%)	t (min)	Yield (%)
Et_2S	120	0	120	96	20	100
PhSMe	1200	20	1200	100	90	100
PhSPh	1920	88[a]	1920	71[a]	1200	95[b]
2-Fu–CH_2SMe	600	5	600	95	30	100
p-OH–C_6H_4SMe	900	5	900	100	15	100

[a]At 70°C.
[b]At 50°C.

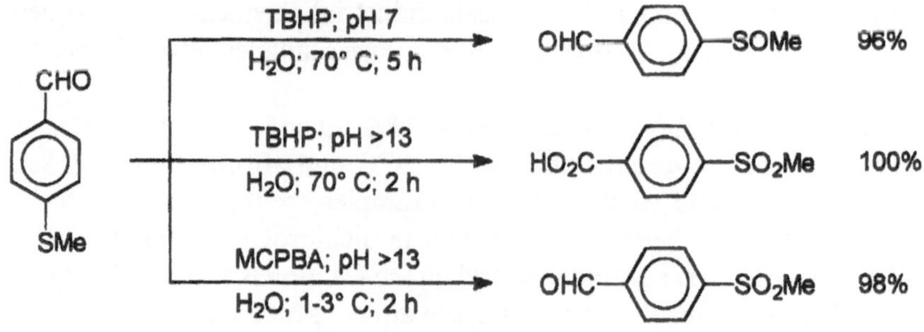

Scheme 6.6

results are obtained by using functionalized, β-CD crystalline complexes and peracetic acid [52d] and oxo–metal complexes [52c] as oxidants.

6.5   Reductions in water: introduction

The reductive methods have been greatly improved and developed in the last 10 years with respect to the type of bonds that may be reduced and with regard to the regio- and stereoselectivity of the process. In this context reductions in aqueous media represent one of the most innovative results of research.

Reductions by hydrides, which at one time seemed impossible to carry out in water, are now a reality, and several water-soluble catalysts, which allow high yields and high selectivities, have been synthesized. Even the hydrogenation of aromatic rings can be performed in water under mild conditions.

Some recent and innovative examples of reductions performed by chemical reagents in aqueous media are illustrated below. Enzymatic reductions in water have recently been reviewed [54].

6.6   Reduction of carbon–carbon double and triple bonds

The synthesis and use of water-soluble hydrogenation catalysts have received considerable attention [55]. Water solubilization has been performed by incorporating highly polar functionalities such as amino, carboxylic, hydroxide and sulfonate into phosphine ligands.

Asymmetric hydrogenations of prochiral 2-acetamidoacrylates are carried out under a hydrogen atmosphere at room temperature in neat water in the presence of water-soluble chiral Rh(I) and Ru(II) complexes with (R)-BINAP(SO$_3$NA)$_4$ (BINAP = 2,2'-bis(diphenylphosphino-1,1'-binaphthyl)] [56, 57] (Table 6.7). The ruthenium catalyst, more stable than the corresponding rhodium analog, gives higher enantioselectivity and, in some

Table 6.7 Hydrogenation of 2-acetamidoacrylates by rhodium- and ruthenium-sulfonated BINAP complexes in water

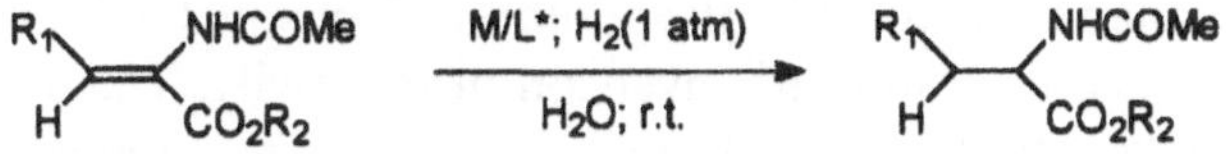

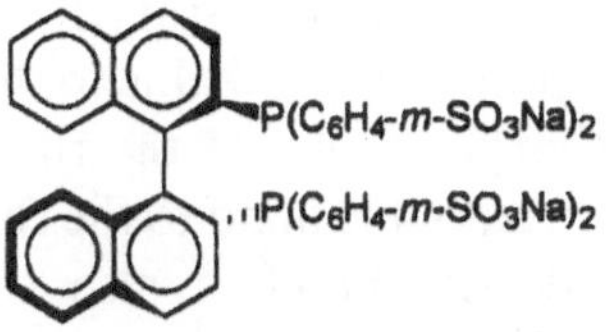

L* = BINAP(So₃Na)₄

R₁	R₂	M	ee (%)
H	H	Ru	68.5
H	Me	Ru	75.9
Ph	H	Ru	87.7
H	H	Rh	70.4
H	Me	Rh	69.0

circumstances, is superior to the analogous nonsulfonated complex. Enantioselection favors the (S)-enantiomer and the *ee* in neat water is comparable with that obtained in alcoholic solution. A detailed investigation [58] of the reaction using hydrogen in the presence of deuterium oxide has shown a regiospecific incorporation of deuterium at the α-position of the ester group, which suggests that water acts as a solvent as well as a reactant.

The system Zn/NiCl₂ 9:1 in 2-methoxyethanol (ME)–water chemoselectively reduces the double bond of α,β-unsaturated carbonyl compounds [59]. US irradiation increases the reaction rate. An example is reported below:

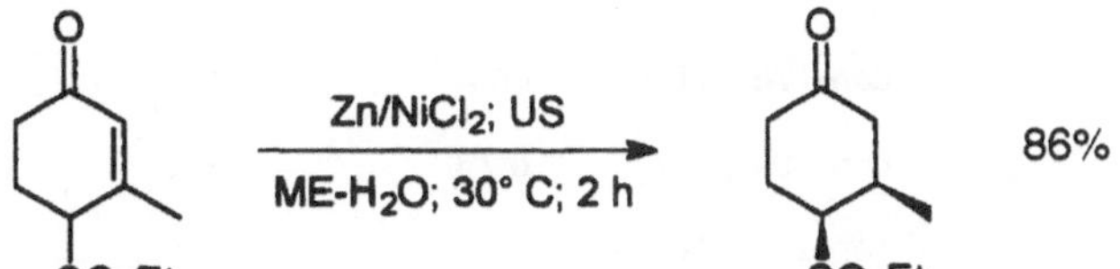

Water in sufficient amount is necessary to have a satisfactory reaction rate. (–)-Carvone is selectively reduced to (+)-dihydrocarvone and carvotanacetone in 95% and 83% yields, respectively, by a simple variation of the experimental conditions [60]. The first step of the reaction probably involves the reduction of the Ni(II) to a low-valent form adsorbed on the zinc surface. Hydrogen is simultaneously produced in the medium but it does not seem to participate in the main reaction pathway.

A regioselective hydrogenation of the internal double bond of alkadienoic acids has been achieved with phosphino-rhodium catalyst in benzene–water at 1 atm hydrogen [61]. For instance half-hydrogenation of 3,8-nonadienoic acid (Scheme 6.7) in anhydrous benzene with $RhCl[P(p\text{-tolyl})_3]_3$ gives a prevalence of 3-nonenoic acid, while the addition of an equal volume of water to the reaction medium causes an inversion of selectivity. The use of aqueous KOH strongly retards the hydrogenation rate. The presence of the free carboxylic acid group is therefore necessary for a successful reaction. The inversion of selectivity is probably ascribable to the fact that water causes the coordination of the carboxylic group to the rhodium, which facilitates the coordination of the internal double bond to the rhodium center.

The carbon–carbon double bond is also reduced by the samarium diodide–H_2O system (see section 6.8.5).

An example of carbon–carbon triple bond reduction in water is the reaction of disubstituted electron-deficient alkynes with water-soluble monosulfonated and trisulfonated triphenylphosphine [62] (Scheme 6.8). The water acts either as solvent or as reactant, and the amount of phosphine controls the *cis/trans* ratio of alkenes because it catalyzes the *cis–trans* olefin isomerization.

6.7 Reduction of carbonyl and nitro functionalities

The reductions of ketones, aldehydes, carboxylic acids and acyl derivatives are fundamental reactions in organic synthesis and the use of aqueous media has allowed high regio- and stereoselective processes to be performed under mild conditions.

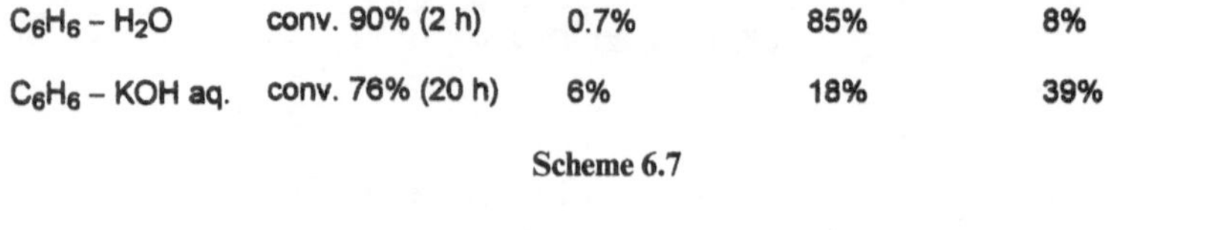

C_6H_6	conv. 78% (4 h)	66%	10%	20%
$C_6H_6 - H_2O$	conv. 90% (2 h)	0.7%	85%	8%
$C_6H_6 - KOH$ aq.	conv. 76% (20 h)	6%	18%	39%

Scheme 6.7

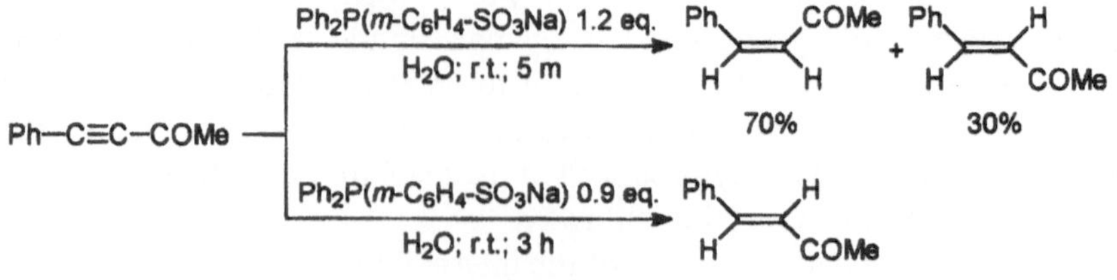

Scheme 6.8

Carbonyl compounds are quantitatively reduced regio- and stereo-selectively by $NaBH_4$ at room temperature in aqueous solutions containing glycosidic amphiphiles such as methyl-β-D-galactoside, dodecanoyl-β-D-maltoside, L-arabinose and sucrose [63]. α,β-Unsaturated ketones give 1,2-reduction to corresponding allylic alcohols and cyclohexanones furnish the thermodynamically more stable alcohol. The observed stereodifferentiation can be attributed to hydrophobic interactions between amphiphilic carbohydrates and lipophilic substrates.

High enantioselective hydrogenation of the carbonyl functionality of β-ketoesters has been observed by using Ru–BINAP catalysts at high temperature and high hydrogen pressure in organic solvents [64]. Recently [65] non-biarylphosphine-based catalyst systems have been prepared that allow high enantioselective hydrogenation of β-ketoesters under low hydrogen pressure in aqueous media. An example is the ruthenium catalyst derived from (R,R)-1,2-bis(trans-2,5-diisopropylphospholano)ethane ((R,R)-i-Pr-BPE-Ru), which allows β-hydroxyesters to be obtained with high conversion and high ee under mild conditions:

R₁, R₂ = Me, Et, i-Pr, t-Bu

ee > 98%

(R,R)-i-Pr-BPE

Carrying out the reaction in organic solvents (i-PrOH, DCM, THF), lower conversions were observed.

The reductions of benzaldehyde and p-tolualdehyde with various types of Raney alloys (Ni–Al, Co–Al, Cu–Al, Fe–Al) in 10% aqueous NaOH afford the corresponding benzyl alcohols in 17–83% yields [66]. Benzoic and p-toluic acids were obtained as by-products, which suggests that the Cannizzaro reaction is competitive. In aqueous $NaHCO_3$ and under US irradiation, the reaction is accelerated and the alcohol is the highly prevalent product.

The effect of hydrotropes on the crossed Cannizzaro reaction rate of benzaldehydes with aqueous formaldehyde was investigated [67]. In the presence of polyethyleneglycol-200 the reaction rate of m-phenoxybenzaldehyde increases more than six-fold and the m-phenoxybenzyl alcohol is also obtained with higher selectivity. The enhancement of the reaction rate of m-bromobenzaldehyde is lower than that observed for m-phenoxybenzaldehyde but the reaction selectivity is higher. p-Anisaldehyde and p-chlorobenzaldehyde show an intermediate behavior. The cationic surfactants cetylpyridinium chloride and cetyltrimethylammonium bromide (CTABr) favor the Cannizzaro reaction instead of the crossed Cannizzaro one.

Monoprotected 2-alkylresorcinols have been obtained in good yield via ortho-hydroxy-assisted reduction of hydroxyphenones [68]:

R = Me, Ph   92-95%   54-80%

The first step of the process consists of the bis-carbonatation of phenone with methylchloroformate and the second step is the reduction of bis-carbonate with sodium borohydride in THF–H$_2$O medium. The entire process can be carried out in one pot. The reduction step does not occur if no water is used. A possible mechanism for the reductive transformation is proposed.

Carboxylic acid derivatives are resistant to reduction with NaBH$_4$ and various catalysts have been used to facilitate the reaction [69]. A variety of aromatic and aliphatic carboxylic acid imidazolides are readily reduced in H$_2$O, H$_2$O–dioxan and H$_2$O–THF solution to primary alcohols by NaBH$_4$ [69]:

R = H, Br, CN, NO$_2$

The reduction is very fast, the reaction yield is high and other functional groups present in the molecule are not reduced. The presence of water is essential for the success of the reaction. When the reaction is carried out in THF only, the alcohol is obtained in poor yield and the main product is the acyl derivative, coming from the reaction with the starting imidazolide.

Lanthanoid reagents have received much attention in the last decade especially for their use in aqueous media. Yb, Sm and Ce metals have a high reduction potential and exhibit a powerful reducing property in acidic media for many organic functionalities [70]. Carboxylic acids, esters, amides and nitriles are rapidly and easily reduced to the corresponding alcohol or primary amine with Sm or Yb metal in 10% HCl at room temperature in MeOH–H$_2$O under an argon atmosphere. Some results are reported in Table 6.8. Aromatic compounds give higher yields than aliphatic ones. By using Y, La, Pr, Nd and Gd, benzoic acid is reduced to benzylic alcohol under the same conditions in 48, 82, 79, 66 and 69% yield, respectively. The reduction probably proceeds *via* single electron transfer of active low-valence lanthanoid species (LnCl or LnCl$_2$).

Many reagents are known to reduce the nitro group of nitro aromatics, but only a limited number of reagents give chemoselective nitro reduction and are suitable for a large-scale reaction. By using the iron–ammonium chloride

Table 6.8 Reduction of organic functionalities with Sm–HCl and Yb–HCl systems

Compound	Time (min)	Product	Yield (%)	
			Sm	Yb
$PhCO_2H$	5–10	$PhCH_2OH$	99	99
$PhCO_2Me$	10	$PhCH_2OH$	87	69
$PhCONH_2$	10	$PhCH_2NH_2$	64	61
$PhCN$	5–10	$PhCH_2NH_2$	99	88
$o\text{-}AcNH\text{-}C_6H_4\text{-}CO_2H$	10	$o\text{-}NH_2\text{-}C_6H_4\text{-}CH_2OH$	87	92
$o\text{-}NO_2\text{-}C_6H_4\text{-}CO_2H$	10	$o\text{-}NH_2\text{-}C_6H_4\text{-}CH_2OH$	59	27
		$o\text{-}NH_2\text{-}C_6H_4\text{-}CO_2H$	20	48

system in neutral aqueous medium, the reduction is fast and chemoselective, and the corresponding anilines are obtained with good yield [71]:

R_1; R_2; R_3 = H, NH_2, Br, Cl, NO_2, OH, CH_2CN, CH_2CO_2H, NHAc

Nitroalkenes can be converted into a variety of functionalities, and the transformation of conjugated nitroalkenes into carbonyl compounds is of particular interest. A simple, cheap and practicable procedure has been developed using the $NaBH_4/H_2O_2$ system [72]:

R_1; R_2 = Me, $c\text{-}C_6H_{11}$, $Me(CH_2)_n$, $-(CH_2)_n-$

Reductions of carbonyl and nitro functionalities via inclusion complexes and by SmI_2–H_2O systems are illustrated in the pertinent sections 6.8.2 and 6.8.5.

6.8   Other reductions

6.8.1   Reduction of the aromatic ring

The hydrogenation of benzenoids to cyclohexane derivatives is a very useful synthetic reaction and various methodologies have been reported.

Recently, aromatics have been reduced in aqueous medium at 50 atm of H_2 and at room temperature with ruthenium trichloride stabilized by trioctylamine ($RuCl_3$/TOA) [73]:

$R_1 = OMe, CO_2Me$; $R_2 = H$, Me, NH_2 cis/trans = 6-15

The aqueous medium increases the reaction rate by a factor of 12 with respect to the organic solvent. The *cis* isomer is always the major product, as is usually observed in heterogeneous catalytic hydrogenation.

Lanthanoid metal–HCl systems are used to reduce aromatic nuclei of heterocyclic compounds [74]. Pyridines, quinolines and isoquinolines are reduced with Sm–20% HCl system with excellent yields (Table 6.9). Surprisingly the amino group is eliminated from aminopyridine, and piperidine is the major reaction product.

The aromatic ring is also reduced with the SmI_2–H_2O system (see section 6.8.5).

6.8.2   Reduction via inclusion complexes

Reduction via inclusion complexes has received considerable attention because it allows both the diastereo- and enantioselectivity of the reaction to be controlled. Cyclodextrins and water-soluble cyclophanes are the most frequently used hosts in aqueous medium. β-CD/ketone (1:1) complexes suspended in $NaBH_4$ aqueous alkaline solution give the corresponding alcohols quantitatively but with modest *ee* [75]. An improvement is obtained by using crystalline α-, β- and γ-CD inclusion complexes of achiral amine-boranes at 0°C in heterogeneous aqueous medium [76]. Under these conditions, 1-phenylethanone and 4-phenyl-2-butanone give the (*S*) and (*R*) alcohol with 91% and 89% *ee*, respectively.

Table 6.9 Reduction of pyridines with Sm–HCl system

R	A (%)	B (%)	C (%)
H	94	2	traces
2-Ph	96		
3-Me	85	11	traces
4-NH_2	60 (R = H)	33 (R = H)	3 (R = H)

Aqueous suspensions of ferrocenyl ketone β-CD inclusion complexes are reduced with NaBH₄ [77]. The presence of salting-out agents (LiCl, NaCl, KCl) increases the enantioselectivity of the reaction (Table 6.10).

Enantioface differentiating reduction of arylglyoxylic acids can be performed with NaBH₄ in aqueous buffer media using a modified CD, 6-deoxy-6-amino-β-cyclodextrin [78]. The reaction proceeds with high yields but the arylglycolic acids obtained had low *ee*. The use of water-soluble chiral paracyclophanes does not increase the enantioselectivity of the reaction.

The ring-opening reaction rate of epoxides with NaBH₄ is very slow in aqueous media, but in the presence of CDs the reaction proceeds smoothly with high regioselectivity, and racemic epoxides are kinetically resolved [79]:

$$\text{Ph} \overset{O}{\triangle} + \text{NaBH}_4 \xrightarrow[\text{r.t.; 48-72 h}]{\text{CD; H}_2\text{O}} \text{Ph} \overset{OH}{\underset{}{\diagup}} + \text{Ph} \diagup\!\diagdown\!\diagup \text{OH} \quad 17\text{-}83\%$$

ee: 0–46%
selectivity   35–94%

6.8.3   Reductive coupling reactions

The coupling of ketones to give pinacols is an old reaction that can be promoted with a variety of reducing agents. The reaction takes place via reduction of the carbonyl to a radical anion followed by radical coupling to give pinacol dianion, which is subsequently protonated by the medium or upon quenching. The reaction has frequently been carried out under aqueous conditions.

Clerici and Porta have extensively investigated the reaction promoted by aqueous TiCl₃ in acidic media [80]. Under these conditions, electron-withdrawing substituted aliphatic, aromatic and heteroaromatic carbonyl compounds give 1,2 diols by a homo-coupling reaction:

Table 6.10 Asymmetric reduction of ferrocenyl ketones using β-cyclodextrin inclusion complexes

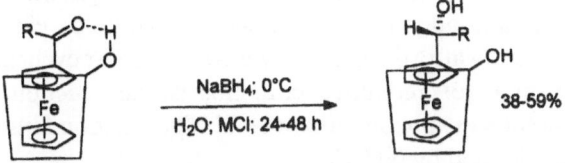

R	M	ee (%)	R	M	ee (%)
Me	–	34	Ph	–	49
Me	Li	32	Ph	Li	43
Me	Na	52	Ph	Na	82
Me	K	40	Ph	K	84

$$R_1-\underset{\displaystyle \overset{\|}{O}}{C}-R_2 \quad \xrightarrow[\text{THF; r.t.; 1-7 h}]{\text{TiCl}_3\text{-H}_2\text{O; H}^+} \quad R_1-\underset{R_2}{\overset{HO}{\underset{|}{C}}}-\underset{R_2}{\overset{OH}{\underset{|}{C}}}-R_1 \quad 61\text{-}75\%$$

$R_1 = \text{Ph, 2-Py}$; $R_2 = \text{CN, CO}_2\text{Me}$

and by a cross-coupling reaction:

$$R_1-\overset{\displaystyle \overset{\|}{O}}{C}-R_2 \; + \; R_3-\overset{\displaystyle \overset{\|}{O}}{C}-R_3 \quad \xrightarrow[\text{THF; r.t.; 0.5-10 h}]{\text{TiCl}_3\text{-H}_2\text{O}} \quad R_1-\underset{R_2}{\overset{HO}{\underset{|}{C}}}-\underset{R_3}{\overset{OH}{\underset{|}{C}}}-R_3 \quad 54\text{-}95\%$$

$R_1 = \text{Ph, Me, 2-Py}$; $R_2 = \text{CN, CO}_2\text{H, CO}_2\text{CH}_3, \text{Me}$; $R_3 = \text{Me}$

with acetone, acetaldehyde or benzaldehyde used in excess or as reaction solvent [80a,b].

The latter reaction is particularly important: the addition of an alkyl radical to carbonyl carbon does not give carbon–carbon bond formation because the intermediate alkoxy radical undergoes fast β-bond cleavage. Under aqueous titanium trichloride conditions, the intermediate alkoxy radical is very quickly reduced by Ti(III), making the addition step irreversible and allowing the formation of a carbon–carbon bond.

The TiCl₃–H₂O-promoted cross-coupling reaction has been extended to the addition of a carbonyl compound to α,β-unsaturated aldehydes [80c]. The addition occurs selectively to the carbon atom of the carbonyl group, and highly functionalized allylic pinacols are obtained with good yields.

The couple Zn–Cu also promotes the pinacol reductive cross-coupling between saturated and α,β-unsaturated carbonyl compounds in aqueous heterogeneous media. The reaction between α,β-unsaturated cycloalkenones and acetone is accelerated by US irradiation and is significantly influenced by the Zn/Cu and Zn/substrate ratios [81].

Aqueous TiCl₃ also promotes a chemoselective reduction of α,β-dicarbonyl compounds to give ketyl radicals that add to the carbonyl carbon of aldehydes affording α,β-dihydroxyketones. The reaction yields are excellent and the diastereoselection depends on the steric hindrance of the groups bonded to the reagents [80e]. When the ketyl radical is generated from methyl phenylglyoxalate and added to β-ketoesters and β-keto acids, the intermediate pinacols rapidly afford lactones by intramolecular cyclization [80d].

Another example of reductive coupling is the reaction of aldimines promoted by indium in aqueous ethanol [82]. The reaction fails in MeCN and DMF but is accelerated by NH₄Cl:

$$R_1-CH{=}NR_2 \quad \xrightarrow[\text{NH}_4\text{Cl; 75}^\circ \text{ C; 16 h}]{\text{In; H}_2\text{O-EtOH}} \quad R_1-\underset{\underset{R_2}{\overset{|}{\underset{|}{NH}}}}{\overset{|}{CH}}-\underset{\underset{R_2}{\overset{|}{\underset{|}{NH}}}}{\overset{|}{CH}}-R_1 \quad 40\text{-}100\%$$

$R_1;R_2 = \text{Ph, } p\text{- and } o\text{-X-C}_6\text{H}_4 \; (X = \text{OMe, Cl, Me})$

6.8.4 Reductive dehalogenation

The hydrogenolysis of the carbon–halogen bond is of importance in organic synthesis especially for aromatic compounds [83].

Water-soluble aryl chlorides undergo hydrogenolysis in aqueous alkaline media in the presence of $PdCl_2$ with NaH_2PO_2 as hydrogen source [84]:

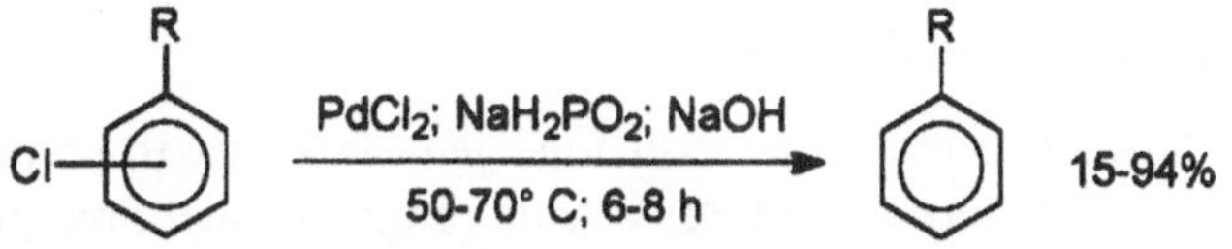

The protocol gives high yields of *para*-substituted chlorobenzenes, low yields of the *meta*-substituted ones and is inefficient for nitrogen-containing heterocycles.

A prominent example of aqueous radical reduction, reported by Breslow and Light, uses the water-soluble tris[3-(2-methoxyethoxy)propyl]stannane in the presence of 4,4'-azobis(4-cyanovaleric acid) (ACVA) or sunlamp as initiators [85]. In aqueous $NaHCO_3$ the *m*-bromobenzoic acid is reduced to benzoic acid in 88% yield.

Collum and Rai [86] use the more easily available [bis(potassium propanoate)*n*(hydroxy)]stannane, which, in the presence of $NaBH_4$ and ACVA, affords reductions and free-radical cyclizations of aryl and alkenyl bromides:

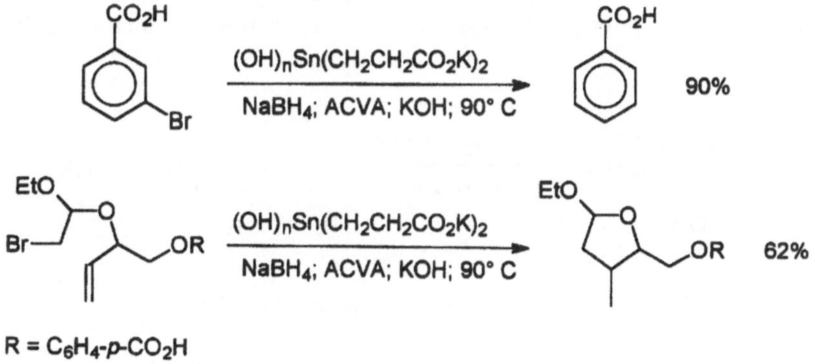

$R = C_6H_4\text{-}p\text{-}CO_2H$

Reductions with $NaBH_4/D_2O$ and $NaBD_4/D_2O$ show that the hydrogen atom is derived from $NaBH_4$. By simply using the $NaBH_4$–ACVA system, aliphatic bromides are not reduced and the reaction rate of aryl bromides is lowered.

Water-soluble and water-insoluble organohalides are hydrogenolyzed in high yields by *n*-Bu_3SnH in water in the presence of radical initiators [87]. For water-insoluble substrates the presence of a surfactant, such as cetyltrimethyl-ammonium bromide, sodium dodecylsulfate and triton-X-100, is necessary.

The hydrogenolysis of halopyridines can be conveniently carried out with 15% aqueous $TiCl_3$ in the presence of acetic acid [80a]. Aqueous titanium trichloride quantitatively removes also the cyano group from cyanopyridines and promotes a facile and quantitative reduction of heterocyclic N-oxides [80a].

For the reductive dehalogenation catalyzed by SmI$_2$ see the next subsection.

6.8.5   Reduction catalyzed by samarium diiodide

Samarium diiodide is a versatile reducing agent. It is easy to prepare, can be used in aqueous media and reduces a wide range of functional groups [88]. Reductions with SmI$_2$ are generally carried out in the presence of proton sources (water or low-molar-mass alcohols) and electron donors (hexamethylphosphoramide (HMPA), 1,3-dimethyl-3,4,5,6-tetrahydro-2(1H)-pyrimidone (DMPU), inorganic bases) [88].

Hasegawa and Curran have shown that water serves not only as a proton source, but can also accelerate certain reductions [89]. An example is reported in Table 6.11. The presence of water is essential for the rapid conversion of A to B, and 'quenching' dry reactions with water only serves to promote this conversion.

At room temperature the SmI$_2$–H$_2$O system rapidly reduces carboxylic acids, esters, amides, nitriles, chlorides, alkenes, nitro compounds and heteroaromatic rings [90]. Some examples are reported in Table 6.12.

Hanessian recently reported a facile and direct preparation of 2-deoxy-aldonolactones from their unprotected oxygenated progenitors by using SmI$_2$ in THF–H$_2$O [91]. The α-deoxygenation of D-arabino-1,4-lactone is illustrated below:

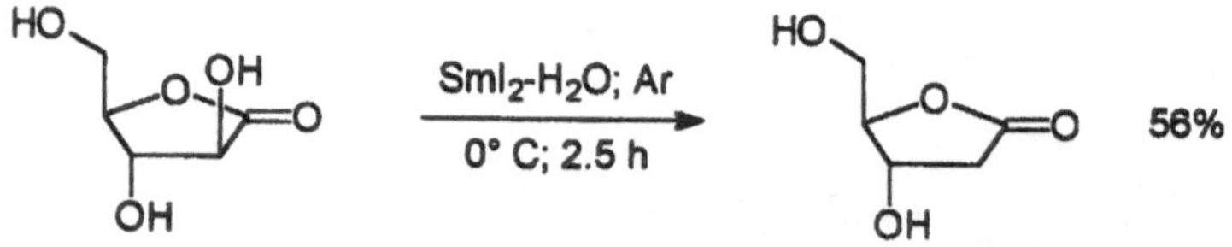

Table 6.11 Effect of water on the reduction of 1,3-diphenylacetone by SmI$_2$

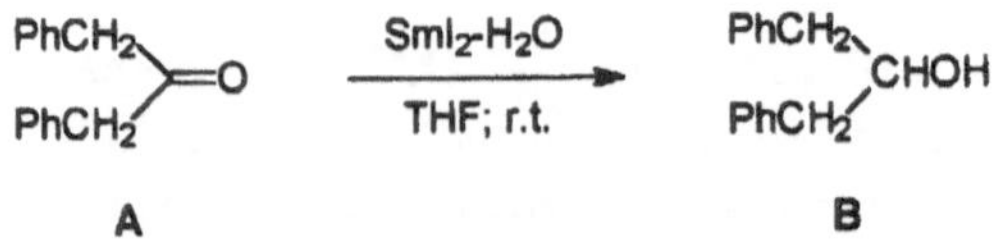

H$_2$O	Time (min)	Quenching	A/B
no	10	H$_2$O–HCl	6/94
no	10	H$_2$O–NH$_4$Cl	26/74
no	10	dry air	100/0
yes	10	dry air	1/99

Table 6.12 Reduction of aromatic compounds with SmI_2–H_2O system

Compounds	Time(s)	Product	Yield (%)
$PhCO_2H$	60	$Ph-CH_2OH$	89
$PhCO_2Me$	3	$Ph-CH_2OH$	93
$PhCONH_2$	3	$Ph-CH_2OH$	94
PhCN	5	$Ph-CH_2NH_2$	88
p-NO_2-C_6H_4-CO_2Me	120	p-NH_2-C_6H_4-CO_2Me	89
o-Cl-C_6H_4-CO_2Me	30	o-Cl-C_6H_4-CH_2OH	53
$Ph-CH=CH-CONH_2$	30	$Ph-CH_2CH_2-CONH_2$	99
pyridine	240	piperidine	94
2-NH_2-pyridine	600	piperidine	94
2-Cl-pyridine	600	piperidine	79
2-CN-pyridine	2	pyridine	96
2-$CONH_2$-pyridine	3	2-Me-pyridine	78
2-CO_2H-pyridine	600	2-Me-piperidine	68

References

[1] Hudlicky, M., *Oxidation in Organic Chemistry*, ACS Monograph 186; American Chemical Society, Washington, DC, 1990.

[2] Vogel, A.I., *Textbook of Practical Organic Chemistry Including Qualitative Organic Analysis*, 4th edn; Longmans, London, 1988.

[3] Minisci, F., *Chim. Ind. (Milan)*, 1990, **72**, 512.

[4] (a) Labinger, J.A.; Herring, A.M.; Bercaw, J.E., *J. Am. Chem. Soc.*, 1990, **112**, 5628; (b) Luinsta, A.; Labinger, J.A.; Bercaw, J.E., *J. Am. Chem. Soc.*, 1993, **115**, 3004; (c) Sen, A.; Lin, M.; Kao, L.C.; Hutson, C.A., *J. Am. Chem. Soc.*, 1992, **114**, 6385.

[5] Saudagar, S.R; Samant, S.D., *Ultrasonic Sonochem.*, 1995, **2**, 515.

[6] Ganin, E.; Amer, I., *Synth. Commun.*, 1995, **25**, 3149.

[7] Bartlett, P.D., *Rec. Chem. Prog.*, 1957, **18**, 111.

[8] Yamades, S.; Kondou, C.; Minato, T., *J. Org. Chem.*, 1996, **61**, 616.

[9] Fringuelli, F.; Germani, R.; Pizzo, F.; Savelli, G., *Tetrahedron Lett.*, 1989, **30**, 1427.

[10] Fringuelli, F.; Pizzo, F.; Germani, R.; Savelli, G., *Org Prep. Proced. Int.*, 1989, **21**, 757.

[11] Fringuelli, F.; Germani, R.; Pizzo, F.; Savelli, G., *Synth. Commun.*, 1989, **19**, 1939.

[12] Kende, A.S.; Delair, P.; Blass, B.E., *Tetrahedron Lett.*, 1994, **44**, 8123.

[13] Neumann, R.; Miller, H., *J. Chem. Soc., Chem. Commun.*, 1995, 2277.

[14] Yatabe, J.; Sugizaki, O.; Kageyama, T., *Nippon Kagau Kaishi*, 1992, 1446.

[15] Zheng, T.C.; Richardson, D.E., *Tetrahedron Lett.*, 1995, **36**, 833.

[16] Turk, H.; Ford, T., *J. Org. Chem.*, 1991, **56**, 1253.

[17] (a) Notari, B., *Stud. Surf. Sci. Catal.*, 1991, **67**, 243; (b) Sheldon, R.A., *J. Mol. Catal.*, 1980, **7**, 107.

[18] Otsuka, K.; Yoshinaka, M.; Yamanaka, I., *J. Chem. Soc., Chem. Commun.*, 1993, 611.

[19] Katagiri, T.; Obara, F.; Toda, S.; Furuhashi, K., *Synlett*, 1994, 507.

[20] Guy, A.; Doussot, J.; Garreau, R.; Godefroy-Falguieres, A., *Tetrahedron: Asymmetry*, 1992, **3**, 247.

[21] See references of footnote 3 of reference [24].

[22] See references of footnote 4 of reference [24].

[23] (a) Prat, D.; Lett, R., *Tetrahedron Lett.*, 1986, **27**, 707; (b) Prat, D.; Delpech, B.; Lett, R., *Tetrahedron Lett.*, 1986, **27**, 711.

[24] Fringuelli, F.; Germani, R.; Pizzo, F.; Santinelli, F.; Savelli, G., *J. Org Chem.*, 1992, **57**, 1198.

[25] Fringuelli, F.; Pizzo, F.; Germani, R., *Synlett*, 1991, 475.

[26] Fringuelli, F.; Germani, R.; Pizzo F.; Suarez, M.J., unpublished results.

[27] Ye, D.; Fringuelli, F.; Piermetti, O.; Pizzo, F., *J. Org. Chem.*, 1997, **62**, 3748.

[28] Muzart, J., *Synthesis*, 1995, 1325.

[29] Reed, K.L.; Gupton, J.T.; Solarz, T.L., *Synth. Commun.*, 1989, **19**, 3579.

[30] Payne, G.B.; Williams, P.H., *J. Org Chem.*, 1959, **24**, 54.

[31] Kirshenbaum, K.S.; Sharpless, K.B., *J. Org Chem.*, 1985, **50**, 1979.

[32] (a) Curci, R.; Fiorentino, M.; Troisi, L.; Edwards, J.O.; Pater, R.H., *J. Org. Chem.*, 1980, **45**, 4758; (b) Corey, P.F.; Ward, F.E., *J. Org. Chem.*, 1986, **51**, 1925.

[33] Oglioruso, M.A.; Wolfe, J.F., *Synthesis of Carboxylic Acids, Esters and Their Derivatives*, Patai, S.; Rappoport, Z., Eds; Wiley, New York, 1991.

[34] Dodd, R.H.; Le Hyaric, M., *Synthesis*, 1993, 295.

[35] Fringuelli, F.; Pellegrino, R.; Piermatti, O.; Pizzo, F., *Synth. Commun.*, 1994, **24**, 2665.

[36] Kabalka, G.W.; Reddy, N.K.; Narayana, C., *Tetrahedron Lett.*, 1992, **33**, 865.

[37] Fringuelli, F.; Germani, R.; Pizzo, F.; Savelli, G., *Gazz. Chim. Ital.*, 1989, **119**, 249.

[38] Thomas, A.F.; Rey, F., *Tetrahedron*, 1992, **48**, 1927.

[39] Yang, D.T.C.; Evans, T.T.; Yamazaki, F.; Narayana, C.; Kabalka, G.W., *Synth. Commun.*, 1993, **23**, 1183.

[40] Balicki, R.; Kaczmarek, L., *Synth. Commun.*, 1993, **23**, 3149.

[41] Heaney, H., *Aldrichim. Acta*, 1993, **26**, 35.

[42] McKillop, A.; Kemp, D., *Tetrahedron*, 1989, **45**, 3299.

[43] Kabalka, G.W.; Deshpande, S.M.; Wadgaonkar, P.P.; Chatla, N., *Synth. Commun.*, 1990, **20**, 1445.

[44] Reed, K.L.; Gupton, J.T.; Solarz, T.L., *Synth. Commun.*, 1990, **20**, 563.

[45] Baudoin, B.; Ribeill, Y.; Vicker, N., *Synth. Commun.*, 1993, **23**, 2833.

[46] Webb, K.S.; Seneviratne, V., *Tetrahedron Lett.*, 1995, **36**, 2377.

[47] (a) Albini, A.; Pietra, S., *Heterocyclic N-Oxides*; CRC Press, Boca Raton, FL, 1991; (b) Albini, A., *Synthesis*, 1993, 263.

[48] Robke, G.J.; Behrman, E., *J. Chem. Res. (S)*, 1993, 412.

[49] Webb, K.S., *Tetrahedron Lett.*, 1994, **35**, 3457.

[50] McKillop, A.; Tarbin, J.A., *Tetrahedron*, 1987, **43**, 1753.

[51] Fringuelli, F.; Pellegrino, R.; Pizzo, F., *Synth. Commun.*, 1993, **23**, 3157.

[52] (a) Rao, K.R.; Sattur, P.B., *J. Chem. Soc., Chem. Commun.*, 1989, 342; (b) Czarnk, A.W., *J. Org. Chem.*, 1984, **49**, 924; (c) Sakuraba, H.; Natori, K.; Tanaka, Y., *J. Org. Chem.*, 1991, **56**, 4124; (d) Bonchio, M.; Carofiglio, T.; Di Furia, F.; Fornasier, R., *J. Org. Chem.*, 1995, **60**, 5986.

[53] Takahashi, K.; Hattori, K., *J. Incl. Phenom. Mol. Recogn. Chem.*, 1994, **17**, 1.

[54] (a) Azerad, R., *Bull. Soc. Chim. Fr.*, 1995, **132**, 17; (b) Fang, J.M.; Lin, C.H.; Bradshaw, C.W.; Wong, C.H., *J. Chem. Soc., Perkin Trans. 1*, 1995, 967.

[55] Kalck, K.; Monteil, F.; *Adv. Organomet. Chem.*, 1992, **34**, 219.

[56] Wan, K.; Davis, M.E., *Tetrahedron: Asymmetry*, 1993, **4**, 2461.

[57] Wan, K.; Davis, M.E., *J. Chem. Soc., Chem. Commun.*, 1993, 1262.

[58] Bakos, J.; Karaivanov, R.; Laghmari, M.; Sinou, D., *Organometallics*, 1994, **13**, 2951.

[59] Petrier, C.; Luche, J.L., *Tetrahedron Lett.*, 1987, **28**, 2347.

[60] Petrier, C.; Luche, J.L., *Tetrahedron Lett.*, 1987, **28**, 2351.

[61] Okano, T.; Kaji, M.; Isotani, S.; Kiji, J., *Tetrahedron Lett.*, 1992, **33**, 5547.

[62] Larpent, C.; Meignan, G., *Tetrahedron Lett.*, 1993, **34**, 4331.

[63] Denis, C.; Laignel, B.; Plusquellec, D.; Le Marouille, J.Y.; Botrel, A., *Tetrahedron Lett.*, 1996, **37**, 53.

[64] (a) Kitamura, M.; Takunaga, M.; Ohkuma, T.; Noyori, R., *Org. Synth.*, 1992, **71**, 1; (b) Heiser, B.; Broger, E.A.; Crameri, Y., *Tetrahedron: Asymmetry*, 1991, **2**, 51; (c) Noyori, R.; Takaya, H., *Acc. Chem. Res.*, 1990, **23**, 345; King, S.A.; Thompson, A.S.; King, A.O.; Verhoeven, T.R., *J. Org. Chem.*, 1992, **57**, 6689.

[65] Burk, M.J.; Harper, T.G.P.; Kalberg, C.S., *J. Am. Chem. Soc.*, 1995, **117**, 4423.

[66] Tsukinoki, T.; Ishimoto, K.; Tsuzuki, H.; Mataka, S.; Tashiro, M., *Bull. Chem. Soc. Jpn.*, 1993, **66**, 3419.

[67] Sane, P.V.; Sharma, M.M., *Synth. Commun.*, 1987, **17**, 1331.

[68] Mitchell, D.; Doecke, C.W.; Hay, L.A.; Koening, T.M.; Wirth, D.D., *Tetrahedron Lett.*, 1995, **36**, 5335.

[69] Sharmara, R.; Voynov, G.H.; Ovaska, T.V.; Marquez, V.E., *Synlett*, 1995, 839.

[70] Kamochi, Y.; Kudo, T., *Chem. Pharm. Bull.*, 1994, **42**, 402.

[71] Ramadas, K.; Srinivasan, N., *Synth. Commun.*, 1992, **22**, 3189.

[72] Ballini, R.; Bosica, G., *Synthesis*, 1994, 723.
[73] Fache, F., Lehuede, S.; Lemaire, M., *Tetrahedron Lett.*, 1995, **36**, 885.
[74] Kamochi, Y.; Kudo, T., *Chem. Pharm. Bull.*, 1995, **43**, 1422.
[75] Fornasier, R.; Reniero F.; Scrimin, P.; Tonellato, U., *J. Org Chem.*, 1985, **50**, 3209.
[76] Sakuraba, H.; Inomata, N.; Tanaka, Y., *J. Org. Chem.*, 1989, **54**, 3482.
[77] Kawajiri, Y.; Matohashi, N., *J. Chem. Soc., Chem. Commun.*, 1989, 1336.
[78] Hattori, K.; Takahashi, K.; Sakai, N., *Bull. Chem. Soc. Jpn.*, 1992, **65**, 2690.
[79] Hu, Y.; Uno, M.; Harada, A.; Takahashi, S., *Bull. Chem. Soc. Jpn.*, 1991, **64**, 1884.
[80] (a) Clerici, A.; Porta, O., *Tetrahedron*, 1982, **38**, 1293; (b) Clerici, A.; Porta, O., *J. Org. Chem.*, 1982, **47**, 2852; (c) Clerici, A.; Porta, O., *J. Org Chem.*, 1983, **48**, 1690; (d) Clerici, A.; Porta, O.; Zago, P., *Tetrahedron*, 1986, **42**, 572; (e) Clerici, A.; Porta, O., *J. Org. Chem.*, 1989, **54**, 3872.
[81] Delair, P.; Luche, J.L., *J. Chem. Soc., Chem. Commun.*, 1989, 398.
[82] Kalyanam, H.; Rao, G.V., *Tetrahedron Lett.*, 1993, **34**, 1647.
[83] Pinder, A.R., *Synthesis*, 1980, 425.
[84] Davydov, D.V.; Beletskaya, I.P., *Russ. Chem Bull.*, 1993, **42**, 572.
[85] (a) Light, J.; Breslow, R., *Tetrahedron Lett.*, 1990, **31**, 2957; (b) Light, J.; Breslow, R., *Org. Synth.*, 1993, **72**, 199.
[86] Rai, R.; Collum, D.B., *Tetrahedron Lett.*, 1994, **35**, 6221.
[87] Maitra, U.; Das Sarma, K., *Tetrahedron Lett.*, 1994, **35**, 7861.
[88] (a) Molander, G.A., *Chem. Rev.*, 1992, **92**, 26; (b) Soderquist, J.A., *Aldrichim. Acta.*, 1991, **24**, 15.
[89] Hasegawa, E.; Curran, D.P., *J. Org. Chem.*, 1993, **58**, 5008.
[90] (a) Kamochi, Y.; Kudo, T., *Heterocycles*, 1993, **36**, 2383; (b) Kamochi, Y.; Kudo, T., *Chem. Lett.*, 1993, 1495.
[91] Hanessian, S.; Girard, C., *Synlett*, 1994, 861.

7 Base-catalyzed aldol- and Michael-type condensations in aqueous media

F. FRINGUELLI, O. PIERMATTI and F. PIZZO

7.1 Introduction

The aldol reaction is one of the most important ways to construct carbon–carbon bonds in organic synthesis. Nature itself seems to prefer this reaction in its biosynthetic processes, for example, in the prebiotic formation of saccharides [1]. Strictly speaking, the aldol reaction is the self-coupling of an aldehyde, having at least one active hydrogen in the α-position, to give a β-hydroxyaldehyde called an aldol (aldol addition), which sometimes dehydrates (aldol condensation).

The aldol reaction and other venerable processes such as the Knoevenagel, Claisen–Schmidt, Perkin, Darzen, Tollens and Wittig reactions are base-catalyzed (sometimes acid-catalyzed too) reactions between an active methylene compound and an aldehyde or a ketone. In the last decade the term 'aldol-type reaction' has been used to indicate that the initial addition step is mechanistically the same for all these reactions.

When the aldol-type addition is carried out under classical conditions (protic solvents and basic or acid catalysis), by-products such as dimers, polymers and α,β-unsaturated carbonyl compounds are frequently formed. Variants such as the Mukaiyama reaction [2] and the use of lanthanide triflates as catalysts [3] generally reduce these side-reactions. A significant contribution has also come from investigations on the use of water as the reaction medium.

The aldol-type reaction occurs in some processes (i.e. Weiss–Cook and Hantzsch reactions), in tandem with the Michael reaction. Therefore, some significant applications of this last reaction in aqueous media will be illustrated.

7.2 Claisen–Schmidt reaction

The cross-aldol addition of a ketone to an aldehyde (Claisen–Schmidt reaction) is profitably carried out by using the silyl enol ether of the ketone in an organic solvent in the presence of $TiCl_4$ (Mukaiyama reaction) [2], but this protocol is not suitable for acid-sensitive substrates. High pressure may be employed in place of the catalyst, but longer reaction times are required [4].

Lubineau [5] has shown that the reaction of the trimethylsilyl enol ether of cyclohexanone with benzaldehyde occurs in water in heterogeneous phase, at room temperature and atmospheric pressure in the absence of catalyst, with the same stereoselectivity as under pressure (Table 7.1). In organic solvents (PhMe, tetrahydrofuran (THF), dichloromethane (DCM) CH_3CN), no reaction was observed at room temperature and atmospheric pressure in the absence of catalyst. Methanol lowers the yield and diastereoselectivity. When using t-butyldimethylsilyl enol ethers, the competitive hydrolysis of the silyl enol ether to cyclohexanone is reduced as it requires higher reaction temperatures. The reaction is favored by an electron-withdrawing substituent in the *para* position of the phenyl ring of benzaldehyde (Table 7.1). It has been hypothesized [5] that the acceleration in water is due to steric compression of the six-membered transition state in a cavity of the water structure.

Cationic surfactants such as the cetyltrimethylammonium compounds CTACl, CTABr, (CTA)$_2$SO$_4$ and CTAOH favor, under weakly alkaline conditions, the Claisen–Schmidt condensation of acetophenones with benzaldehydes (Scheme 7.1), allowing the synthesis of biologically interesting compounds, such as chalcones and flavonols [6, 7], in water only.

A substantial enhancement of the condensation rate of these substances is observed by using hydrotropes such as sodium butylmonoglycosulfate (NaBMGS) and the sodium salts of aromatic sulfonic acids in alkaline media [8]. In the absence of NaOH, the reaction does not proceed, which indicates that the hydrotropes do not catalyze the reaction and that the effect is simply that of the solubilization of the reactants at higher concentrations. The selectivity of the reaction of p-nitrobenzaldehyde with acetone can be controlled in aqueous media by working in the presence of zinc nitrate hexahydrate and N,N-dimethylaminoethanol [9]. The amino alcohol–zinc combination favors aldol addition at pH 9.1–9.5 and aldol condensation at pH 11.4. The same reaction catalyzed by metal complexes bearing ligands of the α-amino acids

Table 7.1 Condensation of silyl enol ethers of cyclohexanone with benzaldehydes under various conditions

R	R_1	Medium	T (°C)	t (h)	Condition	Yield (%)	*syn/anti*
Me	H	CH_2Cl_2	r.t.	360	stirring	0	–
Me	H	CH_2Cl_2	20	2	TiCl$_4$	82	25/75
Me	H	CH_2Cl_2	60	216	10 kbar	90	75/25
Me	H	H_2O	20	120	stirring	23	85/15
Me	H	H_2O–THF	55	24	stirring	76	74/26
t-Bu	H	H_2O–THF	100	16	stirring	84	57/43
Me	NO$_2$	H_2O–THF	55	36	sonication	82	70/30
Me	OMe	H_2O–THF	55	36	sonication	29	70/30

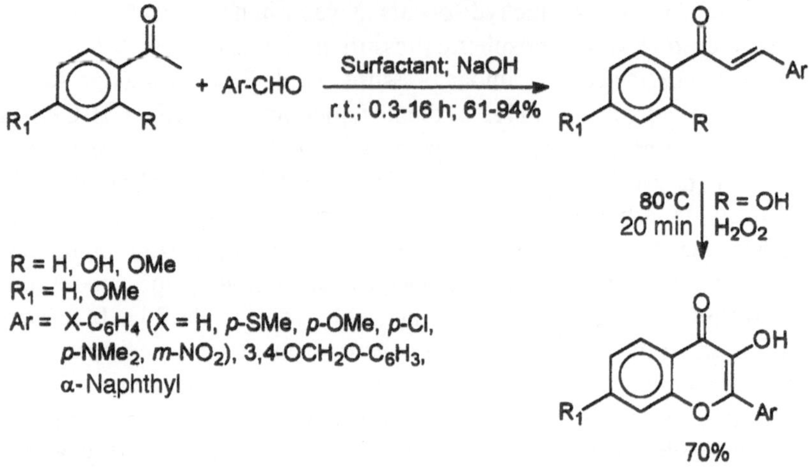
Surfactant; NaOH
r.t.; 0.3-16 h; 61-94%
+ Ar-CHO
80°C
20 min
R = OH
H2O2
R = H, OH, OMe
R1 = H, OMe
Ar = X-C6H4 (X = H, p-SMe, p-OMe, p-Cl, p-NMe2, m-NO2), 3,4-OCH2O-C6H3, α-Naphthyl
70%

Scheme 7.1

has been studied in the presence of cyclodextrins (CDs) in aqueous media in an attempt to obtain optically active β-hydroxyketones, but the results have been discouraging [10, 11].

The intramolecular Claisen–Schmidt addition rate of 6-keto-6-*p*-*t*-butylphenyl hexanal in the presence of β-CD is minimal at neutral pH but is strongly accelerated in alkaline medium [12]. β-CD molecules bearing one or two imidazole groups catalyze the cyclization by a factor of 20–50 at pH 7, but the enantio-induction is minimal.

The addition reaction of cyclohexanone to benzaldehyde carried out in water only gives a high yield of ketol as a 1:1 *threo–erythro* mixture, while in the presence of CTACl, the bis-condensation product is obtained quantitatively [6b]:

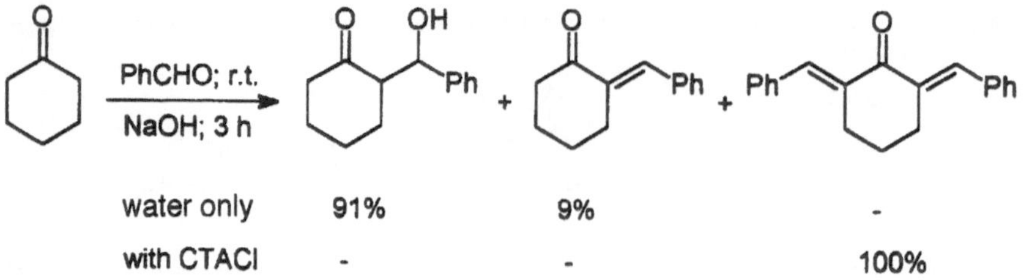
PhCHO; r.t.
NaOH; 3 h
water only
91%
9%
-
with CTACl
-
-
100%

Acetone and cyclopentanone react similarly with benzaldehyde to give bis-adducts even when an excess of ketone is used [7].

The aqueous medium influences not only the reaction rate but also the stereoselection of the aldol addition. One significant example [13] is the intramolecular cyclization of ketoaldehyde depicted in Scheme 7.2. In organic solvents there is a preference for *syn* or *anti* adduct depending on the presence of coordinating cations (K^+, Na^+, Li^+, $MgBr^+$) or a complexing agent such as

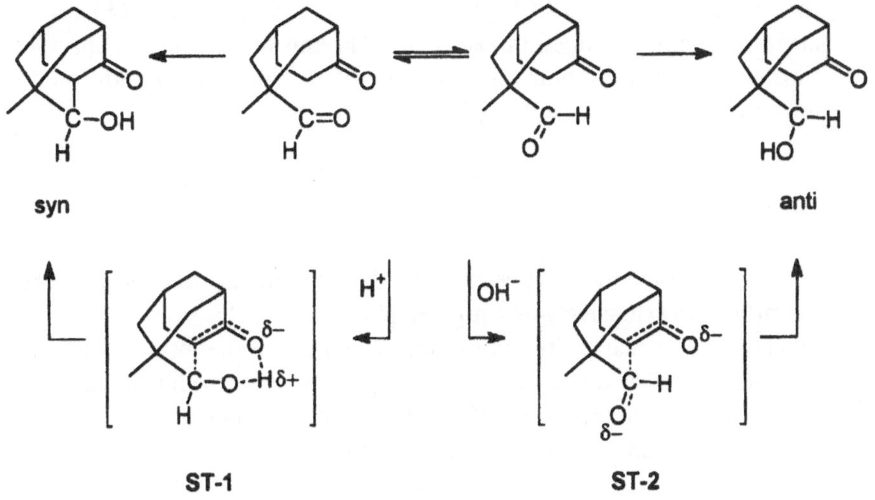

syn   **anti**

ST-1   **ST-2**

Scheme 7.2

kryptofix 222, respectively. In aqueous alkaline medium a prevalent *anti*-selectivity is observed, which can be explained by the strong hydration of the metal cation by the water molecule. Enolate does not coordinate the metal cation and cyclization occurs through the open transition state **ST-2**. Under acidic aqueous conditions, the *syn* adduct prevails, which is formed through the transition state **ST-1** stabilized by an intramolecular hydrogen bond.

7.3   Vinylogous aldol reaction

The γ-hydrogen of α,β-unsaturated ketones, nitriles and esters is an 'active' hydrogen, and electrophilic addition at the γ-position ('vinylogous aldol addition' when the electrophile is an aldehyde) competes with electrophilic addition at the α′-position.

The reaction of isophorone with benzaldehyde in water only at room temperature gives only vinylogous aldol addition but with low conversion:

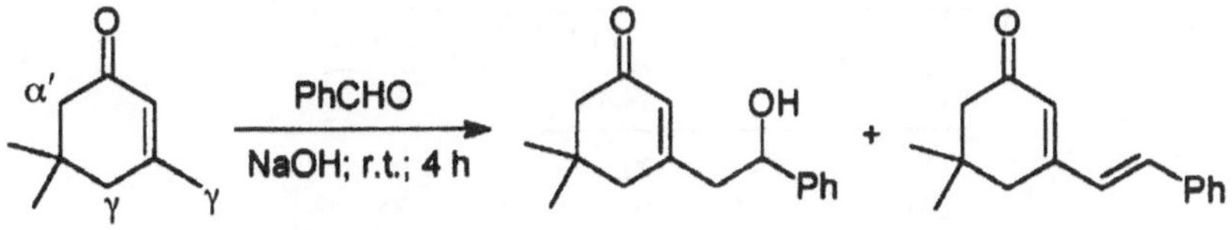

water only	24%	-
CTACl	-	80%
TBACl	27%	58%

In the presence of CTACl, the condensation product, (E)-benzylideneisophorone, is obtained in excellent yield and no dimers are observed. By using tetrabutylammonium chloride (TBACl), a mixture of addition and condensation adducts is observed [6a].

7.4 Knoevenagel reaction

The Knoevenagel reaction between benzaldehydes and acetonitriles in water has recently been extensively investigated [6a, 14].

Salicylaldehydes react with malononitrile in heterogeneous aqueous alkaline medium at room temperature to give α-hydroxybenzylidenemalononitriles, which are converted directly in high yield to 3-cyanocoumarins by acidification and heating of the reaction mixture [14]:

R = H, OH, OMe   75-95%

By using substituted acetonitriles, the addition reaction sometimes requires the presence of catalytic amounts of CTABr (Table 7.2). Comparison with reactions carried out in homogeneous alcoholic medium shows that the aqueous reaction gives better yields [14] (Table 7.2).

The condensation of benzaldehyde with aryl acetonitriles does not occur in

Table 7.2 Synthesis of 3-substituted coumarins in aqueous and ethanolic media

R	Yield (%) of coumarin	
	Water	Ethanol
CN	90	70
CO₂Et	66	35
NO₂	87	80
Ph	90[a]	traces
2-Py	98	55
2-Th	95[a]	20

[a]In the presence of 0.1 mol/equiv of CTABr.

water only but requires the presence of catalytic amounts of CTACl or TBACl to have high yield of aryl cinnamonitriles [6a]:

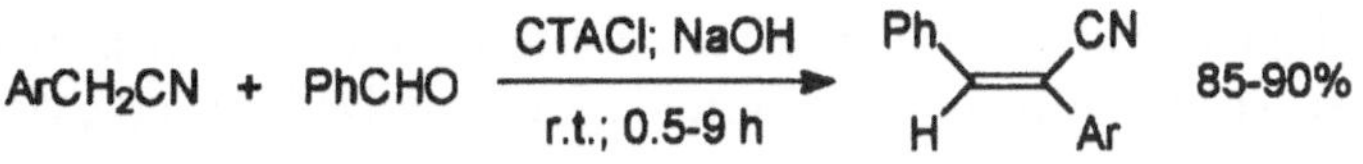

$$ArCH_2CN \; + \; PhCHO \xrightarrow[\text{r.t.; 0.5-9 h}]{\text{CTACl; NaOH}} \underset{\text{H}}{\overset{\text{Ph}}{\bigg\backslash}}C=C\underset{\text{Ar}}{\overset{\text{CN}}{\bigg/}} \quad 85\text{-}90\%$$

Ar = Ph, $p\text{-}NO_2\text{-}C_6H_4$, $PhSO_2$

The presence of a cationic surfactant is also necessary to achieve a selective condensation between indene and benzaldehyde in water. A catalytic amount of CTACl favors the bis-condensation, while a large excess of surfactant allows the mono-adduct to be isolated quantitatively [6a].

Malononitrile reacts with acetone in water in the presence of KF–alumina to give 2-aza[2.2.2]octane in high yield via Knoevenagel condensation followed by double cyclization [15]. The bicyclo adduct heated directly or refluxed in tetralin releases 2-methylpropene (retro Diels–Alder reaction), giving the corresponding pyridine derivative.

Aqueous solutions of hydrotropes (e.g. NaBMGS, sodium methylcellosolve sulfate) have been used in the Hantzsch dihydropyridine synthesis, a tandem Knoevenagel and Michael reaction, in which acetoacetic ester reacts with benzaldehydes and methyl aminocrotonate or aqueous ammonia at room temperature or heated with microwaves (MW) [16, 17]:

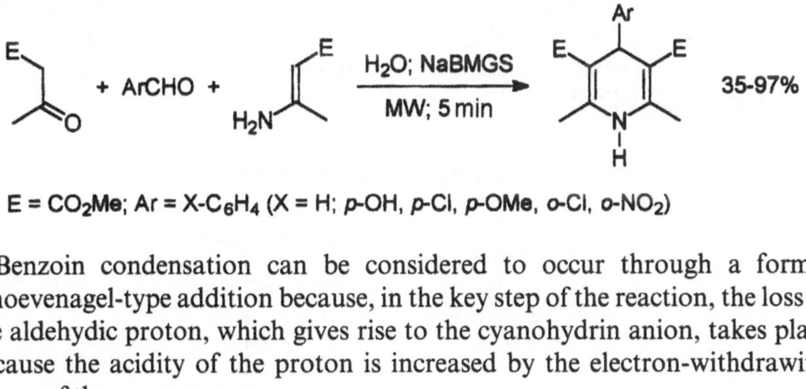

E = CO_2Me; Ar = $X\text{-}C_6H_4$ (X = H; $p\text{-}OH$, $p\text{-}Cl$, $p\text{-}OMe$, $o\text{-}Cl$, $o\text{-}NO_2$)

Benzoin condensation can be considered to occur through a formal Knoevenagel-type addition because, in the key step of the reaction, the loss of the aldehydic proton, which gives rise to the cyanohydrin anion, takes place because the acidity of the proton is increased by the electron-withdrawing power of the cyano group:

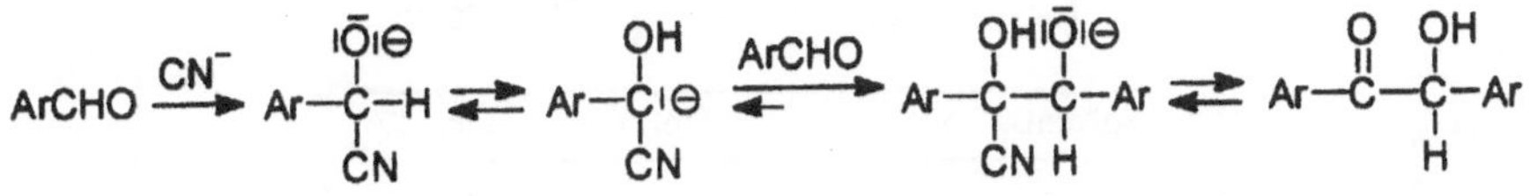

Breslow [18] reports that the reaction rate of cyanide-catalyzed aqueous benzoin condensation is affected both positively and negatively by various salts depending upon the salting-out (e.g. LiCl) and salting-in (e.g. $LiClO_4$) properties. These results are consistent with the hypothesis that the hydrophobic packing of the reactants in the transition state is responsible for the acceleration of the reaction in water.

The coupling reaction between acrylic derivatives and aldehydes (Baylis–Hillman reaction) is catalysed by tertiary amines (e.g. reaction of Table 7.3). The addition of a zwitterionic form, derived from the coupling of acrylate with amine, to the aldehyde is a key step of the reaction and is a formal Knoevenagel-type addition.

Lubineau and co-workers [19] found that the reaction of benzaldehyde with acrylonitrile, catalyzed by 1,4-diazabicyclo[2.2.2]octane (DABCO), is greatly accelerated when carried out in aqueous media. In water only under heterogeneous conditions, the reaction is 5–24 times faster than in organic solvents such as MeOH, THF and PhMe (Table 7.3), and is even faster when lithium or sodium iodide is added. Surprisingly, the reaction was slower with lithium chloride.

7.5 Tollens reaction

An advantage of using an aqueous medium is that the protection of functional groups such as OH is sometimes unnecessary. An example is the one- or two-carbon homologation of 1,3-dihydroxyacetone and glycero-tetrulose by reaction with formaldehyde in the presence of bases [20] (Scheme 7.3).

Aqueous formaldehyde has also been used to prepare α-substituted methyl vinyl ketones from monoalkylated 1,3-diketones under heterogeneous liquid–solid conditions in highly concentrated aqueous solution of K_2CO_3 at room temperature [21]:

Table 7.3 Coupling reaction of benzaldehyde with acrylonitrile

Solvent	Time (h)	Yield (%)
H_2O	7–8	90–98
MeOH	34	90–98
THF or PhMe	168	15–30
H_2O (4 M LiI or NaI)	2–3	92–93
H_2O–CsI	11	92
H_2O–LiCl	24	82

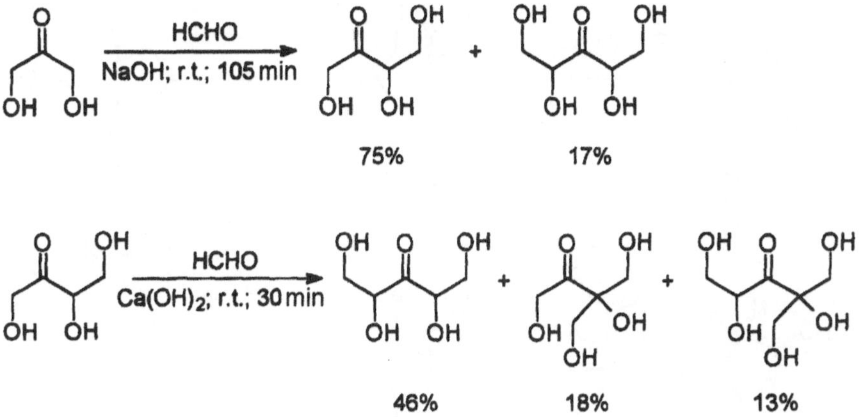

Scheme 7.3

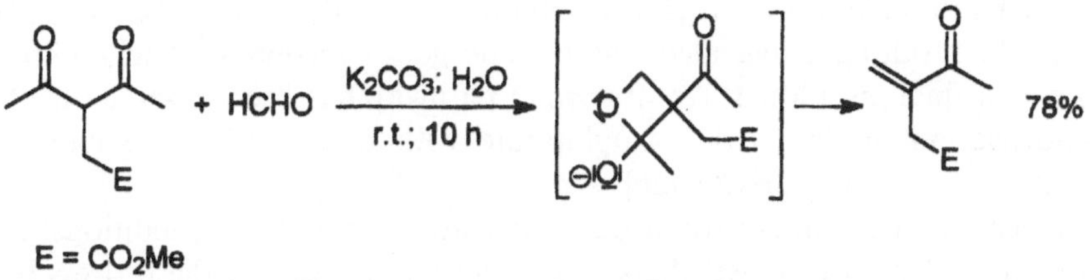

E = CO$_2$Me

Formally, the process consists of an initial Tollens addition followed by an intramolecular acetalyzation and a final expulsion of acetate.

Tollens addition between HCOH and CH$_3$CHO and the intermolecular aldol addition of CH$_3$CHO have been used as reaction models to study, by quantum-mechanical methods, the importance of water in aldol-like reactions carried out in aqueous media [22]. Water accelerates the addition process because it coordinates the reactants, making the geometry of the initial complex more suitable for the reaction, and stabilizes the transition state of the reaction. Water therefore acts as a catalyst.

7.6 Weiss–Cook reaction

In 1968 Weiss and Edwards discovered [23] that the reaction of dimethyl 3-oxoglutarate with glyoxal in 2:1 molar ratio in aqueous acidic solution gives *cis*-bicyclo[3.3.0]octane-3,7-dione-2,4,6,8-tetracarboxylate, which is then converted to *cis*-bicyclo[3.3.0]octane-3,7-dione by acid-catalyzed hydrolysis followed by spontaneous decarboxylation (Scheme 7.4).

The reaction mechanism [24, 25] is a double Knoevenagel reaction that gives an α-β-unsaturated γ-hydroxycyclopentenone, which reacts with

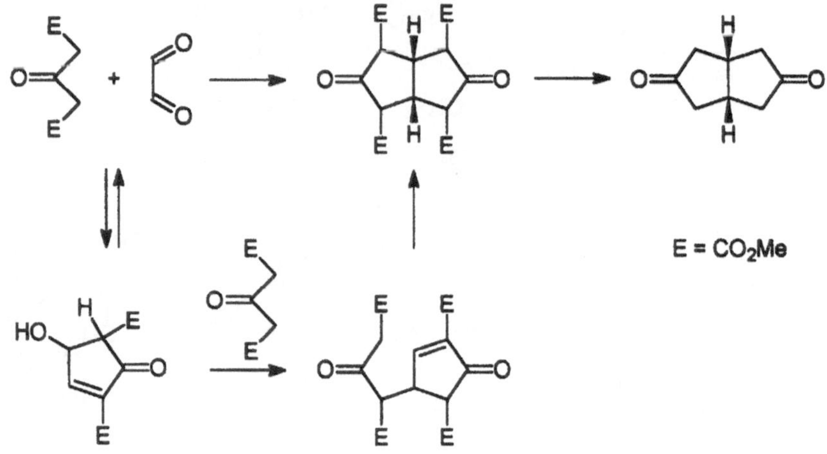

Scheme 7.4

another molecule of dimethyl 3-oxoglutarate according to a Michael addition. The resulting dehydrated adduct undergoes a second Michael addition to give a final *cis* adduct. An abnormal pathway has been observed in the condensation of dimethyl 3-oxoglutarate with dimethyl-2,3-dioxobutanedionate in aqueous NaHCO$_3$ [26].

When the reaction is carried out under aqueous alkaline conditions, the rate and the yield increase, the side-reactions decrease and the process is a good vehicle for the formation of multicyclic polyquinane systems [27, 28].

Paquette [29] has recently shown that the (Z)-cyclododec-7-ene-1,2-dione:

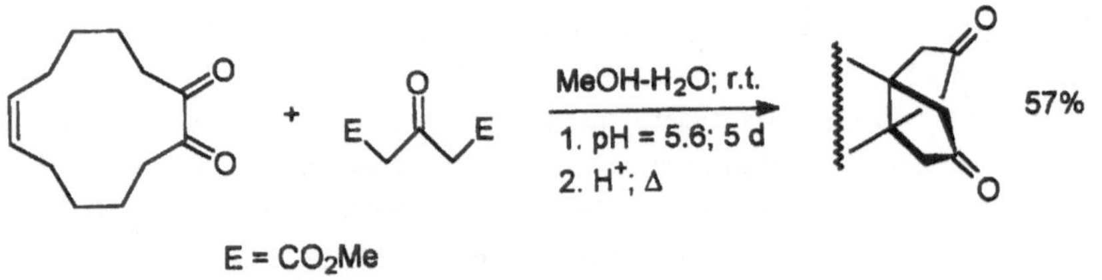

and its alkyne equivalent undergo a Weiss–Cook reaction normally, giving diquinane products in high yield. Sterically congested eight- and twelve-1,2-cycloalkadienones show no tendency to react.

7.7 Michael reaction

The Michael reaction is currently catalyzed by bases. Since strong bases sometimes cause side-reactions, the use of other catalysts such as transition-metal complexes [30], alumina [31, 32], lanthanides [33], phase-transfer catalysts [34], lithium iodide [35] and cesium fluoride [36] have been investigated.

The Michael reaction is accelerated under high pressure and therefore should be facilitated by water as solvent in agreement with the observation that the aqueous medium promotes reactions between hydrophobic molecules having a negative activation volume.

Lubineau and Augé [37] found that nitromethane reacts with methyl vinyl ketone (Scheme 7.5) in water under neutral conditions in the absence of catalyst to give a 4:1 mixture of Michael adducts. Under neat conditions or in solvents such as THF, DCM and PhMe the reaction does not occur in the absence of catalyst. The reaction is about four times faster when it is carried out in the presence of glucose or saccharose, which are known to increase the hydrophobic interaction.

The aqueous medium is also important in the reaction of N-methylglycine with N-ethylmaleimide and with formaldehyde [38] (Scheme 7.6). In refluxing toluene with azeotropic removal of water, paraformaldehyde generates the azomethine ylide by reaction with the amino acid via decarboxylation of the

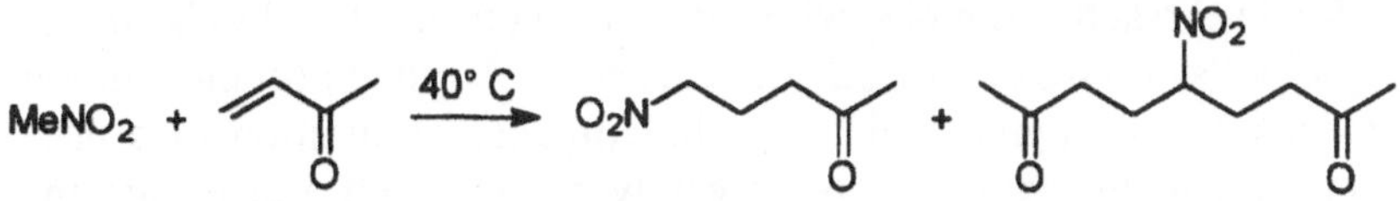

Solvent	Time (h)	A/B	Yield
H$_2$O	32	4:1	quantitative
MeOH	120	1:1	quantitative

Scheme 7.5

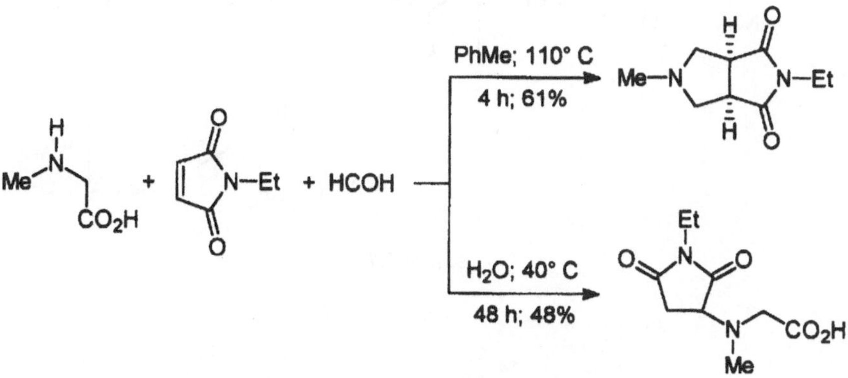

Scheme 7.6

iminium salt. The ylide is trapped by the efficient dipolarophile and bicyclic pyrrolidine is obtained. The ylide is protonated in water, which prevents cycloaddition, and thus only a substituted succinimide results from the Michael addition of the amino acid to the electronically deficient double bond of the maleimide. In the absence of formaldehyde, 1,4-addition is favored in water (40°C, 1 day, 60%) with respect to classical organic solvents such as DMF (21%), THF (traces) and PhMe (0%).

The Michael reaction of α,β-unsaturated ketones, such as methyl vinyl ketone and 3-penten-2-one, with β-dicarbonyl compounds has been investigated in aqueous solution in the presence of CTABr and other cationic surfactants [39]. The reaction yield depends on the temperature, concentration, nucleophile precursor, surfactant and structure of the substrate.

An example of asymmetric Michael addition carried out in water is the reaction of aromatic thiols with 2-cyclohexanone and maleic acid esters via formation of their crystalline CD complexes [40]. The best chiral inductions (*ee* ca. 30%, yield 50–93%) were obtained by the combination of the crystalline β-CD complex of benzenethiol with 2-cyclohexenone and octyl maleate, respectively. The opposite combination gives very low *ee*.

A Michael reaction coupled with aldol-type condensation has been used in the one-pot synthesis of allylrethrone (Scheme 7.7), an important component of an insecticidal pyrethroid [32]. The conjugative addition of 5-nitro-1-pentene to methyl vinyl ketone is catalyzed by Al_2O_3 and occurs in the absence of solvent. An intramolecular aldol-type condensation is then carried out in alkaline aqueous medium after the conversion of the nitro group into a carbonyl by the Nef reaction.

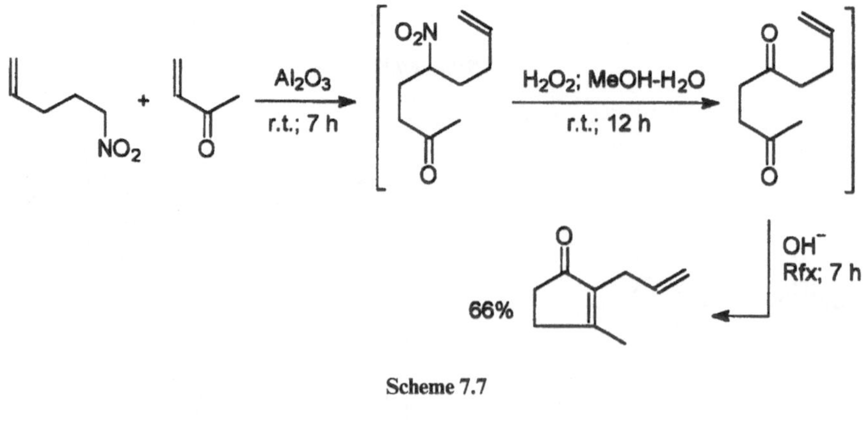

Scheme 7.7

References

[1] (a) Decker, P.; Schweer, H., *Origins of Life*, 1984, **14**, 335; (b) Decker, P.; Schweer, H.; Pohlmann, R.J., *Chromatography*, 1982, **244**, 281; (c) Harsch, G.; Bauer, H.; Voelter, W., *Liebigs Ann. Chem.*, 1984, 623.

[2] Mukaiyama, T.; Banno, K,; Narasaka, K., *J. Am. Chem. Soc.*, 1974, **96**, 7503.

[3] Kobayashi, S., *Synlett*, 1994, 689.

[4] Yamamoto, Y.; Maruyama, K.; Matsumoto, K., *J. Am. Chem. Soc.*, 1983, **105**, 6963.

[5] (a) Lubineau, A., *J. Org. Chem.*, 1986, **51**, 2142; (b) Lubineau, A.; Meyer, E., *Tetrahedron*, 1988, **44**, 6065.

[6] (a) Fringuelli, F.; Pani, G.; Piermatti, O.; Pizzo, F., *Tetrahedron*, 1994, **50**, 11499; (b) Fringuelli, F.; Pani, G.; Piermatti, O.; Pizzo, F., *Life Chem. Rep.*, 1995, **13**, 133.

[7] Nivalkar, K.R.; Mudaliar, C.D.; Mashraqui, S.H., *J. Chem. Res. (S)*, 1992, 98.

[8] Sadvilcar, V.G.; Samant, S.D.; Gaikar, V.G., *J. Chem. Tech. Biotechnol.*, 1995, **62**, 405.

[9] Buonora, P.T.; Rosauer, K.G.; Dai, L., *Tetrahedron Lett.*, 1995, **36**, 4049.

[10] Watanabe, K.; Yamada, Y.; Goto, K., *Bull. Chem. Soc. Jpn.*, 1985, **58**, 1401.

[11] Zhang, Y.; Xu, W., *Synth. Commun.*, 1989, **19**, 1291.

[12] Desper, J.M.; Breslow, R., *J. Am. Chem. Soc.*, 1994, **116**, 12081.

[13] Denmark, S.E.; Lee, W., *Tetrahedron Lett.*, 1992, **33**, 7729.

[14] Brufola, G.; Fringuelli, F.; Piermatti, O.; Pizzo, F., *Heterocycles*, 1996, **43**, 1257.

[15] Nakano, Y.; Niki, S.; Kinouchi, S.; Miyamae, H.; Igarashi, M., *Bull. Chem. Soc. Jpn.*, 1992, **65**, 2934.

[16] Sadvilkar, V.G.; Khadilkar, B.M.; Gaikar, V.G., *J. Chem. Tech. Biotechnol.*, 1995, **63**, 33.

[17] Khadilkar, B.M.; Gaikar, V.G.; Chitnavis, A.A., *Tetrahedron Lett.*, 1995, **36**, 8083.

[18] Kool, E.T.; Breslow, R., *J. Am. Chem. Soc.*, 1988, **110**, 1596.

[19] Augé, J.; Lubin, M.; Lubineau, A., *Tetrahedron Lett.*, 1994, **35**, 7947.

[20] Shigemasa, Y.; Yokoyama, K.; Sashiwa, H.; Saimoto, H., *Tetrahedron Lett.*, 1994, **35**, 1263.

[21] Ben Ayed, T.; Amri, H., *Synth. Commun.*, 1995, **25**, 3813.

[22] Coitiño, E.L.; Tomasi, J.; Ventura, O.N., *J. Chem. Soc., Faraday Trans.*, 1994, **90**, 1745.

[23] Weiss, U.; Edwards, J.M., *Tetrahedron Lett.*, 1968, 4885.

[24] Yang-Lan, S.; Mueller-Johnson, M.; Oehldrich, J.; Wichman, D.; Weiss, U.; Cook, J.M., *J. Org Chem.*, 1976, **41**, 4053.

[25] Mitschka, R.; Oehldrich, J.; Takahashi, K.; Cook, J.M.; Weiss, U.; Silverton, J.V., *Tetrahedron*, 1981, **37**, 4521.

[26] Deslongchamps, G.; Mink, D.; Boyle, P.D.; Singh, N., *Can. J. Chem.*, 1994, **72**, 1162.

[27] Fu, X.; Cook, J.M.; *Aldrichim. Acta*, 1992, **25**, 43.

[28] Gupta, A.K.; Fu, X.; Snyder, J.P.; Cook, J.M., *Tetrahedron*, 1991, **47**, 3665.

[29] Detert, H.; Lanter, J.C.; Paquette, L.A., *J. Org. Chem.*, 1995, **60**, 353.

[30] Corsico Coda, A.; Desimoni, G.; Righetti, P.; Tacconi, G., *Gazz. Chim. Ital.*, 1984, **114**, 417.

[31] Ranu, B.C.; Bahar, S., *Tetrahedron*, 1992, **48**, 1327.

[32] Ballini, R., *Synthesis*, 1993, 687.

[33] Van Westrenen, J.; Roggen, R.M.; Hoefnagel, M.A.; Peters, J.A.; Kieboom, A.P.G.; Van Bekkum, H., *Tetrahedron*, 1990, **46**, 5741.

[34] Kryshtal, G.V.; Kulganeck, V.V.; Kucherov, V.F.; Yanovskaya, L.A., *Synthesis*, 1979, 107.

[35] Antonioletti, R.; Bonadies, F.; Monteagudo, E.S.; Scettri, A., *Tetrahedron Lett.*, 1991, **32**, 5373.

[36] Boyer, J.; Corriv, R.J.P.; Perz, R.; Réyé, C., *J. Chem. Soc., Chem. Commun.*, 1981, 122.

[37] Lubineau, A.; Augé, J., *Tetrahedron Lett.*, 1992, **33**, 8073.

[38] Lubineau, A.; Bouchain, G.; Queneau, Y., *J. Chem. Soc. Perkin Trans. 1*, 1995, 2433.

[39] Bassetti, M.; Cerichelli, G.; Floris, B., *Gazz. Chim. Ital.*, 1991, **121**, 527.

[40] Sakuraba, H.; Tananoka, Y.; Toda, F., *J. Incl. Phenom. Mol. Recogn. Chem.*, 1991, **11**, 195.

8 Water-stable rare-earth Lewis-acid catalysis in aqueous and organic solvents

S. KOBAYASHI

8.1 Introduction

Lewis-acid-catalyzed carbon–carbon bond-forming reactions have been of great interest in organic synthesis because of their unique reactivities and selectivities, and for the mild conditions used [1]. While various kinds of Lewis-acid-promoted reactions have been developed and many have been applied in industry, these reactions must be carried out under strict anhydrous conditions. The presence of even a small amount of water stops the reaction, because most Lewis acids immediately react with water, rather than with the substrates, and decompose or deactivate. This fact has restricted the use of Lewis acids in organic synthesis.

On the other hand, the utility of aqueous reactions is now generally recognized [1]. It is desirable to perform the reactions of compounds containing water of crystallization or other water-soluble compounds in aqueous media, because tedious procedures to remove water are necessary when the reactions are carried out in organic solvents. Moreover, aqueous reactions of organic compounds avoid the use of harmful organic solvents.

We started our research efforts to develop a new type of Lewis acid that could be used in aqueous media. The chemistry of metal trifluoromethanesulfonates (triflates) [2], some of which have been used as Lewis acids, has been studied in our laboratories [3]. One of the most successful examples is the use of chiral diamine-coordinated tin(II) triflate as a chiral Lewis-acid catalyst in asymmetric aldol and related reactions [3c–e]. Metal triflates have several unique properties compared with the corresponding metal halides, and, in the course of our investigations to search for other metal triflates, we first focused on lanthanide triflates [4].

Lanthanide compounds were expected to act as strong Lewis acids because of their hard character and to have strong affinity toward carbonyl oxygens [5]. Among these compounds, lanthanide triflates were expected to be one of the strongest Lewis acids because of the electron-withdrawing trifluoromethanesulfonyl group. On the other hand, their hydrolysis was postulated to be slow, based on their hydration energies and hydrolysis constants [6]. In fact, while most metal triflates are prepared under strict anhydrous conditions, lanthanide triflates are reported to be prepared in aqueous solution [4,

7]. The large radius of lanthanide(III) and the specific coordination number also attracted us, and we started to investigate the use of lanthanide triflates as Lewis-acid catalysts for the reaction of amines with nitriles under anhydrous conditions [7a]. After starting this study, we found that scandium triflate (Sc(OTf)$_3$) and yttrium triflate (Y(OTf)$_3$) are also excellent water-tolerant Lewis acids, and we have developed new synthetic reactions using these rare-earth triflates as catalysts.

In this chapter, we describe the unique properties of a new type of Lewis acid, rare-earth triflates, and useful synthetic reactions using these Lewis acids in both aqueous and organic solvents.

8.2 Aldol reactions

8.2.1 Aldol reactions in aqueous media

The titanium tetrachloride-mediated aldol reaction of silyl enol ethers with aldehydes was first reported in 1973 [8, 9]. The reaction is notably distinguished from conventional aldol reactions carried out under basic conditions: it proceeds in a highly regioselective manner to afford cross-aldols in high yields [10]. Since this pioneering effort, several efficient activators such as trityl salts [11], Clay montmorillonite [12] and fluoride anions [13] have been developed to realize high yields and selectivities. Lanthanide(III) chlorides and some organolanthanide compounds have been reported to catalyze aldol reactions of ketene silyl acetals with aldehydes [14]. Now the aldol reaction is considered to be one of the most important carbon–carbon bond-forming reactions in organic synthesis. These reactions are usually carried out under strictly non-aqueous conditions. The presence of even a small amount of water causes lower yields, probably due to the rapid decomposition or deactivation of the promoters and the hydrolysis of the silyl enol ethers. Furthermore, the promoters cannot be recovered and reused because they decompose under usual quenching conditions.

In 1986, Lubineau *et al.* reported the water-promoted aldol reaction of silyl enol ethers with aldehydes [15]. While the report that the aldol reactions proceeded without catalyst in water was unique, the yields and the substrate range were not satisfactory.

In our laboratories, the hydroxymethylation reaction of silyl enol ethers with commercial formaldehyde (methanal) solution was first attemped by using lanthanide triflates. Formaldehyde is a versatile reagent, being one of the most highly reactive C(1) electrophiles in organic synthesis [16]. Dry gaseous formaldehyde, required for many reactions, has some disadvantages because it must be generated before use from solid polymer paraformaldehyde by way of thermal depolymerization, and it self-polymerizes easily. (Snider and Yamamoto separately developed formaldehyde–organoaluminum complex as a source of formaldehyde in several reactions [17].) On the other hand, commercial formaldehyde solution, which is an aqueous solution containing 37% formaldehyde and 8–10% methanol, is cheap, easy to handle and stable even at room temperature. However, the use of this reagent is strongly restricted owing to the existence of a large amount of water. For example, the titanium tetrachloride (TiCl₄)-promoted hydroxymethylation reaction of silyl enol ethers was carried out by using trioxane as a formaldehyde source under strict anhydrous conditions [9]. (The TMSOTf-mediated aldol-type reaction of silyl enol ethers with dialkoxymethanes has also been reported [18].) Formaldehyde water solution could not be used because TiCl₄ and the silyl enol ether reacted with water rather than with formaldehyde in that water solution.

The effects of lanthanide triflates in the reaction of the silyl enol ether of propiophenone (1) with commercial formaldehyde solution were examined [19]. In most cases, the reactions proceeded smoothly to give the corresponding adducts in high yields (Table 8.1). The reactions were most effectively carried out in commercial formaldehyde solution–THF medium under the influence of a catalytic amount of Yb(OTf)₃.

Several examples of the hydroxymethylation reaction of silyl enol ethers with commercial formaldehyde solution are shown in Table 8.1, and the following characteristic features of this reaction are noted. (i) In every case, the reactions proceeded smoothly under extremely mild conditions (almost neutral) to give the corresponding hydroxymethylated adducts in high yields. Sterically hindered silyl enol ethers also worked well and the diastereoselectivities were high. (ii) Di- and poly-hydroxymethylated products were not observed [20]. (iii) The absence of equilibrium (double-bond migration) in silyl enol ethers allowed for a regiospecific hydroxymethylation reaction. (iv) Only a catalytic amount of Yb(OTf)₃ was required to complete the reaction. The amount of the catalyst was examined by taking the reaction of the silyl enol ether derived from propiophenone (1) with commercial formaldehyde solution as a model, and the reaction was found to be catalyzed with good yield by even 1 mol% of Yb(OTf)₃: 1 mol% gave 90% yield; 5 mol%, 90% yield; 10 mol%, 94% yield; 20 mol%, 94% yield; 100 mol%, 94% yield.

Next, the use of lanthanide triflates in the activation of aldehydes other than formaldehyde was investigated [21]. The model reaction of 1-trimethylsiloxycyclohexene (2) with benzaldehyde under the influence of a catalytic amount of Yb(OTf)₃ (10 mol%) was examined (Table 8.2). The reaction proceeded smoothly in H_2O–THF (1:4), but the yields were low when water or THF was used alone. Among several lanthanide triflates screened, neodymium triflate (Nd(OTf)₃), gadolinium triflate (Gd(OTf)₃), Yb(OTf)₃ and lutetium triflate (Lu(OTf)₃) were quite effective, while the yield of the desired aldol adduct was lower in the presence of lanthanum triflate (La(OTf)₃), praseodymium triflate (Pr(OTf)₃) or thulium triflate (Tm(OTf)₃) (Table 8.2).

Table 8.1 Reaction of silyl enol ethers with commercial formaldehyde solution catalyzed by Yb(OTf)$_3$

Entry	Silyl enol ether	Product	Yield (%)
1	OSiMe$_3$ a) — Ph, **1**	O — Ph ... OH	94
2	OSiMe$_3$ b)	O ... OH	85
3	OSiMe$_3$ — Ph	O — Ph ... OH	77
4	OSiMe$_3$ **2**	O ... OH	82
5	OSiMe$_3$	O ... OH (3:2)	86
6	OSiMe$_3$	O ... OH	92
7	Me$_3$SiO a) — Ph	O — Ph ... OH	92
8	Me$_3$SiO a) — Ph, O, SEt	O, O — Ph, SEt, OH (O — Ph, O, O) c)	88
9	OSiMe$_3$, O, SEt	O, OH, O, SEt, H (O, O, O, H) d)	83
10	OSiMe$_3$, O, S^{t}Bu, H	O, OH, O, S^{t}Bu, H + O, OH, O, S^{t}Bu, H (9:1)	90

^{a}$Z/E \geq 98/2$.
^{b}$Z/E = 1/4$.
^{c}The mixture of the hydroxythioester and the lactone (2:1) was obtained. The other diastereomers were not observed.
^{d}The mixture of the hydroxythioester and the lactone (3:1) was obtained. Less than 3% yields of the other diastereomers were observed.

Table 8.2 Effect of lanthanide triflates on the reaction:

$$PhCHO + \underset{\mathbf{2}}{\overset{OSiMe_3}{\bigcirc}} \xrightarrow[\substack{H_2O\text{-}THF\ (1:4) \\ rt,\ 20\ h}]{Ln(OTf)_3\ (10\ mol\%)} Ph\overset{OH}{\underset{}{\wedge}}\overset{O}{\bigcirc}$$

$Ln(OTf)_3$	Yield (%)	$Ln(OTf)_3$	Yield (%)
$La(OTf)_3$	8	$Dy(OTf)_3$	73
$Pr(OTf)_3$	28	$Ho(OTf)_3$	47
$Nd(OTf)_3$	83	$Er(OTf)_3$	52
$Sm(OTf)_3$	46	$Tm(OTf)_3$	20
$Eu(OTf)_3$	34	$Yb(OTf)_3$	91
$Gd(OTf)_3$	89	$Lu(OTf)_3$	88

The effect of ytterbium salts was also investigated (Table 8.3) [22]. While the Yb salts with less-nucleophilic counter-anions such as OTf^- or ClO_4^- effectively catalyzed the reaction, only low yields of the product were obtained when the Yb salts with more-nucleophilic counter-anions such as Cl^-, OAc^-, NO_3^- and SO_4^{2-} were employed. The Yb salts with less-nucleophilic counter-anions are more cationic and the high Lewis acidity promotes the desired reaction.

Several examples of the present aldol reaction of silyl enol ethers with aldehydes are listed in Table 8.4. In every case, the aldol adducts were obtained in high yields in the presence of a catalytic amount of $Yb(OTf)_3$, $Gd(OTf)_3$, or $Lu(OTf)_3$ in aqueous media. Diastereoselectivities were generally good to moderate. One feature in the present reaction is that water-soluble aldehydes, for instance, acetaldehyde, acrolein and chloroacetaldehyde, can be reacted with silyl enol ethers to afford the corresponding cross-aldol adducts in high

Table 8.3 Effect of Yb salts on the reaction:

$$PhCHO + \underset{\mathbf{2}}{\overset{OSiMe_3}{\bigcirc}} \xrightarrow[\substack{H_2O\text{-}THF \\ (1:4) \\ rt,\ 19\ h}]{\substack{Yb\ salt \\ (10\ mol\%)}} Ph\overset{OH}{\underset{}{\wedge}}\overset{O}{\bigcirc}$$

Yb salt	Yield (%)
$Yb(OTf)_3$	91[a]
$Yb(ClO_4)_3$	88[b]
$YbCl_3$	3
$Yb(OAc)_3 \cdot 8H_2O$	14
$Yb(NO_3)_3 \cdot 5H_2O$	7
$Yb_2(SO_4)_3 \cdot 5H_2O$	trace

[a] *syn/anti* = 73/27.
[b] *syn/anti* = 76/24.

Table 8.4 Lanthanide-triflate-catalyzed aldol reactions in aqueous media:

$$R^1CHO + \overset{OSiMe_3}{\underset{R^2}{\bigvee}}R^3 \xrightarrow[\text{H}_2\text{O-THF, rt}]{\substack{Yb(OTf)_3 \\ (10\ mol\%)}} R^1\overset{OH\quad O}{\underset{R^3}{\bigvee}}R^2$$

Entry	Aldehyde	Silyl enol ether	Yield (%)
1	PhCHO	(cyclohexene OSiMe₃) **2**	91[a]
2	PhCHO	(OSiMe₃)	89[b]
3	PhCHO	(OSiMe₃)	93[c]
4	PhCHO	Ph—(OSiMe₃) **1**	81[d]
5	CH₃CHO	**1**	93[e,f]
6	⌄CHO	**1**	82[e,g]
7	Cl⌄CHO	**1**	95[h]
8	Ph—C(O)—CHO•H₂O	**1**	67[i]
9	(benzene)—OH / CHO	**1**	81[j,k]
10	(pyridine N)—CHO	**1**	87[j,l]

[a]*syn/anti* = 73/27.
[b]*syn/anti* = 63/37.
[c]*syn/anti* = 71/29.
[d]*syn/anti* = 53/47.
[e]Gd(OTf)₃ was used instead of Yb(OTf)₃.
[f]*syn/anti* = 46/54.
[g]*syn/anti* = 60/40.
[h]*syn/anti* = 45/55.
[i]*syn/anti* = 27/73.
[j]Lu(OTf)₃ was used instead of Yb(OTf)₃.
[k]*syn/anti* = 55/45.
[l]*syn/anti* = 42/58.

yields (entries 5–7 in Table 8.4). Some of these aldehydes are commercially supplied as water solutions and are appropriate for direct use. Phenylglyoxal monohydrate also worked well (entry 8). It is known that water often interferes with the aldol reactions of metal enolates with aldehydes, and that, in the cases where such water-soluble aldehydes are employed, some troublesome purifications including dehydration are necessary. Moreover, salicylaldehyde (entry 9) and 2-pyridinecarboxaldehyde (entry 10) could be successfully employed. The former has a free hydroxy group that is incompatible with metal enolates or Lewis acids, and the latter is generally difficult to use under the influence of Lewis acids because of the coordination of the nitrogen atom to the Lewis acids, resulting in the deactivation of the acids.

8.2.2   Recovery and reuse of the catalyst

Lanthanide triflates are more soluble in water than in organic solvents such as dichloromethane. Very interestingly, almost 100% of lanthanide triflates was quite easily recovered from the aqueous layer after the reaction was completed and it could be reused. For example, first use (20 mol% of $Yb(OTf)_3$) in the reaction of 1 with formaldehyde water solution gave 94% yield; second use, 91% yield; third use, 93% yield. The reactions are usually quenched with water and the products are extracted with an organic solvent (for example, dichloromethane). Lanthanide triflate is in the aqueous layer and so only removal of the water is needed to give the catalyst, which can be used in the next reaction (Figure 8.1). It is noteworthy that lanthanide triflates are expected to solve some severe environmental problems induced by Lewis-acid-promoted reactions in industrial chemistry [23].

8 2.3   Reaction mechanism

Although hydrolysis of lanthanide compounds is expected to be very slow [6], there still remains a possibility that a small number of protons exist in the aqueous medium according to the following equation:

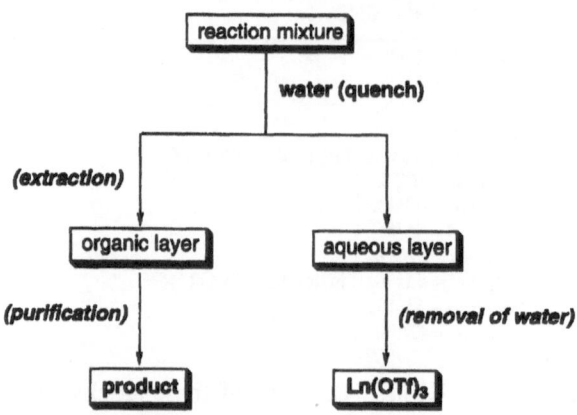

Figure 8.1 Recovery of the catalyst.

$$Yb(OTf)_3 + m\,H_2O \; \longrightarrow \; \rightleftharpoons \; Yb(H_2O)_m{}^{3+} + 3OTf$$

$$\Updownarrow$$

$$Yb(H_2O)_{m-1}(OH)^{2+} + H^+$$

Various pH aqueous solutions (pH = 1–6) of trifluoromethanesulfonic acid were prepared independently, and the model reaction of the silyl enol ether derived from cyclohexanone (**2**) with benzaldehyde was tested. Six independent experiments were performed (pH = 1, 2, 3, 4, 5, 6). In the pH 5 and 6 solutions, only a trace amount of the product was detected on TLC, the yields were less than 5%, and hydrolysis of **2** was also observed. In the pH 1–4 solutions, the silyl enol ether immediately hydrolyzed to give the original ketone, and no aldol adduct was obtained. From these experiments, the protons that may be produced from the hydrolysis of the lanthanide triflates were found not to be an active catalytic species in the present aldol reaction of silyl enol ethers with aldehydes: the pH values of Yb(OTf)$_3$ solutions were measured as 5.90 (1.6 × 10^{-2} M, THF–H$_2$O, 4:1) and 6.40 (8.0 × 10^{-2} M, H$_2$O).

In the present aldol reactions, the amount of water influenced yields and diastereoselectivities strongly. The effects of the amount of water on the yield in the model reaction of silyl enol ether **2** with benzaldehyde in the presence of 10 mol% of Yb(OTf)$_3$ in THF are shown in Figures 8.2 and 8.3. The best yields in Figure 8.2 were obtained when the ratio of water was 10–20%. When the amount of water increased, the yield began to decrease. The reaction system became two phases when the amount of water increased, and the yield decreased. Only 18% of the product was isolated in 100% water solution. On

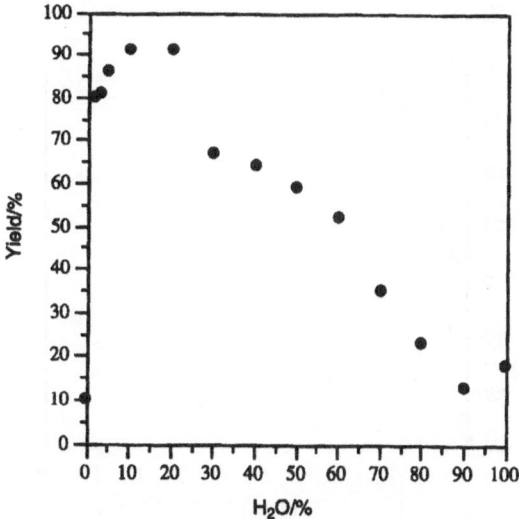

Figure 8.2 Effect of water on yield.

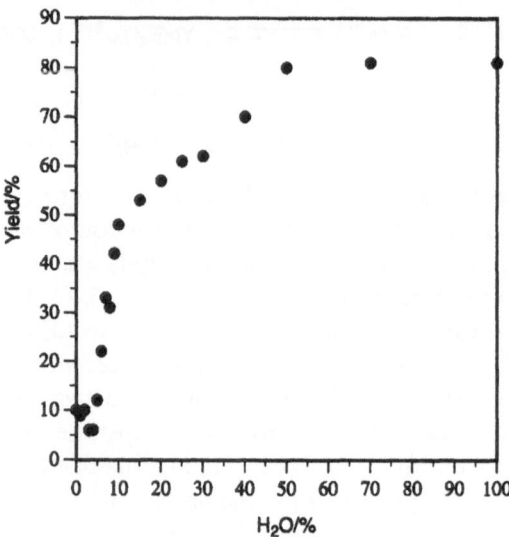

Figure 8.3 Effect of water on yield.

the other hand, in Figure 8.3 when water was not added or 1–5 eq. of water were added to Yb(OTf)$_3$, the yield of the desired aldol adducts was also low (ca. 10% yield). The yield improved as water was increased to 6–10 eq.; and when more than 50 eq. of water were added, the yield improved to more than 80%.

As for the diastereoselectivities, the amount of water also has an important role in deciding the course of the reactions (Figure 8.4). In the absence of water, the reaction proceeded with *anti* preference. The selectivity changed as

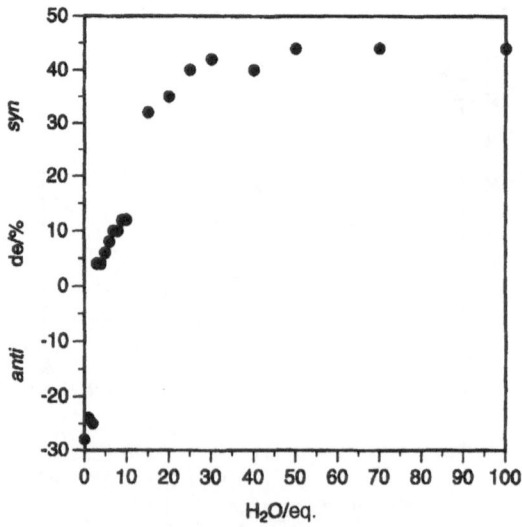

Figure 8.4 Effect of water on diastereoselectivity.

the amount of water increased and the *syn* aldol began to be obtained when more than 3 eq. of water were added to $Yb(OTf)_3$. The selectivity improved in accordance with the amount of water, and almost the same selectivities were obtained when more than 15 eq. of water were added.

These phenomena can be explained as follows. First, in the absence of water or in the presence of a small amount of water, THF predominantly coordinates to $Yb(OTf)_3$ and the activity of THF-coordinated $Yb(OTf)_3$ as a Lewis acid is low. The reaction proceeds slowly via the cyclic six-membered transition state with *anti* preference [24]. On the other hand, when the moles of water are gradually increased, water is prone to coordinate to $Yb(OTf)_3$ instead of tetrahydrofuran (THF), and $Yb(OTf)_3$ dissociates to form the active Yb cation. The solid-state structure of $Yb(OTf)_3 \cdot 9H_2O$ has been investigated [7d]. When $Yb(OTf)_3 \cdot 9H_2O$ (prepared according to the literature) was used, the aldol reactions proceeded, but faster hydrolysis of the silyl enol ethers was observed. At this stage, intramolecular and intermolecular exchange reactions of water molecules occur frequently [25]. There is a chance for an aldehyde to coordinate to Yb^{3+} instead of to water molecules and the aldehyde is activated:

$$Yb(H_2O)_m^{3+} + 3\,OTf^- \underset{H_2O}{\overset{RCHO}{\rightleftharpoons}} \quad Yb(H_2O)_{m-1}^{3+} \cdots \underset{H}{\overset{O}{\underset{R}{\Vert}}} + 3\,OTf^-$$

A silyl enol ether attacks this activated aldehyde to produce the aldol adduct. This ytterbium-catalyzed aldol reaction would proceed via the acyclic transition state to give *syn* aldols [26]. When the amount of water is further increased, a competitive reaction, hydrolysis of the silyl enol ether, precedes the desired aldol reaction.

8.2.4   Aldol reactions in organic solvents

Next, the application of ketene silyl acetals was tried in the above aqueous reactions of silyl enolates with aldehydes. Ketene silyl acetals are useful ester enolate equivalents that can be isolated [27, 28], and the aldol-type reaction of ketene silyl acetals with aldehydes is among the most important and mildest methods of carbon–carbon bond formation [29]. Disappointingly, no aldol adduct was obtained when the ketene silyl acetal derived from methyl 2-methylpropionate (3) was employed as a representative ketene silyl acetal (structure 3 is shown later in Table 8.10). In aqueous media, hydrolysis of the ketene silyl acetal preceded the desired aldol reaction.

The reaction was then carried out in organic solvents. First, ketene silyl acetal 3 was treated with benzaldehyde in the presence of 10 mol% of $Yb(OTf)_3$ in dichloromethane. The reaction proceeded smoothly at –78°C to afford the corresponding aldol-type adduct in a 94% yield. The same reaction at room temperature also went quite cleanly without side-reactions and the desired adduct was obtained in a 95% yield. In other organic solvents such as toluene, THF, acetonitrile and DMF, $Yb(OTf)_3$ worked well, and it was found that other lanthanide triflates ($Ln(OTf)_3$: Ln = La, Ce, Pr, Nd, Sm, Eu, Gd, Dy, Ho, Er, Tm and Lu) also catalyzed the above aldol reaction effectively (85–95% yields).

Several examples of the present aldol reactions in organic solvent using $Yb(OTf)_3$, $Eu(OTf)_3$, $Gd(OTf)_3$, or $Ho(OTf)_3$ as a catalyst are examined (Scheme 8.1) [30]. Silyl enolates derived from not only esters but also thioesters and ketones reacted with aldehydes to give the corresponding adducts in high yields. Furthermore, acetals reacted smoothly with silyl enolates to afford the corresponding aldol-type adducts in high yields. It should be noted that the catalysts could be easily recovered from the aqueous layer after the reactions were quenched with water and could be reused, and that the yields of the second run were almost comparable to those of the first run in every case.

Thus, the lanthanide-triflate-catalyzed aldol reactions of silyl enolates with aldehydes or acetals were successfully carried out not only in aqueous but also in organic solvents. The extreme mildness of the reaction conditions, the simple procedures, the successful use of both aqueous and organic solvents, and the striking feature of the reusability of these catalysts are especially noteworthy.

8.2.5   Aldol reactions in water–ethanol–toluene and continuous use of the Ln(OTf)₃ catalyst

Quite recently, the aldol reactions of silyl enol ethers with aldehydes were found to proceed smoothly in a new solvent system, water–ethanol–toluene [31]. The reactions proceeded much faster in the above solvent than in THF–water. Furthermore, the new solvent system realized continuous use of the catalyst by a very simple procedure.

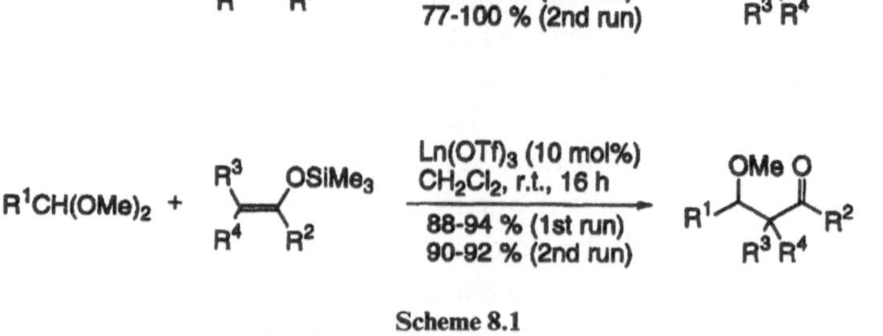

Scheme 8.1

Several water–organic solvent systems were examined in the model reaction of 1-phenyl-1-trimethylsiloxypropene with 2-pyridinecarboxalde-hyde under the influence of 10 mol% Yb(OTf)$_3$ (Table 8.5). While the reaction proceeded sluggishly in a water–toluene system, the adduct was obtained in a good yield when ethanol was added to this system. The yield increased in accordance with the amount of ethanol, and it was noted that the reaction proceeded much faster in the water–ethanol–toluene system than in the orig-inal water–THF system.

Next, we examined the reuse of the catalyst. Although the water–ethanol–toluene (1:7:4) system was one phase, it easily became two phases by adding toluene after the reaction was completed. The product was isolated from the organic layer by a usual workup. On the other hand, the catalyst remained in the aqueous layer, which was used directly in the next reaction without removing water. It is noteworthy that the yields of the second, third and fourth runs were comparable to that of the first run (Scheme 8.2).

Several examples of the present aldol reactions of silyl enol ethers with aldehydes in water–ethanol–toluene are listed in Table 8.6. 3-Pyridine-carboxaldehyde as well as 2-pyridinecarboxaldehyde (examples of nitrogen-containing aldehydes), salicylaldehyde (an example of an aldehyde

Table 8.5 Effect of solvents on the reaction:

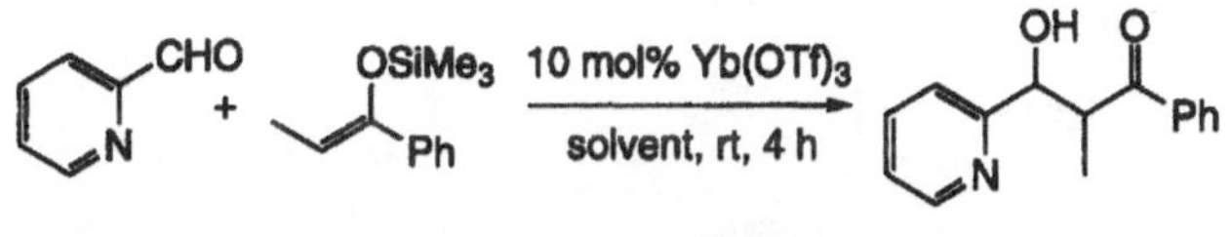

Solvent	Yield (%)	syn/anti
H$_2$O–toluene (1:4)	0	–
H$_2$O–EtOH–toluene (1:3:4)	30	37/63
(1:5:4)	41	39/61
(1:7:4)	70	41/59
(1:10:4)	96	40/60
H$_2$O–THF (1:4)	12	43/57

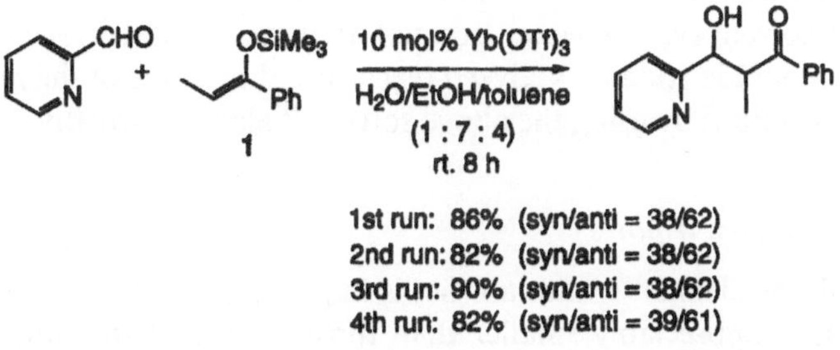

1st run: 86% (syn/anti = 38/62)
2nd run: 82% (syn/anti = 38/62)
3rd run: 90% (syn/anti = 38/62)
4th run: 82% (syn/anti = 39/61)

Scheme 8.2

Table 8.6 Yb(OTf)$_3$-catalyzed aldol reactions of silyl enol ethers with aldehydes in water–ethanol–toluene:

Aldehyde	Silyl enol ether	Product	Yield (%)
PhCHO	**2**		89[a]
PhCHO			95
			87
			82
			96
HCHO aq.	**1**		90

[a]*syn/anti* = 74/26.

containing a free hydroxyl group) and formaldehyde water solution worked well. As for silyl enol ethers, not only ketone enol ethers but also silyl enolates derived from thioesters were used. In every case, the adducts were obtained in high yields in the presence of 10 mol% Yb(OTf)$_3$.

A new solvent system, water–ethanol–toluene, facilitates reuse of the catalyst in lanthanide-triflate-catalyzed reactions. Shortening of the reaction time in these reactions by using the above solvent is also noteworthy.

8.2.6 Scandium triflates

Although the element scandium (Sc) is in group 3 and lies above La and Y, its radius is appreciably smaller than those of any of the other rare-earth elements, and the chemical behavior of scandium is known to be intermediate between that of aluminum and that of the lanthanides [32]. Scandium is uncommon probably due to the lack of rich sources and to difficulties in separation, and its use in organic synthesis is rather limited in spite of its promising properties. In the course of our investigations to search for novel Lewis-acid catalysts, especially metal triflates, we focused on the element scandium.

Sc(OTf)$_3$ is prepared from the corresponding oxide (Sc$_2$O$_3$) and trifluoro-methanesulfonic acid (TfOH) [4]; 1.5 mol of Sc$_2$O$_3$ was added to an aqueous solution of TfOH (ca. 50% v/v), and the mixture was heated at 100°C for 1 h:

$$Sc_2O_3 + 6\,TfOH \longrightarrow 2\,Sc(OTf)_3 + 3\,H_2O$$
(ca 50% TfOH/H$_2$O)

After filtration to remove the unreacted oxide, the water was evaporated under reduced pressure. The resulting white powder was dried by heating *in vacuo* at 200°C.

Scandium triflate was found to be an effective catalyst in the aldol reactions [33]. The activities of various triflate catalysts were evaluated in the aldol reaction of 1-trimethylsiloxycyclohexene (2) with benzaldehyde in dichloromethane (Table 8.7). While the reaction scarcely proceeded at −78°C in the presence of Yb(OTf)$_3$ or Y(OTf)$_3$, the aldol adduct was obtained in an 81% yield in the presence of Sc(OTf)$_3$. Obviously, Sc(OTf)$_3$ is more active than Y(OTf)$_3$ or Yb(OTf)$_3$ in this case. (However, Y(OTf)$_3$ and Yb(OTf)$_3$ did catalyze the aldol reaction of 2 with benzaldehyde at room temperature in dichloromethane to afford the aldol adduct in 58% yield (*syn/anti* = 32/68) and 81% yield (*syn/anti* = 29/71), respectively.)

This interesting and promising result prompted us to examine the use of Sc(OTf)$_3$ as a Lewis-acid catalyst in the aldol reactions of silyl enolates with carbonyl compounds. Several examples of the Sc(OTf)$_3$-catalyzed aldol reactions of silyl enolates with aldehydes were examined. Silyl enolates derived from ketones, thioesters and esters reacted smoothly with aldehydes in the presence of 5 mol% of Sc(OTf)$_3$ to afford the aldol adducts in high yields. Sc(OTf)$_3$ was also found to be an effective catalyst in the aldol-type

Table 8.7 Effect of catalysts on the reaction:

Entry	Catalyst	Yield (%)
1	Sc(OTf)$_3$	81
2	Y(OTf)$_3$	trace
3	Yb(OTf)$_3$	trace

reaction of silyl enolates with acetals. The reactions proceeded smoothly at −78°C or room temperature to give the corresponding aldol-type adducts in high yields without side-reaction products. It should be noted that aldehydes were more reactive than acetals. (It was reported that TMSOTf selectively activated acetals rather than aldehydes in the aldol-type reaction of silyl enolates [34, 35]. Selective activation of acetals or aldehydes under certain nonbasic conditions are now under investigation in our group [36].) For example, while 3-phenylpropionaldehyde reacted with the ketene silyl acetal of methyl 2-methylpropionate (**3**) at −78°C to give the aldol adduct in an 80% yield (Scheme 8.3), no reaction occurred at −78°C in the reaction of the same ketene silyl acetal with 3-phenylpropionaldehyde dimethylacetal. The acetal reacted with the ketene silyl acetal at 0°C to room temperature to give the aldol-type adduct in a 97% yield (Scheme 8.3).

Sc(OTf)₃ can behave as a Lewis-acid catalyst even in aqueous media. Sc(OTf)₃ was stable in water and was effective in the aldol reactions of silyl enolates with aldehydes in aqueous media. The reactions of the usual aromatic and aliphatic aldehydes such as benzaldehyde and 3-phenylpropionaldehyde with silyl enolates were carried out in both aqueous and organic solvents, and water-soluble formaldehyde and chloroacetaldehyde were directly treated as water solutions with silyl enolates to afford the aldol adducts in good yields. Moreover, the catalyst could be recovered almost quantitatively from the aqueous layer after the reaction was completed. The recovered catalyst was also effective in the second reaction, and the yield of the second run was comparable to that of the first run:

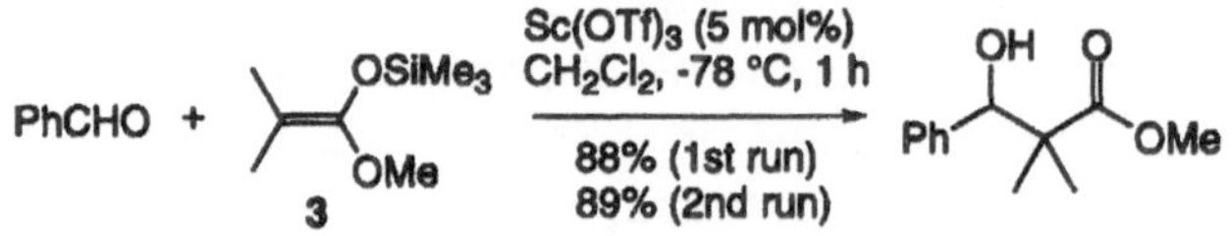

Although some of these unique characteristics are comparable to those of the lanthanide triflates, Sc(OTf)₃ was more effective than lanthanide triflates in these reactions.

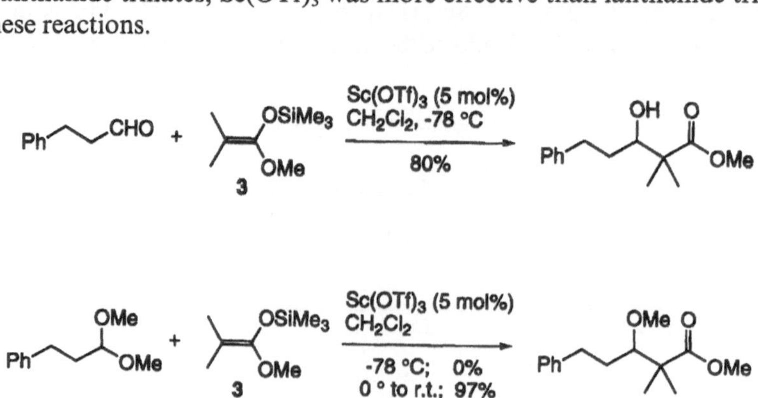

Scheme 8.3

8.3 Allylation reactions

Synthesis of homoallylic alcohols by the reaction of allyl organometallics with carbonyl compounds is one of the most important processes in organic synthesis [37]. The allylation reactions of carbonyl compounds were found to proceed smoothly under the influence of 5 mol% of Sc(OTf)$_3$ [38] by using tetraallyltin as an allylating reagent. (For the reactions of carbonyl compounds with tetraallyltin, see references [39a–d]. Quite recently Yamamoto *et al.* reported allylation reactions of aldehydes with tetraallyltin in the presence of hydrochloric acid [39e]. Lewis-acid-promoted allylation reactions of carbonyl compounds with allyltrialkyltin have also been reported [40].) Several examples are shown in Table 8.8. The reactions proceeded smoothly in the presence of only a catalytic amount of Sc(OTf)$_3$ under extremely mild conditions [11] and the adducts, homoallylic alcohols, were obtained in high yields. Ketones could also be used in the reaction (entries 4 and 5 in Table 8.8). In most cases, the reactions were successfully carried out in aqueous media. It is noteworthy that unprotected sugars reacted directly to give the adducts in high yields (entries 7–9). The allylated adducts are intermediates for the synthesis of higher sugars [41]. Moreover, an aldehyde containing water of crystallization such as phenylglyoxal mono-hydrate reacted with tetraallyltin to give the diallylated adduct in high yield (entry 10). Under the present reaction conditions, salicylaldehyde and 2-pyridinecarboxaldehyde reacted with tetraallyltin to afford the homoallylic alcohols in good yields (entries 11 and 12). Under general Lewis-acid condi-tions, these compounds react with the Lewis acids rather than with the nucleophile. Furthermore, several kinds of solvents could be used. The reactions also proceeded under non-aqueous conditions. Water-sensitive substrates under Lewis-acid conditions could be reacted in an appropriate organic solvent (entries 13 and 14).

Yb(OTf)$_3$ is also effective in the present allylation reactions. For example, 3-phenylpropionaldehyde reacted with tetraallyltin in the presence of 5 mol% of Yb(OTf)$_3$ to afford the adduct in a 90% yield.

In addition, the water–ethanol–toluene system could be successfully applied to the present allylation reactions. An example of the allylation reaction of tetraallyltin with aldehyde is shown below, and in this case also continuous use of the catalyst was realized:

1st run: 90%; 2nd run: 95%; 3rd run: 96%; 4th run: 89%

Table 8.8 Sc(OTf)$_3$-catalyzed allylation reactions of carbonyl compounds with tetraallytin[a]

Entry	Carbonyl compound	Product	Solvent	Yield (%)
1	Ph～CHO	Ph～⟶OH～	H$_2$O–THF (1:9) H$_2$O–EtOH (1:9) H$_2$O–CH$_3$CN (1:9) EtOH CH$_3$CN	92 96 96 86 94
2	PhCHO	Ph～OH～	H$_2$O–THF (1:9) CH$_3$CN	94 82
3	Ph～CHO	Ph～OH～	H$_2$O–THF (1:9) CH$_3$CN	98 94
4	Ph～C(O)～	Ph～OH～	CH$_2$Cl$_2$	78
5	Ph–C(O)–CO$_2$Me	MeO$_2$C OH Ph～	H$_2$O–THF (1:9)	87
6	Ph～C(O)	Ph～OH～	CH$_2$Cl$_2$	82
7	D-arabinose	AcO～OAc OAc / OAc OAc b)	H$_2$O–THF (1:4) H$_2$O–EtOH (1:4) H$_2$O–CH$_3$CN (1:9)	81[c] 89[d] 93[e]
8	2-deoxy-D-ribose	AcO～OAc OAc / OAc ～H$_2$O b)	H$_2$O–THF (1:9)	89[f]
9	2-deoxy-D-glucose	AcO～OAc OAc OAc / OAc b)	H$_2$O–THF (1:9)	88[f]
10	Ph–C(O)–CHO•H$_2$O	Ph OH ～⟶ OH	CH$_3$CN	78[g]
11	(2-hydroxyphenyl)CHO	(2-hydroxyphenyl)CH(OH)～	H$_2$O–THF (1:9) CH$_3$CN	quant. 90
12	(pyridin-2-yl)CHO	(pyridin-2-yl)CH(OH)～	H$_2$O–THF (1:9) CH$_3$CN	quant. 84

Table 8.8 *cont'd*

Entry	Carbonyl compound	Product	Solvent	Yield (%)
13	[structure: benzene ring with CHO and O-CH₂-OMe substituents]	[structure: benzene ring with CH(OH)-CH₂-CH=CH₂ and O-CH₂-OMe]	CH₃CN	87
14	[structure: benzene ring with CHO and CH(OEt)₂]	[structure: benzene ring with CH(OH)-CH₂-CH=CH₂ and CH(OEt)₂]	CH₃CN	90

[a]Carried out at 25°C except for entries 8 and 9 (60°C).
[b]The products were isolated after acetylation.
[c]*syn/anti* = 28/72.
[d]*syn/anti* = 27/73.
[e]*syn/anti* = 26/74.
[f]*syn/anti* = 50/50
[g]Diastereomer ratio = 88/12. Relative configuration assignment was not made.

8.4 Diels–Alder reactions

Next, the function of rare-earth triflates as catalysts in the Diels–Alder reaction was examined. Although many Diels–Alder reactions have been carried out at higher reaction temperatures without catalysts, heat-sensitive compounds cannot be employed in complex and multistep syntheses. While Lewis-acid catalysts allow the reactions to proceed at room temperature with satisfactory yields, they are often accompanied by diene polymerization and excess amounts of the catalyst are often needed to catalyze carbonyl-containing dienophiles [42].

Lanthanide triflates were also found to be efficient catalysts in the Diels–Alder reactions of carbonyl-containing dienophiles with cyclopentadiene [43]. A catalytic amount of Yb(OTf)$_3$ was enough to promote the reactions to give the corresponding adducts in high yields, and the catalyst could be easily recovered and reused.

In the Diels–Alder reactions, Sc(OTf)$_3$ was clearly more effective than the lanthanide triflates as a catalyst [44]. While in the presence of 10 mol% of Y(OTf)$_3$ or Yb(OTf)$_3$, only a trace amount of the adduct was obtained in the Diels–Alder reaction of methyl vinyl ketone (MVK) with isoprene, the reaction proceeded quite smoothly to give the adduct in a 91% yield in the presence of 10 mol% of Sc(OTf)$_3$.

Several examples of the Sc(OTf)$_3$-catalyzed Diels–Alder reactions are shown in Table 8.9. In every case, the Diels–Alder adducts were obtained in high yields with endoselectivities.

Table 8.9 Sc(OTf)₃-catalyzed Diels–Alder reactions

Dienophile	Diene	Product	Yield (%)	*endo/exo*
			95	87/13
			89	100/0
			90	–
			86	–
			97	84/16
			96	89/11
			83	>95/5
			91	–
			73	–
			83	100/0
			89	94/3
			92	–

The present Diels–Alder reactions proceeded even in aqueous media:

Sc(OTf)$_3$ (10 mol%)

THF : H$_2$O (9 : 1)

93% yield, endo/exo=100/ 0

(Indeed, some Diels–Alder reactions in water without catalyst have been reported [45].) Thus, naphthoquinone reacted with cyclopentadiene in THF–H$_2$O (9:1) at room temperature to give the corresponding adduct in a 93% yield (endo/exo = 100/0). Recovery and reuse of the catalyst were also possible in this reaction. After the reaction was completed, the aqueous layer was concentrated to give the catalyst. The recovered catalyst was effective in subsequent Diels–Alder reactions, and it should be noted that the yields of the second and even the third runs were comparable to that of the first run.

8. 5   Mannich-type reactions

8. 5.1   Reactions of imines with silyl enolates

The Mannich and related reactions provide one of the most fundamental and useful methods for the synthesis of β-aminoketones or β-aminoesters. Although the classical protocols include some severe side-reactions, new modifications using preformed iminium salts and imines have been developed [46]. Among them, reactions of imines with enolate components, especially silyl enolates, provide useful and promising methods leading to β-amino-ketones or β-aminoesters. The first report using a stoichiometric amount of TiCl$_4$ as a promoter appeared in 1977 [47] and, since then, some efficient catalysts have been developed [48].

In aqueous media, water coordinates rare-earth triflates under equilibrium conditions, and thus activation of carbonyl compounds using a catalytic amount of the Lewis acid has been performed. It was expected that, based on the same consideration, the catalytic activation of imines would be possible by using rare-earth triflates.

We tested the reactions of imines with silyl enolates in the presence of 5 mol% of lanthanide triflates (Ln(OTf)$_3$) and scandium triflate (Sc(OTf)$_3$), and selected examples are shown in Table 8.10 [49]. In most cases, the reactions proceeded smoothly in the presence of 5 mol% of Yb(OTf)$_3$ (a representative lanthanide triflate) to afford the corresponding β-aminoester derivatives in good to high yields. Yttrium triflate (Y(OTf)$_3$) was also effective, and the yield was improved when Sc(OTf)$_3$ was used as a catalyst instead of Yb(OTf)$_3$. Not only silyl enolates derived from esters, but also

Table 8.10 Reactions of imines with silyl enolates:

Imine	Silyl enolate	Ln	Yield (%)
	3	Yb	97[a]
		Y	81
		Yb	95
	3	Yb	86
		Y	78
		Yb	65
		Sc	80
		Yb	80[b]
	3	Yb	95
		Yb	67[c]
	3	Yb	88
	3	Yb	88
	3	Yb	60
	3	Yb	47

[a]Second use = 96% yield; third use = 98% yield.
[b]*syn/anti* = 18/82.
[c]*syn/anti* = 21/79.

that derived from a thioester, worked well to give the desired β-aminoesters and thioester in high yields. In the reactions of the silyl enolate derived from benzyl propionate, *anti* adducts were obtained in good selectivities. In addition, the catalyst could be recovered after the reaction was completed and could be reused.

8.5.2   One-pot synthesis of β-aminoesters from aldehydes using Ln(OTf)₃ as catalyst

While the Lewis-acid-catalyzed reactions of imines with silyl enolates are one of the most efficient methods for the preparation of β-aminoesters, many imines are hygroscopic, unstable at high temperatures and difficult to purify by distillation or column chromatography. It is desirable from a synthetic point of view that imines, generated *in situ* from aldehydes and amines, immediately react with silyl enolates and provide β-aminoesters in a one-pot reaction. However, most Lewis acids cannot be used in this reaction because they decompose or deactivate in the presence of the amines and water that exist during imine formation. Judging from the unique properties of rare-earth triflates, we planned to use them as catalysts for the above one-pot preparation of β-aminoesters from aldehydes.

A general scheme of the one-pot synthesis of β-aminoesters from aldehydes is shown below [50]:

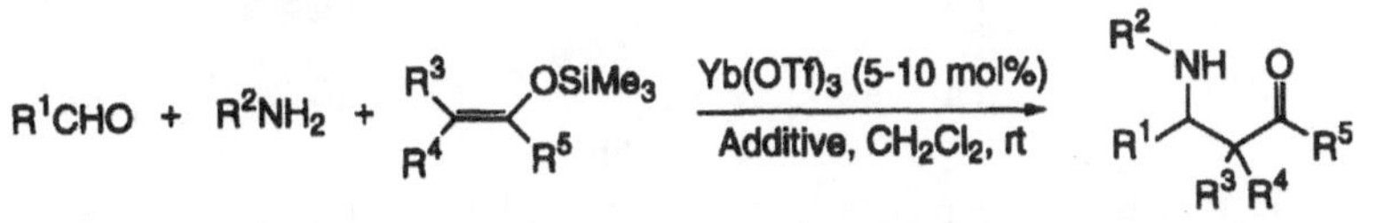

In the presence of a catalytic amount of Yb(OTf)₃ and an additive (a dehydrating reagent such as MS 4A or MgSO₄), an aldehyde was treated with an amine and then with a silyl enolate in the same vessel. Several examples are shown in Tables 8.11 and 8.12, and the following characteristic features of this reaction are noted.

(i) In every case, β-aminoesters were obtained in high yields. Silyl enolates derived from esters as well as thioesters reacted smoothly to give the adducts. No adducts between aldehydes and the silyl enolates were observed in any reaction.

(ii) A silyl enol ether derived from a ketone also worked well to afford the β-aminoketone in a high yield (Table 8.11, entry 10).

(iii) Aliphatic aldehydes reacted with amines and silyl enolates to give the corresponding β-aminoesters in high yields. In some reactions of imines, it is known that aliphatic enolizable imines prepared from aliphatic aldehydes gave poor results.

(iv) Phenylglyoxal monohydrate also worked well in this reaction. The imine derived from phenylglyoxal is unstable, and a troublesome treatment is known to be required for its use [51].

(v) The catalyst could be recovered after the reaction was completed and could be reused (first run, 91%; second run, 92%, in the reaction of benzaldehyde, *p*-anisidine and silyl enolate 3 (Table 8.11, entry 3)).

Table 8.11 One-pot synthesis of β-aminoesters from aldehydes

Entry	R^1	R^2	Silyl enolate	Additive[a]	Yield (%)
1	Ph	Ph	(OSiMe₃ / OMe structure) **3**	MS4A / MgSO₄	90 / 89
2	Ph	Bn	**3**	MS4A	85
3	Ph	p-MeOPh	**3**	MgSO₄	91[b]
4	Ph	o-MeOPh	**3**	MS4A	96
5	Ph	Ph	(OSiMe₃ / SEt structure)	MS4A	90
6	Ph	Bn	(OSiMe₃ / SEt structure)	MS4A	62[b]
7	Ph	p-MeOPh	(OSiMe₃ / SEt structure)	MS4A	79 / 84[b], 87[b,c]
8	Ph	C₄H₉	**3**	MS4A	89
9	PhCO[d]	Ph	**3**	MgSO₄	82
10	PhCO[d]	Ph	(OSiMe₃ / Ph structure)	MgSO₄	87
11	PhCH=CH	p-MeOPh	**3**	MgSO₄	92[e]
12	Ph(CH₂)₂	Bn	**3**	MgSO₄	83[f]
13	C₄H₉	Bn	**3**	MgSO₄	77[f]
14	C₈H₁₇	Bn	**3**	MgSO₄	81[f]
15	C₈H₁₇	Ph₂CH	**3**	MgSO₄	89[g]

[a]MS4A or MgSO₄ was used. Almost comparable yields were obtained in each case.
[b]CH₃CN was used as a solvent.
[c]Sc(OTf)₃ was used instead of Yb(OTf)₃.
[d]Monohydrate.
[e]C₂H₅CN, –78°C.
[f]–78°C to 0°C.
[g]0°C.

(vi) As for the diastereoselectivity of this reaction, good results were obtained after examination of the reaction conditions. While *anti* adducts were produced preferentially in the reactions of benzaldehyde, *syn* adducts were obtained with high selectivities in the reactions of aliphatic aldehydes (Table 8.12).

(vii) The high yields of the present one-pot reactions depend on the unique properties of lanthanide triflates as Lewis-acid catalysts. Although TiCl₄ and TMSOTf are known to be effective for the activation of imines [47, 52], the

Table 8.12 Diastereoselective one-pot synthesis of β-aminoesters from aldehydes:

Entry	R^1	R^2	Silyl enolate	Yield (%)	*syn/anti*[a]
1	Ph	Bn		90	1/13.3
2	Ph	Bn		78	1/9.0
3	Ph(CH₂)₂	Ph₂CH		88	8.1/1
4	C₄H₉	Ph₂CH		90	8.1/1
5	(CH₃)₂CHCH₂	Ph₂CH		86	7.3/1

[a]Determined by ^{1}H NMR anslysis.

use of even stoichiometric amounts of TiCl$_4$ and TMSOTf instead of lanthanide triflate in the present one-pot reactions gave only trace amounts of the product in both cases (Table 8.13).

(viii) One-pot preparation of a β-lactam from an aldehyde, an amine and a silyl enolate has been achieved based on the present reaction:

Table 8.13 Effect of catalysts on the reaction:

Entry	Activator		Yield (%)
1	Yb(OTf)₃	10 mol%	89
2	TiCl₄	100 mol%	trace
3	TMSOTf	100 mol%	trace

The reaction of the aldehyde, the amine and **2** was carried out under standard conditions, and Hg(OCOCF$_3$)$_2$ was then added to the same pot. The desired β-lactam was isolated in a 78% yield.

Thus, the one-pot synthesis of β-aminoesters from aldehydes has been achieved by using rare-earth triflate catalysis. The high efficiency using simple starting materials and a catalytic amount of a reusable catalyst is especially noteworthy.

8.5.3   Aqueous Mannich-type reactions

As mentioned in the previous sections, silyl enolates are excellent enolate components in the Mannich-type reactions with imines. Alternatively, it was found that vinyl ethers also reacted with imines smoothly in the presence of a catalytic amount of Ln(OTf)$_3$, to afford the corresponding β-aminoketones. In addition, the reactions proceeded smoothly by the combination of aldehydes, amines and vinyl ethers in aqueous media [53].

A general scheme for the new Mannich-type reaction is shown below:

$$R^1CHO + R^2NH_2 + \underset{R^3}{\overset{OMe}{\diagup\!\!\diagdown}} \xrightarrow[\text{THF - H}_2\text{O (9:1)}]{\text{Yb(OTf)}_3(10 \text{ mol\%})} \underset{R^1}{\overset{R^2\diagdown_{NH}\quad O}{\diagdown\!\!\diagup\!\!\diagdown R^3}}$$

The procedure is very simple. In the presence of 10 mol% of Yb(OTf)$_3$, an aldehyde, an amine and a vinyl ether were combined in a solution of THF–water (9:1) at room temperature to afford β-aminoketone.

Selected examples of the present reaction are shown in Table 8.14. In all cases, β-aminoketones were obtained in good yields. Several characteristic features are noteworthy in this reaction. The procedure is very simple, consisting of simply mixing an aldehyde, an amine, a vinyl ether and a small amount of lanthanide triflate in aqueous solution. The catalyst could be recovered after the reaction was completed and could be reused (first run, 93%; second run, 83%; third run, 87%, in the reaction of phenylglyoxal monohydrate, p-chloroaniline and 2-methoxypropene). Commercially available formaldehyde and chloroacetaldehyde water solutions were used directly and the corresponding β-aminoketones were obtained in good yields. Phenylglyoxal monohydrate, methyl glyoxylate, an aliphatic aldehyde and an α,β-unsaturated aldehyde also worked well to give the corresponding β-aminoesters in high yields. In some Mannich reactions with preformed iminium salts and imines, it is known that yields are often low because of the instability of the imines derived from these aldehydes, or troublesome treatments are known to be required for their use [51]. Other lanthanide triflates can also be used. In the reaction of phenylglyoxal monohydrate, p-chloroaniline and 2-methoxypropene, 90% (Sm(OTf)$_3$), 94% (Tm(OTf)$_3$) and 91% (Sc(OTf)$_3$) yields were obtained.

A possible mechanism for the present reaction is shown in Scheme 8.4. It

Table 8.14 Synthesis of β-aminoketones in aqueous media

R¹	R²	R³	Yield (%)
H	p-ClPh	Me	92
H	p-Ans[a]	Me	76
H	p-Ans	Ph	quant.
Ph	p-ClPh	Me	90
Ph	p-Ans	Me	74
Ph(CH₂)₂	p-ClPh	Me	55
ClCH₂	p-ClPh	Me	59
PhCH=CH	p-ClPh	Me	73
PhCO	p-ClPh	Me	93
PhCO	Ph	Me	90
PhCO	p-Ans	Me	75
PhCO	p-Ans	Ph	85
MeO₂C	p-Ans	Me	67
MeO₂C	p-Ans	Ph	58

[a]p-Ans = p-anisidine.

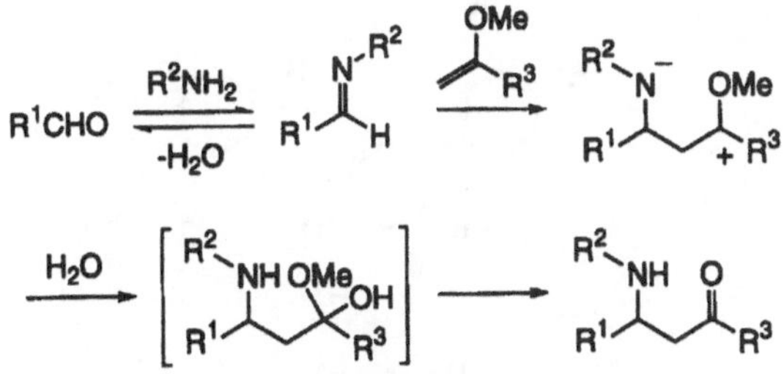

Scheme 8.4

should be noted that dehydration accompanied by imine formation and successive addition of a vinyl ether proceed smoothly in aqueous solution, and that the first aqueous Mannich-type reaction catalyzed by a lanthanide triflate has been developed: Grieco *et al.* reported *in situ* generation and trapping of immonium salts under Mannich-like conditions [54, 55]. Use of lanthanide triflate, a water-tolerant Lewis acid, is key and essential in this reaction.

8. 6 Imino Diels–Alder reactions: synthesis of pyridine and quinoline derivatives

8.6.1 Reactions of imines with dienes or alkenes

The imino Diels–Alder reaction is among the most powerful synthetic tools for constructing *N*-containing six-membered heterocycles, such as pyridines and quinolines [56]. Although Lewis acids often promote these reactions, more than stoichiometric amounts of the acids are required owing to the strong coordination of the acids to nitrogen atoms [56]. We intended to use rare-earth triflates as a catalyst in this reaction.

In the presence of 10 mol% of ytterbium triflate (Yb(OTf)$_3$, a representative lanthanide triflate), N-benzylideneaniline (**4a**) was treated with 2-trimethyl-siloxy-4-methoxy-1,3-butadiene (Danishefsky's diene, **5** [57]) in acetonitrile at room temperature. The imino Diels–Alder reaction proceeded smoothly to afford the corresponding tetrahydropyridine derivative in a 93% yield (Table 8.15). The adduct was obtained quantitatively when Sc(OTf)$_3$ was used as a catalyst. Imines **4b** and **4c** also reacted smoothly with **5** to give the corresponding adducts in high yields. Next, the reaction of **4a** with cyclopenta-

Table 8.15 Syntheses of pyridine derivatives catalyzed by Ln(OTf)$_3$:

R^1	Diene	Ln(OTf)$_3$ (mol%)	Product	Yield (%)
H (**4a**)	OSiMe$_3$	Yb(OTf)$_3$ (10)	**6a**	93 (99)[a]
OMe (**4b**)			**6b**	82
Cl (**4c**)	OMe **5**		**6c**	92
H (**4a**)		Sc(OTf)$_3$ (20)	**7a+8a**	54[b]
OMe (**4b**)			**7b+8b**	71[c]
Cl (**4c**)			**7c+8c**	50[d]
H (**4a**)		Yb(OTf)$_3$ (10)	**9a**	69 (91)[e,f]
OMe (**4b**)			**9b**	38[f]
Cl (**4c**)			**9c**	85[f]

[a] 10 mol% Sc(OTf)$_3$ was used. The reaction was carried out at 0°C.
[b] **7a** 37%, **8a** 17%.
[c] **7b** 8%, **8b** 63%.
[d] **7c** 37%, **8c** 13%.
[e] 20 mol% Sc(OTf)$_3$ was used.
[f] *cis/trans* = 94/6.

diene was performed under the same reaction conditions. It was found that the reaction course changed in this case and that a tetrahydroquinoline derivative was obtained in a 69% yield. In this reaction, the imine worked as an azadiene toward one of the double bonds of cyclopentadiene as a dienophile [51, 58]. In the reactions of 2,3-dimethylbutadiene, mixtures of tetrahydropyridine and tetrahydroquinoline derivatives were obtained.

Other examples and effects of lanthanide triflates are shown in Tables 8.16 and 8.17, respectively [59]. A vinyl sulfide, a vinyl ether and a silyl enol ether worked well as dienophiles to afford tetrahydroquinoline derivatives in high yields [60]. Fujiwara *et al.* reported a similar type of reaction [61]. As for the lanthanide triflates, heavy lanthanides such as Er, Tm and Yb gave better results.

8.6.2   Three-component coupling reactions of aldehydes, amines and dienes/alkenes

One synthetic problem in the imino Diels–Alder reactions is the imines' stability under the influence of Lewis acids. It is desirable that the imines acti-

Table 8.16 Syntheses of quinoline derivatives catalyzed by Ln(OTf)$_3$:

R^1	R^2		Alkene	Product	Yield (%)	*cis/trans*
H	H	(4a)		10a	75	57/43
OMe	H	(4b)	SPh	10b	0	–
Cl	H	(4c)		10c	quant.	nd[a]
H	OMe	(4d)		10d	70	nd[a]
H	H	(4a)		11a	96	nd[a]
OMe	H	(4b)	OEt	11b	77	67/33
Cl	H	(4c)		11c	95	nd[a]
OMe	H	(4b)	Ph, OSiMe$_3$	12	quant.	83/17[b]

[a]Not determined.
[b]Relative configuration assignment was not made.

Table 8.17 Effects of Ln(OTf)$_3$ on the reaction:

Ln	Yield (%)
Sc	94
Y	60
La	45
Pr	57
Nd	60
Sm	63
Eu	63
Gd	76
Dy	63
Ho	64
Er	97
Tm	92
Yb	85
Lu	72

vated by Lewis acids are immediately trapped by dienes or dienophiles [54, 55]. In 1989, Weinreb *et al.* reported a convenient procedure for the imino Diels–Alder reaction of an aldehyde and a 1,3-diene with *N*-sulfinyl *p*-toluenesulfonamide via *N*-sulfonyl imine produced *in situ*, by using a stoichiometric amount of BF$_3$•OEt$_2$ as a promoter [62].

Bearing in mind the usefulness and efficiency of one-pot procedures, three-component coupling reactions between aldehydes, amines and alkenes via imine formation and imino Diels–Alder reactions were examined by using lanthanide triflate as a catalyst.

In the presence of 10 mol% of ytterbium triflate and magnesium sulfate, benzaldehyde was treated with aniline and **5** successively in acetonitrile at room temperature. The three-component coupling reaction proceeded smoothly to afford the corresponding tetrahydropyridine derivative in an 80% yield. It is noteworthy that Yb(OTf)$_3$ kept its activity and effectively catalyzed the reaction even in the presence of water and the amine. (When typical Lewis acids such as BF$_3$•OEt$_2$ and ZnCl$_2$ (100 mol%) were used instead of the rare-earth triflates under the same reaction conditions, lower yields were observed (23% and 12%, respectively).) Use of Sc(OTf)$_3$ slightly improved the yield. Other examples of the three-component coupling reaction are shown in Table 8.18. In the reaction between benzaldehyde, anisidine and cyclopentadiene under the same reaction conditions, the

Table 8.18 One-pot syntheses of pyridine and quinoline derivatives:

$$R^1CHO + \underset{H_2N}{\overset{R^2}{\bigcirc}} + \text{diene or alkene} \xrightarrow[\text{CH}_3\text{CN, rt.}]{\substack{\text{Yb(OTf)}_3 \\ (5\text{-}20 \text{ mol\%})}} \text{pyridine or quinoline derivatives}$$

R^1	R^2	Diene or alkene	Product	Yield (%)	*cis/trans*
Ph	H	**5**	**6a**	80 (83)[a]	–
		(cyclopentene)	**9a**	56	94/6
		SPh (vinyl)	**10a**	70	nd[b]
	Cl		**10c**	quant.	nd[b]
	H	OEt (vinyl)	**11a**	60	79/21
PhCO	H	**2**	**13**	76	–
	H	(cyclopentene)	**14a**	94 (97)[c]	96/4 (96/4)
	OMe		**14b**	94	94/6
	Cl		**14c**	quant.	96/4
MeO$_2$C	H		**15a**	82	99/1
	Cl		**15c**	84	99/1
	Cl	SPh (vinyl)	**16**	65	nd[b]
H[d]	Cl	(cyclopentene)	**17**	90[c]	–

[a]Sc(OTf)$_3$ (10 mol%) was used.
[b]Not determined.
[c]The reactions were carried out in aqueous solution (H$_2$O–EtOH–toluene = 1:9:4).
[d]Commercial formaldehyde water solution was used.

reaction course changed and the tetrahydroquinoline derivative was obtained in a 56% yield. A vinyl sulfide, a vinyl ether and a silyl enol ether worked well as dienophiles to afford tetrahydroquinoline derivatives in high yields. Phenylglyoxal monohydrate reacted with amines and **5** or cyclopentadiene to give the corresponding tetrahydropyridine or quinoline derivatives in high yields. As mentioned in the previous section, the imine derived from phenylglyoxal is known to be highly hygroscopic and its purification by distillation or chromatography is very difficult owing to its instability [51]. Moreover, the three-component coupling reactions proceeded smoothly in aqueous solution, and commercial formaldehyde water solution could be used directly. Most lanthanide triflates tested were effective in the three-component coupling reactions (Table 8.19). These reactions provide very useful routes for the synthesis of pyridine and quinoline derivatives.

8.6.3   Reaction mechanism

In the reactions of **4a–c** with cyclopentadiene, a vinyl sulfide, or a vinyl ether (**4a–c** work as azadienes), **4c** gave the best yields, while the yields using **4b** were lowest. The HOMO and LUMO energies and coefficients of **4a–c** and protonated **4a–c** are summarized in Table 8.20. These data do not correspond to the differences in reactivity between **4a**, **4b** and **4c** if the reactions are postulated to proceed via concerted [4+2] cycloaddition. On the other hand,

Table 8.19 Effects of $Ln(OTf)_3$ on the reaction:

Ln	Yield (%)
Sc	63
Y	77
La	88
Pr	75
Nd	97
Sm	91
Eu	87
Gd	91
Dy	87
Ho	76
Er	84
Tm	84
Yb	94
Lu	80

Table 8.20 HOMO and LUMO energies and coefficient of **7a–c**[a]:

	HOMO (eV)	Coefficient			
		C(1)	N(2)	C(3)	C(4)
4a	−8.97	0.32	0.35	−0.34	−0.24
4b	−8.68	0.32	0.29	−0.40	−0.24
4c	−8.93	0.30	0.31	−0.35	−0.22
4a–H⁺	−13.19	0.15	0.31	−0.39	−0.19
4b–H⁺	−12.53	0.20	0.18	−0.46	−0.08
4c–H⁺	−12.58	0.16	0.15	−0.37	−0.10
	LUMO (eV)	Coefficient			
		C(1)	N(2)	C(3)	C(4)
4a	−0.69	0.43	−0.38	−0.30	−0.27
4b	−0.64	0.43	−0.38	−0.28	−0.28
4c	−0.85	0.43	−0.35	−0.33	−0.27
4a–H⁺	−5.75	0.63	−0.48	−0.09	−0.21
4b–H⁺	−5.58	0.62	−0.49	−0.07	−0.21
4c–H⁺	−5.77	0.63	−0.47	−0.19	−0.21

[a]Calculated with MOPAC Ver. 6.01 using the PM3 Hamiltonian. (MOPAC Ver. 6: Stewart, J.J.P., *QCPE Bull.*, 1989, **10**, 9. Revised as Ver. 6.01 by Tsuneo Hirano, University of Tokyo, for HITAC and UNIX machines: Hirano, T., *JCPE Newslett.* 1989, **10**, 1.)

the high reactivity of **4c** toward electrophiles compared to **4a** and **4b** may be accepted by assuming a stepwise mechanism.

The reaction of **4a** with 2-methoxypropene was tested in the presence of Yb(OTf)₃ (10 mol%). The main product was tetrahydroquinoline derivative **18a**, and small amounts of quinoline **19a** and β-aminoketone dimethylacetal **20a** were also obtained (Scheme 8.5, top). On the other hand, the three-component coupling reaction between benzaldehyde, aniline and 2-methoxypropene gave only a small amount of tetrahydroquinoline derivative **18a**, and the main products in this case were β-aminoketone **21a** and its dimethylacetal **20a** (Scheme 8.5, bottom). Similar results were obtained in the reaction of **4b** with 2-methoxypropene, and in the three-component coupling reaction between benzaldehyde, anisidine and 2-methoxypropene (Scheme 8.6).

A possible mechanism of these three-component coupling reactions is shown in Scheme 8.7. Intermediate **22** is quenched by water and methanol generated *in situ* to afford **20** and **21**, respectively. While **18** is predominantly

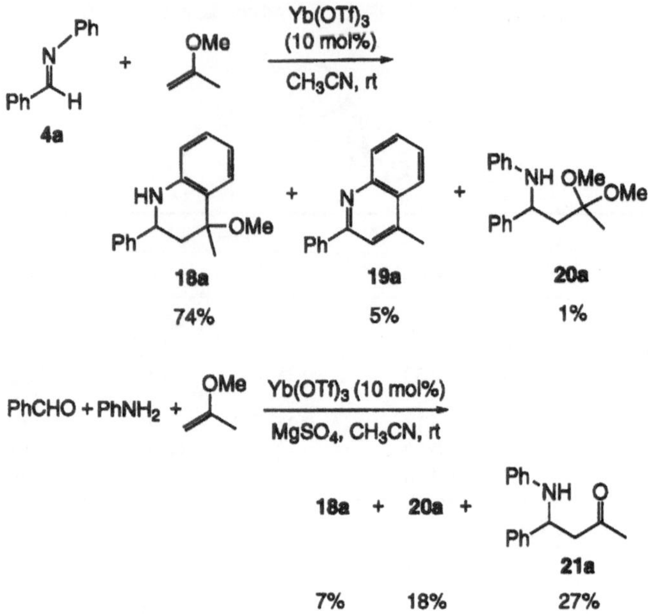

Scheme 8.5

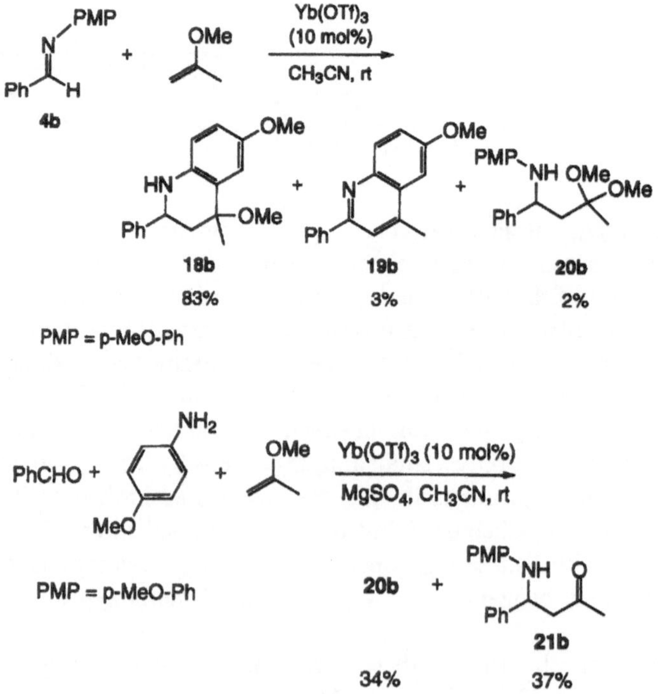

Scheme 8.6

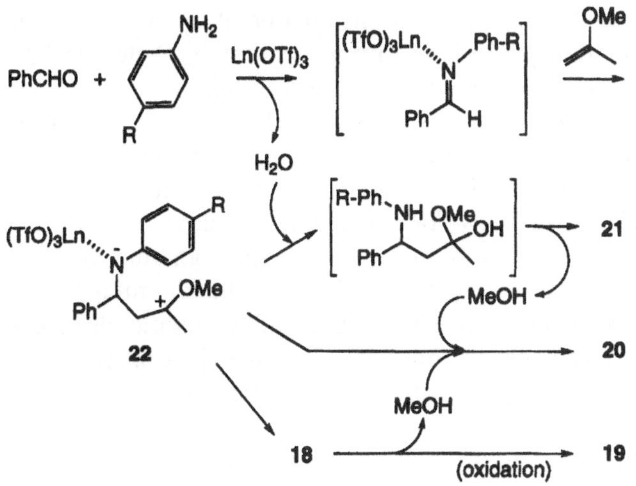

Scheme 8.7

formed from **22** under anhydrous conditions, formation of **20** and **21** predominated in the presence of even a small amount of water. It is noted that these results suggest a stepwise mechanism in these types of imino Diels–Alder reactions [63].

Thus, a new type of Lewis acid, rare-earth triflates, is quite effective for the catalytic activation of imines, and has achieved imino Diels–Alder reactions of imines with dienes or alkenes. The unique reactivities of imines, which work as both dienophiles and azadienes under certain conditions, were also revealed. Three-component coupling reactions between aldehydes, amines and dienes/alkenes were successfully carried out by using lanthanide triflates as catalysts to afford pyridine and quinoline derivatives in high yields. The triflates were stable and kept their activity even in the presence of water and amines. According to these reactions, many substituted pyridines and quino-lines can be prepared directly from aldehydes, amines and dienes/alkenes. A stepwise reaction mechanism in these reactions was suggested from the exper-imental results.

8.7   Asymmetric aza Diels–Alder reactions: synthesis of tetrahydroquinoline derivatives using a chiral lanthanide Lewis acid as catalyst

Recently, asymmetric reactions using chiral Lewis acids have been demon-strated to achieve several highly enantioselective carbon–carbon bond-forming processes using catalytic amounts of chiral sources [64, 65]. However, chiral Lewis-acid-catalyzed asymmetric reactions of nitrogen-containing substrates are rare, probably because most chiral Lewis acids would be trapped by the basic nitrogen atoms to block the catalytic cycle. For example, aza Diels–Alder reactions are one of the most basic and versatile reactions for the synthesis of nitrogen-containing heterocyclic compounds [56, 66]. Although asymmetric versions using chiral auxiliaries or a stoichiometric amount of a chiral Lewis acid have been reported [67], examples using a catalytic amount of a chiral source are unprecedented.

In the previous section, rare-earth triflates were shown to be excellent catalysts for achiral aza Diels–Alder reactions. While stoichiometric amounts of Lewis acids are required in many cases, a small amount of the triflate effectively catalyzes the reactions. On the other hand, chiral lanthanide Lewis acids have been developed to realize highly enantioselective Diels–Alder reactions of 2-oxazolidin-1-one with dienes [68]. Bearing these hopeful results in mind, we started to investigate catalytic asymmetric aza Diels–Alder reactions.

First, the reaction of N-benzylideneaniline with cyclopentadiene was performed under the influence of 20 mol% of a chiral ytterbium Lewis acid prepared from ytterbium triflate (Yb(OTf)$_3$), (R)-(+)-1,1'-binaphthol (BINOL) and trimethylpiperidine (TMP). The reaction proceeded smoothly at room temperature to afford the desired tetrahydroquinoline derivative in a 53% yield. However, no chiral induction was observed. At this stage, it was indicated that bidentate coordination between a substrate and a chiral Lewis acid would be necessary for reasonable chiral induction. N-Benzylidene-2-hydroxyaniline (23a) was then prepared, and the reaction with cyclopentadiene (24a) was examined. It was found that the reaction proceeded smoothly to afford the corresponding 8-hydroxyquinoline derivative (25a) in a high yield. (Some interesting biological activity has been reported in 8-hydroxyquinoline derivatives [69].) The enantiomeric excess of the cis-adduct in the first trial was only 6%. However, the selectivity increased when diazabicyclo-[5.4.0]-undec-7-ene (DBU) was used instead of TMP (Table 8.21). It was also indicated that the phenolic hydrogen of 23a would interact with DBU, which should interact with the hydrogen of (R)-(+)-BINOL [70], to decrease the selectivity. We then examined additives that interact with the phenolic hydrogen of 23a. When 20 mol% of N-methylimidazole (MID) was used, 91% ee of the cis adduct was obtained; however, the chemical yield was low. Other additives were screened, and it was found that the desired tetrahydroquinoline derivative was obtained in a 92% yield with high selectivities ($cis/trans \geq$ 99/1, 71% ee), when 2,6-di-t-butyl-4-methylpyridine (DTBMP) was used.

Other substrates were tested, and the results are summarized in Table 8.22 [71]. Vinyl ethers (24b–d) also worked well to afford the corresponding tetrahydroquinoline derivatives (25b–e) in good to high yields with good to excellent diastereo- and enantioselectivities (entries 1–9 in Table 8.22). Use of 10 mol% of the chiral catalyst also gave the adduct in high yields and selectivities (entries 2 and 6). As for additives, 2,6-di-t-butylpyridine (DTBP) gave the best result in the reaction of imine 23a with ethyl vinyl ether (24b), while higher selectivities were obtained when DTBMP or 2,6-diphenylpyridine

Table 8.21 Effect of additive on the reaction:

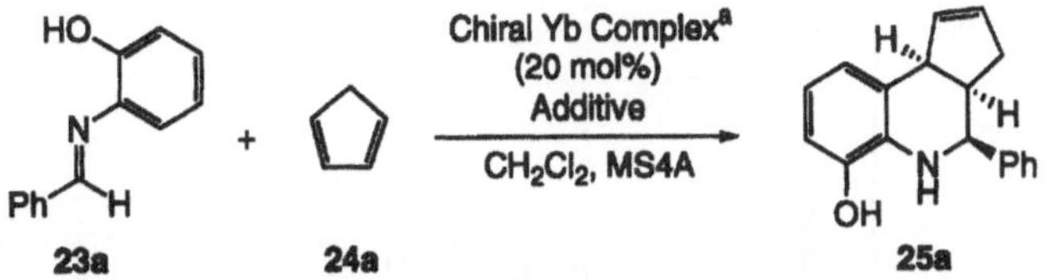

Additive[b] (mol%)	Temp. (°C)	Yield (%)	cis/trans	ee (%) (cis)
–	0	71	98/2	62
–	–15 to 0	48	99/1	68
MID (20)	–15 to 0	21	98/2	91
DTBP (20)	0	49	95/5	31
DTBP (100)	0	67	99/1	61
DMP (100)	0	14	98/2	56
DTBMP (100)	–15	82	>99/1	70
DTBP (100)	–15	92	>99/1	71

[a]Prepared from Yb(OTf)$_3$, (R)–(+)-BINOL and DBU.
[b]MID = 1–methylimidazole, DTBP = 2,6-di-t-butylpyridine, DMP = 2,6-dimethylpyridine, DTBMP = 2,6-di-t-butyl-4-methylpyridine.

(DPP) was used in the reaction of imine **23a** with **23b**. This could be explained by the slight difference in the asymmetric environment created by Yb(OTf)$_3$, (R)-(+)-BINOL, DBU and the additive (see below). While use of butyl vinyl ether (**24c**) decreased the selectivities (entry 7), dihydrofuran (**24d**) reacted smoothly to achieve high levels of selectivity (entries 8 and 9). It was found that the imine (**23c**) prepared from cyclohexanecarboxaldehyde and 2-hydroxyaniline was unstable and difficult to purify. The asymmetric aza Diels–Alder reaction was successfully carried out using the three-component coupling procedure (successively adding the aldehyde, the amine and cyclopentadiene) in the presence of Sc(OTf)$_3$ (instead of Yb(OTf)$_3$), (R)-(+)-BINOL, DBU and DTBMP.

The assumed transition state of this reaction is shown in Scheme 8.8.

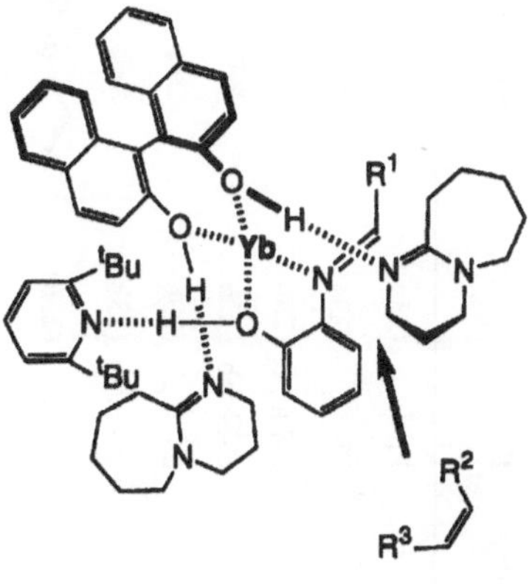

Scheme 8.8

Table 8.22 Asymmetric synthesis of tetrahydroquinoline derivatives:

Entry	R¹		Alkene	Additive[a]	Amount of catalyst (mol%)	Temp. (°C)	Product	Yield (%)	cis/trans	ee (%) (cis)
1	Ph	(23a)	OEt (24b)	DTBP	20	−45	25b	58	94/6	61
2	Ph	(23a)	24b	DTBP	10	−45	25b	52	94/6	77
3	α-Naph	(23b)	24b	DTBP	20	−15	25c	69	>99/1	86
4	α-Naph	(23b)	24b	DPP[b]	20	−15	25c	65	99/1	91
5	α-Naph	(23b)	24b	DTBMP	20	−15	25c	74	>99/1	91
6	α-Naph	(23b)	24b	DTBMP	10	−15	25c	62	98/2	82
7	α-Naph	(23b)	OBu (24c)	DTBMP	20	−15	25d	80	66/34	70
8	α-Naph	(23b)	(24d)	DTBMP	20	−15	25e	90	91/9	78
9	α-Naph	(23b)	24d	DPP[b]	20	−15	25e	67	93/7	86
10	α-Naph	(23b)	(24a)	DTBMP	20	−15	25f	69	>99/1	68
11[c]	c-C$_6$H$_{11}$	(23c)	24a	DTBMP	20	−15	25g	58	>99/1	73

[a]See Table 8.21.
[b]2,6-Diphenylpyridine.
[c]Sc(OTf)$_3$ was used; see text.

(Triflate anions are omitted for clarity.) Yb(OTf)$_3$, (R)-(+)-BINOL and DBU form a complex with two hydrogen bonds, and the axial chirality of (R)-(+)-BINOL is transferred via the hydrogen bonds to the amine parts. The additive would interact with the phenolic hydrogen of the imine, which is fixed by bidentate coordination to Yb(III). Since the top face of the imine is shielded by the amine, the dienophiles approach from the bottom face to achieve high levels of selectivity.

Thus, we have developed catalytic asymmetric aza Diels–Alder reactions of imines with alkenes using a chiral lanthanide Lewis acid, to afford 8-hydroxyquinoline derivatives in high yields with high diastereo- and enantioselectivities. The characteristic points of this reaction are as follows. (i) Asymmetric aza Diels–Alder reactions between achiral azadienes and dienophiles have been achieved using a catalytic amount of a chiral source. (ii) The unique reaction pathway, in which the chiral Lewis acid activates not dienophiles but dienes, is revealed. In most asymmetric Diels–Alder reactions reported using chiral Lewis acids, the Lewis acids activate dienophiles [64, 65]. However, inverse electron-demand asymmetric Diels–Alder reactions of 2-pyrone derivatives have been reported [72]. (iii) A unique lanthanide complex including an azadiene and an additive, which is quite different from the conventional chiral Lewis acids, has been developed.

8.8 Micellar systems

Quite recently, we have found that scandium-triflate-catalyzed aldol reactions of silyl enol ethers with aldehydes were successfully carried out in micellar systems. While the reactions proceeded sluggishly in water, remarkable enhancement of the reactivity was observed in the presence of a small amount of a surfactant. In these systems, versatile carbon–carbon bond-forming reactions proceeded in water without using any organic solvents.

As mentioned in the previous section, lanthanide and scandium triflates (Ln(OTf)$_3$ and Sc(OTf)$_3$) are stable Lewis acids in water, and aldol reactions of silyl enol ethers with aldehydes proceed smoothly in aqueous media in the presence of a catalytic amount of the lanthanide salt. While the reactions were successfully carried out in THF–water or toluene–ethanol–water, lower yields were obtained in pure water. In the course of our investigations to develop new synthetic reactions, especially carbon–carbon bond-forming reactions, in aqueous media, we have found that such reactions proceeded smoothly in micellar systems.

Lewis-acid catalysis in micellar systems was first found in the model reaction of the silyl enol ether of propiophenone (1) with benzaldehyde (Table 8.23). While the reaction proceeded sluggishly in the presence of 0.2 eq. Yb(OTf)$_3$ in water, remarkable enhancement of the reactivity was observed when the reaction was carried out in the presence of 0.2 eq.

Table 8.23 Effect of Ln(OTf)$_3$ and surfactants on the reaction:

PhCHO + [structure: OSiMe$_3$, Ph, **1**] $\xrightarrow[\text{H}_2\text{O}]{\substack{\text{cat. Ln(OTf)}_3 \\ \text{surfactant}}}$ [product: O, OH, Ph, Ph]

Ln(OTf)$_3$ (eq.)	Surfactant (eq.)	Time (h)	Yield (%)
Yb(OTf)$_3$/0.2	–	48	17
Yb(OTf)$_3$/0.2	SDS/0.04	48	12
Yb(OTf)$_3$/0.2	SDS/0.1	48	19
Yb(OTf)$_3$/0.2	SDS/0.2	48	50
Yb(OTf)$_3$/0.2	SDS/1.0	48	22
Sc(OTf)$_3$/0.2	SDS/0.2	17	73
Sc(OTf)$_3$/0.1	SDS/0.2	4	88
Sc(OTf)$_3$/0.1	Triton X-100[a]/0.2	60	89
Sc(OTf)$_3$/0.1	CTAB/0.2	4	trace

[a]The structure of this is:

[structure: H$_3$C-C(CH$_3$)(CH$_3$)-CH$_2$-C(CH$_3$)(CH$_3$)-[benzene ring]-(OCH$_2$CH$_2$)$_x$OH]   Triton X-100

Yb(OTf)$_3$ in an aqueous solution of sodium dodecylsulfate (SDS, 0.2 mol, 35 mM), and the corresponding aldol adduct was obtained in a 50% yield. In the absence of the Lewis acid and the surfactant (water-promoted conditions), only 20% yield of the aldol adduct was isolated after 48 h, while a 33% yield of the aldol adduct was obtained after 48 h in the absence of the Lewis acid in an aqueous solution of SDS. The amounts of the surfactant also influenced the reactivity, and the yield was improved when scandium triflate (Sc(OTf)$_3$) was used as a Lewis-acid catalyst. Judging from the critical micelle concentration, micelles would be formed in these reactions, and it is noteworthy that the Lewis-acid-catalyzed reactions proceeded smoothly in micellar systems. (Although several organic reactions in micelles have been reported, there has been no report on Lewis-acid catalysis in micelles, to the best of our knowledge. In addition, judging from the amount of the surfactant used in the present case, the aldol reaction would proceed not only in the micelles. The precise reaction mechanism is now under investigation [73].) It was also found that the surfactants influenced the yield, and that Triton X-100 was effective in the aldol reaction (but required long reaction times), while only a trace amount of the adduct was detected when using cetyltrimethylammonium bromide (CTAB) as a surfactant. (Hydrolysis of silyl enol ether **1** took place when lower yields were observed in Table 8.23.)

Several examples of the Sc(OTf)$_3$-catalyzed aldol reactions in micellar systems are shown in Table 8.24. Not only aromatic but also aliphatic and α,β-unsaturated aldehydes reacted with silyl enol ethers to afford the corresponding aldol adducts in high yields. Formaldehyde water solution also

Table 8.24 Sc(OTf)$_3$-catalyzed aldol reactions in micellar systems:

R^{1}CHO + [OSiMe$_3$ silyl enol ether with R^2, R^3] $\xrightarrow[\text{H}_2\text{O, rt}]{\begin{array}{c}\text{Sc(OTf)}_3\ (0.1\ \text{eq.})\\ \text{SDS}\ (0.2\ \text{eq.})\end{array}}$ [product O, OH with R^2, R^3, R^1]

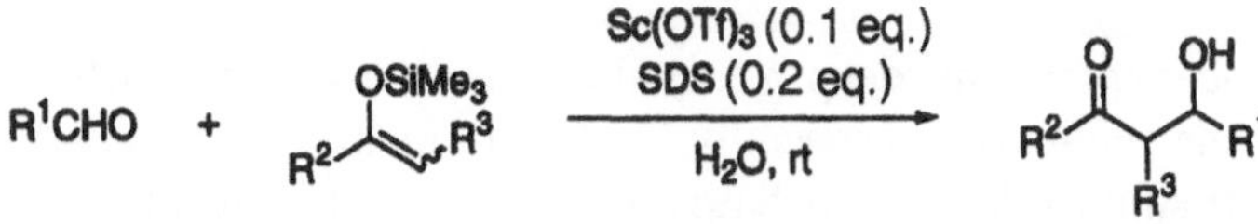

Aldehyde	Silyl enol ether	Yield (%)
PhCHO	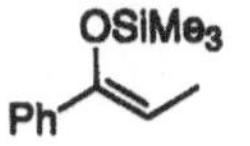 OSiMe$_3$, Ph —— **1**	88[a]
 Ph⌒⌒CHO	**1**	86[b]
 Ph⌒=CHO	**1**	88[c]
HCHO	**1**	82[d]
PhCHO	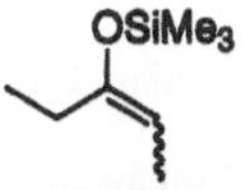 OSiMe$_3$ (trisubstituted enol ether)	88[e]
PhCHO	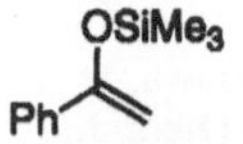 OSiMe$_3$, Ph	75[f,g]
PhCHO	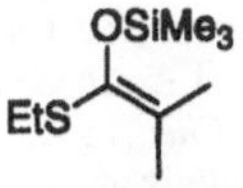 OSiMe$_3$, EtS	94
PhCHO	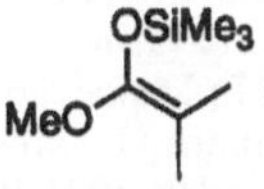 OSiMe$_3$, MeO **3**	84[g]

[a]*syn/anti* = 50/50.
[b]*syn/anti* = 45/55.
[c]*syn/anti* = 41/59.
[d]Commercially available HCHO aq. (3 ml), **1** (0.5 mmol), Sc(OTf)$_3$ (0.1 mmol) and SDS (0.1 mmol) were combined.
[e]*syn/anti* = 57/43.
[f]Sc(OTf)$_3$ (0.2 mol) was used.
[g]Additional silyl enolate (1.5 mol) was charged after 6 h.

worked well. To our great surprise, ketene silyl acetal **3**, which is known to hydrolyze very easily even in the presence of a small amount of water, reacted with an aldehyde in the present micellar system to afford the corresponding aldol adduct in a high yield.

It should be noted that the reactions were successfully carried out in water without using any organic solvents. Use of the reusable scandium catalyst and water as a solvent would result in clean and environmentally friendly systems. Further studies to develop other synthetic reactions in micellar systems and also to clarify the precise mechanism in these reactions are now actively in progress in our laboratories.

Acknowledgements

I thank and express my deep gratitude to my co-workers (whose names appear in the references), especially Dr Iwao Hachiya and Messrs Haruro Ishitani, Mitsuharu Araki and Satoshi Nagayama. Financial support by a Grant-in-Aid for Scientific Research from the Ministry of Education, Science, Sports, and Culture, Japan, is also acknowledged.

References

[1] Schinzer, D., Ed., *Selectivities in Lewis Acid Promoted Reactions*; Kluwer Academic, Dordrecht, 1989.
[2] Reviews: (a) Howells, R.D.; McCown, J.C., *Chem. Rev.*, 1977, **77**, 69; (b) Emde, H.; Domsch, D.; Feger, H.; Frick, U.; Goyz, A.; Hergott, H.H.; Hofmann, K.; Kober, W.; Krageloh, K.; Oesterle, T.; Steppan, W.; West, W.; Simchen, G., *Synthesis*, 1982, 1; (c) Stang, P.J.; Hanack, M.; Subramanian, L.R., *Synthesis*, 1982, 85; (d) Stang, P.J.; White, M.R., *Aldrichim. Acta*, 1983, 15.
See also, R_2BOTf: (e) Inoue, T.; Mukaiyama, T., *Bull. Chem. Soc. Jpn.*, 1980, **53**, 174. R_3SiOTf: (f) Murata, S.; Suzuki, M.; Noyori, R., *J. Am. Chem. Soc.*, 1980, **102**, 3248. $Sn(OTf)_2$: (g) Mukaiyama, T.; Iwasawa, N.; Stevens, R.W.; Haga, T., *Tetrahedron*, 1984, **40**, 1381. $R_2Sn(OTf)_2$: (h) Sato, T.; Otera, J.; Nozaki, H., *J. Am. Chem. Soc.* 1990, **112**, 901. $Al(OTf)_3$: (i) Minowa, N.; Mukaiyama, T., *Chem. Lett.*, 1987, 1719; Olah, G.A.; Farooq, O.; Morteza, S.; Farnia, F.; Olah, J.A., *J. Am. Chem. Soc.*, 1988, **110**, 2560. $Zn(OTf)_2$: (j) Corey, E.J.; Shimoji, K., *J. Am. Chem. Soc.*, 1983, **105**, 1662.
[3] (a) Kobayashi, S.; Tamura, M.; Mukaiyama, T., *Chem. Lett.*, 1988, 91; (b) Mukaiyama, T.; Shimpuku, T.; Takashima, T.; Kobayashi, S., *Chem. Lett.*, 1989, 145; (c) Kobayashi, S.; Uchiro, H.; Fujishita, Y.; Shiina, I.; Mukaiyama, T., *J. Am. Chem. Soc.*, 1991, **113**, 4247; (d) Kobayashi, S.; Uchiro, H.; Shiina, I.; Mukaiyama, T., *Tetrahedron*, 1993, **49**, 1761; (e) Kobayashi, S.; Kawasuji, T., *Synlett*, 1993, 911.
[4] Thom, K.F., US Patent 3615169, 1971; *Chem. Abstr.*, 1972, **76**, 5436a.
[5] Review: Molander, G.A., *Chem. Rev.*, 1992, **92**, 29.
[6] Baes, Jr., C.F.; Mesmer, R.E., *The Hydrolysis of Cations*; Wiley, New York, 1976, p. 129.
[7] (a) Forsberg, J.H.; Spaziano, V.T.; Balasubramanian, T.M.; Liu, G.K.; Kinsley, S.A.; Duckworth, C.A.; Poteruca, J.J.; Brown, P.S.; Miller, J.L., *J. Org. Chem.*, 1987, **52**, 1017. See also: (b) Collins, S.; Hong, Y., *Tetrahedron Lett.*, 1987, **28**, 4391; (c) Almasio, M.-C.; Arnaud-Neu, F.; Schwing-Weill, M.-J., *Helv. Chim. Acta*, 1983, **66**, 1296; (d) Harrowfield, J. M.; Kepert, D. L.; Patrick, J. M.; White, A.H., *Aust. J. Chem.*, 1983, **36**, 483.
[8] Mukaiyama, T.; Narasaka, K.; Banno, T., *Chem. Lett.*, 1973, 1011.
[9] Mukaiyama, T.; Banno, K.; Narasaka, K., *J. Am. Chem. Soc.*, 1974, **96**, 7503.
[10] Mukaiyama, T., *Org. React.*, 1982, **28**, 203.
[11] Kobayashi, S.; Murakami, M.; Mukaiyama, T., *Chem. Lett.*, 1985, 1535.
[12] Kawai, M.; Onaka, M.; Izumi, Y., *Chem. Lett.*, 1986, 1581; *Bull. Chem. Soc. Jpn.*, 1988, **61**, 1237.
[13] Noyori, R.; Yokoyama, K.; Sakata, J.; Kuwajima, I.; Nakamura, E.; Shimizu, M., *J. Am. Chem. Soc.*, 1977, **99**, 1265; Nakamura, E.; Shimizu, M.; Kuwajima, I.; Sakata, J.; Yokoyama, K.; Noyori, R., *J. Org. Chem.* 1983, **48**, 932.
[14] (a) Takai, K.; Heathcock, C.H., *J. Org. Chem.*, 1985, **50**, 3247; (b) Vougioukas, A. E.; Kagan, H.B., *Tetrahedron Lett.*, 1987, **28**, 5513; (c) Gong, L.; Streitwieser, A., *J. Org. Chem.*, 1990, **55**, 6235; (d) Mikami, K.; Terada, M.; Nakai, T., *J. Chem. Soc., Chem. Commun.*, 1993, 343 and references therein.
[15] (a) Lubineau, A., *J. Org. Chem.*, 1986, **51**, 2142; (b) Lubineau, A.; Meyer, E., *Tetrahedron*, 1988, **44**, 6065.
[16] (a) Hajos, Z.G.; Parrish, D.R., *J. Org. Chem.*, 1973, **38**, 3244; (b) Stork, G.; Isobe, M., *J. Am. Chem. Soc.*, 1975, **97**, 4745; (c) Lucast, D.H.; Wemple, J., *Synthesis*, 1976, 724;

(d) Ono, N.; Miyake, H.; Fujii, M.; Kaji, A., *Tetrahedron Lett.*, 1983, **24**, 3477; (e) Tsuji, J.; Nisar, M.; Minami, I., *Tetrahedron Lett.*, 1986, **27**, 2483; (f) Larsen, S.D.; Grieco, P.A.; Fobare, W.F., *J. Am. Chem. Soc.*, 1986, **108**, 3512.

[17] (a) Snider, B. B.; Rodini, D. J.; Kirk, T. C.; Cordova, R., *J. Am. Chem. Soc.*, 1982, **104**, 555; (b) Snider, B.B., in *Selectivities in Lewis Acid Promoted Reactions*, Schinzer, D., Ed.; Kluwer Academic, London, 1989, pp. 147–67; (c) Maruoka, K.; Conception, A.B.; Hirayama, N., Yamamoto, H., *J. Am. Chem. Soc.*, 1990, **112**, 7422; (d) Maruoka, K.; Concepcion, A. B.; Murase, N.; Oishi, M.; Yamamoto, H., *J. Am. Chem. Soc.*, 1993, **115**, 3943.

[18] Murata, S.; Suzuki, M.; Noyori, R., *Tetrahedron Lett.*, 1980, **21**, 2527.

[19] Kobayashi, S., *Chem. Lett.*, 1991, 2187.

[20] Gaut, H; Skoda, J., *Bull. Soc. Chim. Fr.*, 1946, **13**, 308.

[21] Kobayashi, S.; Hachiya, I., *Tetrahedron Lett.*, 1992, 1625.

[22] Kobayashi, S.; Hachiya, I., *J. Org. Chem.*, 1994, **59**, 3590.

[23] Haggin, J., *Chem. Eng. News*, 1994, 18 April, p. 22.

[24] Heathcock, C. H., in *Asymmetric Synthesis*, vol. 3, Part B, Morrison, J.D., Ed.; Academic Press, New York, 1984, p. 111.

[25] Margerum, D.W.; *et al.*, in *Coordination Chemistry*, vol. 2, Martell, A.E., Ed., ACS Monograph No. 174; American Chemical Society, Washington, DC, 1978.

[26] Murata, S.; Suzuki, M.; Noyori, R., *J. Am. Chem. Soc.*, 1980, **102**, 3248.

[27] Colvin, E.W., *Silicon in Organic Synthesis*; Butterworths, London, 1981.

[28] Weber, W.P., *Silicon Reagents for Organic Synthesis*; Springer-Verlag, Berlin, 1983.

[29] Saigo, K.; Osaki, M.; Mukaiyama, T., *Chem. Lett.*, 1975, 989.

[30] Kobayashi, S.; Hachiya, I.; Takahori, T., *Synthesis*, 1993, 371.

[31] Kobayashi, S.; Hachiya, I.; Yamanoi, Y., *Bull. Chem. Soc. Jpn*, 1994, **67**, 2342.

[32] Cotton, F.A.; Wilkinson, G., *Advanced Inorganic Chemistry*, 5th Edn.; Wiley, New York, 1988, p. 973.

[33] Kobayashi, S.; Hachiya, I.; Ishitani, H.; Araki, M., *Synlett*, 1993, 472.

[34] Noyori, R.; Murata, S.; Suzuki, M., *Tetrahedron*, 1981, **37**, 3899.

[35] Mukai, C.; Hashizume, S.; Nagami, K.; Hanaoka, M., *Chem. Pharm. Bull.*, 1990, **38**, 1509. See also reference [25].

[36] Mukaiyama, T.; Ohno, T.; Han, J. S.; Kobayashi, S., *Chem. Lett.*, 1991, 949.

[37] Review: Yamamoto, Y.; Asao, N., *Chem. Rev.*, 1993, **93**, 2207.

[38] Hachiya, I.; Kobayashi, S., *J. Org. Chem.*, 1993, **58**, 6958.

[39] (a) Peet, W.G.; Tam, W., *J. Chem. Soc., Chem. Commun.*, 1983, 853; (b) Daude, G.; Pereyre, M.; *J. Organomet. Chem.*, 1980, **190**, 43; (c) Harpp, D.N.; Gingras, M., *J. Am. Chem. Soc.*, 1988, **110**, 7737.
See also: (d) Fukuzawa, S.; Sato, K.; Fujinami, T.; Sakai, S., *J. Chem. Soc., Chem. Commun.*, 1983, 853; (e) Yanagisawa, A.; Inoue, H.; Morodome, M.; Yamamoto, H., *J. Am. Chem. Soc.*, 1993, **115**, 10356.

[40] (a) Yamamoto, Y.; Yatagai, H.; Naruta, Y.; Maruyama, K., *J. Am. Chem. Soc.*, 1980, **102**, 7107; (b) Pereyre, M.; Quintard, J.-P.; Rahm, A., *Tin in Organic Synthesis*; Butterworths, London, 1987, p. 216.

[41] (a) Schmid, W.; Whitesides, G.M., *J. Am. Chem. Soc.*, 1991, **113**, 6674; (b) Kim, E.; Gordon, D.M.; Schmid, W.; Whitesides, G.M., *J. Org. Chem.*, 1993, **58**, 5500.

[42] (a) Yates, D.; Eaton, P.E., *J. Am. Chem. Soc.*, 1960, **82**, 4436; (b) Hollis, T.K.; Robinson, N.P.; Bosnich, B., *J. Am. Chem. Soc.*, 1992, **114**, 5464.
Review: (c) Carruthers, W., *Cycloaddition Reactions in Organic Synthesis*; Pergamon, Oxford, 1990.

[43] Kobayashi, S.; Hachiya, I.; Takahori, T.; Araki, M.; Ishitani, H., *Tetrahedron Lett.*, 1992, **33**, 6815.

[44] Kobayashi, S.; Hachiya, I.; Takahori, T.; Araki, M.; Ishitani, H., Tetrahedron Lett., 1993, **34**, 3755.

[45] (a) Rideout, D.C.; Breslow, R., *J. Am. Chem. Soc.*, 1980, **102**, 7816; (b) Grieco, P.A.; Garner, P.; He, Z., *Tetrahedron Lett.*, 1983, **24**, 1897.

[46] Kleinman, E.F., *Comprehensive Organic Synthesis*, vol. 2, Trost, B.M., Ed.; Pergamon, Oxford, 1991, p. 893.

[47] Ojima, I.; Inaba, S-I.; Yoshida, K., *Tetrahedron Lett.*, 1977, 3643.

[48] (a) Ikeda, K.; Achiwa, K.; Sekiya, M., *Tetrahedron Lett.*, 1983, **24**, 4707; (b) Mukaiyama, T.; Kashiwagi, K.; Matsui, S., *Chem. Lett.*, 1989, 1397; (c) Mukaiyama, T.; Akamatsu, H.; Han, J.S., *Chem. Lett.*, 1990, 889; (d) Onaka, M.; Ohno, R.; Yanagiya, N.; Izumi, Y., *Synlett*, 1993, 141.
For a stoichiometric use: (e) Dubois, J.-E.; Axiotis, G., *Tetrahedron Lett.*, 1984, **25**, 2143; (f) Colvin, E.W.; McGarry, D.G., *J. Chem. Soc., Chem. Commun.*, 1985, 539; (g) Shimada, S.; Saigo, K.; Abe, M.; Sudo, A.; Hasegawa, M., *Chem. Lett.*, 1992, 1445.
For an enantioselective version: (h) Hattori, K.; Yamamoto, H., *Tetrahedron*, 1994, **50**, 2785.
[49] Kobayashi, S.; Araki, M.; Ishitani, H.; Nagayama, S.; Hachiya, I., *Synlett*, 1995, 233.
[50] Kobayashi, S.; Araki, M.; Yasuda, M., *Tetrahedron Lett.*, 1995, **36**, 5773.
[51] Lucchini, V.; Prato, M.; Scorrano, G.; Tecilla, P., *J. Org. Chem.*, 1988, **53**, 2251.
[52] Guanti, G.; Narisano, E.; Banfi, L., *Tetrahedron Lett.*, 1987, **28**, 4331.
[53] Kobayashi, S.; Ishitani, H., *J. Chem. Soc., Chem. Commun.*, 1995, 1379.
[54] Larsen, S.D.; Grieco, P.A., *J. Am. Chem. Soc.*, 1985, **107**, 1768.
[55] Grieco, P.A.; Parker, D.T., *J. Org. Chem.*, 1988, **53**, 3325 and references therein.
[56] (a) Weinreb, S.M., *Comprehensive Organic Synthesis*, vol. 5, Trost, B.M., Ed.; Pergamon, Oxford, 1991, p. 401; (b) Boger, D.L.; Weinreb, S.M., *Hetero Diels–Alder Methodology in Organic Synthesis*; Academic Press, San Diego, 1987, chs. 2 and 9.
[57] Danishefsky, S.; Kitahara, T., *J. Am. Chem. Soc.*, 1974, **96**, 7807.
[58] (a) Boger, D.L., *Tetrahedron*, 1983, **39**, 2869; (b) Grieco, P.A.; Bahsas, A.J., *Tetrahedron Lett.*, 1988, **29**, 5855 and references therein.
[59] (a) Kobayashi, S.; Ishitani, H.; Nagayama, S.; *J. Chem. Lett.*, 1995, 423; (b) Kobayashi, S.; Ishitani, H.; Nagayama, S., *Synthesis*, 1995, 1195.
[60] For vinyl ethers, see: (a) Joh, T.; Hagihara, N., *Tetrahedron Lett.*, 1967, 4199; (b) Povarov, L.S., *Russ. Chem. Rev.*, 1967, **36**, 656; (c) Worth, D.F.; Perricine, S.C.; Elslager, E.F.; *J. Heterocycl. Chem.* 1970, **7**, 1353; (d) Kametani, T.; Takeda, H.; Suzuki, Y.; Kasai, H.; Honda, T., *Heterocycles*, 1986, **24**, 3385; (e) Cheng, Y.S.; Ho, E.; Mariano, P.S.; Ammon, H.L.; *J. Org. Chem.*, 1985, **56**, 5678 and references therein.
For vinyl sulfides, see: Narasaka, K.; Shibata, T., *Heterocycles*, 1993, **35**, 1039.
[61] Makioka, Y.; Shindo, T.; Taniguchi, Y.; Takaki, K.; Fujiwara, Y., *Synthesis*, 1995, 801.
[62] Sisko, J.; Weinreb, S.M., *Tetrahedron Lett.*, 1989, **30**, 3037.
[63] Nomura, Y.; Kimura, M.; Takeuchi, Y.; Tomoda, S.; *Chem. Lett.*, 1978, 267.
[64] (a) Maruoka, K.; Yamamoto, H., *Catalytic Asymmetric Synthesis*, Ojima, I., Ed.; VCH, Weinheim, 1993, p. 413.
[65] Narasaka, K., *Synthesis*, 1991, 1.
[66] (a) Kametani, T.; Kasai, H., *Stud. Natural Prod. Chem.*, 1989, **3**, 385; (b) Grigos, V.I.; Povarov, L.S.; Mikhailov, B.M., *Izv. Akad. Nauk SSSR, Ser. Khim.*, 1965, 2163; *Chem. Abstr.*, 1966, **64**, 9680.
[67] (a) Waldmann, H., *Synthesis*, 1994, 535; (b) Borrione, E.; Prato, M.; Scorrano, G.; Stiranello, M.J., *Chem. Soc., Perkin Trans. 1*, 1989, 2245; (c) Ishihara, K.; Miyata, M.; Hattori, K.; Tada, T.; Yamamoto, H., *J. Am. Chem. Soc.* 1994, **116**, 10520.
[68] (a) Kobayashi, S.; Hachiya, I.; Ishitani, H.; Araki, M., *Tetrahedron Lett.*, 1993, **34**, 4535; (b) Kobayashi, S.; Ishitani, H., *J. Am. Chem. Soc.*, 1994, **116**, 4083; (c) Kobayashi, S.; Ishitani, H.; Hachiya, I.; Araki, M., *Tetrahedron*, 1994, **50**, 11623; (d) Kobayashi, S.; Araki, M.; Hachiya, I., *J. Org. Chem.*, 1994, **59**, 3758.
[69] (a) Rauckman, B.S.; Tidwell, M.Y.; Johnson, J.V.; Roth, B., *J. Med. Chem.*, 1989, **32**, 1927; (b) Johnson, J.V.; Rauckman, B.S.; Baccanari, D.P.; Roth, B., *J. Med. Chem.*, 1989, **32**, 1942; (c) Ife, R. J.; Brown, T.H.; Keeling, D.J.; Leach, C.A.; Meeson, M. L.; Parsons, M E.; Reavill, D.R.; Theobald, C.J.; Wiggall, K.J., *J. Med. Chem.*, 1992, **35**, 3413; (d) Sarges, R.; Gallagher, A.; Chambers, T.J.; Yeh, L.-A., *J. Med. Chem.*, 1993, **36**, 2828; (e) Mongin, F.; Fourquez, J.-M.; Rault, S.; Levacher, V.; Godard, A.; Trecourt, F.; Queguiner, G., *Tetrahedron Lett.*, 1995, **36**, 8415.
[70] Kobayashi, S.; Ishitani, H.; Araki, M.; Hachiya, I., *Tetrahedron Lett.*, 1994, **35**, 6325.
[71] Ishitani, H.; Kobayashi, S., *Tetrahedron Lett.*, 1996, **37**, 7357.
[72] (a) Posner, G.H.; Carry, J.-C.; Lee, J.K.; Bull, D.S.; Dai, H., *Tetrahedron Lett.*, 1994, **35**, 1321; (b) Markó, I.E.; Evans, G.R., *Tetrahedron Lett.*, 1994, **35**, 2771.
[73] (a) Fendler, J.H.; Fendler, E.J., *Catalysis in Micellar and Macromolecular Systems*,

Academic Press, London, 1975; (b) Holland, P.M.; Rubingh, D.N., Eds., *Mixed Surfactant Systems*; American Chemical Society, Washington, DC, 1992; (c) Cramer, C.J.; Truhlar, D.G., Eds., *Structure and Reactivity in Aqueous Solution*; American Chemical Society, Washington, DC, 1994; (d) Sabatini, D.A.; Knox, R.C.; Harwell, J.H., Eds. *Surfactant-Enhanced Subsurface Remediation*; American Chemical Society, Washington, DC, 1995.

Index